NANKAI TRACTS IN MATHEMATICS

ISSN: 1793-1118

Series Editors: Yiming Long, Weiping Zhang and Lei Fu
Chern Institute of Mathematics

*For the complete list of titles in this series, please visit
http://www.worldscientific.com/series/ntm

GEOMETRY AND ANALYSIS
ON FINSLER SPACES

Nankai Tracts in Mathematics – Vol. 17

GEOMETRY AND ANALYSIS ON FINSLER SPACES

Qiaoling Xia

Hangzhou Dianzi University, China

World Scientific

NEW JERSEY · LONDON · SINGAPORE · BEIJING · SHANGHAI · TAIPEI · CHENNAI

Published by

World Scientific Publishing Co. Pte. Ltd.

5 Toh Tuck Link, Singapore 596224

USA office: 27 Warren Street, Suite 401-402, Hackensack, NJ 07601

UK office: 57 Shelton Street, Covent Garden, London WC2H 9HE

Library of Congress Control Number: 2024059691

British Library Cataloguing-in-Publication Data
A catalogue record for this book is available from the British Library.

Nankai Tracts in Mathematics — Vol. 17
GEOMETRY AND ANALYSIS ON FINSLER SPACES

ISBN 978-981-12-9667-3 (hardcover)
ISBN 978-981-12-9668-0 (ebook for institutions)
ISBN 978-981-12-9669-7 (ebook for individuals)

For any available supplementary material, please visit
https://www.worldscientific.com/worldscibooks/10.1142/13945#t=suppl

Typeset by Stallion Press
Email: enquiries@stallionpress.com

Preface

Finsler geometry is just Riemannian geometry without quadratic restriction. It was originated from Riemann's ground-breaking 'Habilitation' address in the year 1854. There was no essential progress in the general case (i.e., metric geometry of non-quadratic type) until 1918, when P. Finsler studied the variation problems of curves and surfaces with a family of norms (Finsler metrics) in his Phd thesis. For this reason, we call this subject Riemann–Finsler geometry, or Finsler geometry for short.

Since 1918, many geometers had made important and essential contributions to Finsler geometry, such as L. Berwald, E. Cartan, S. S. Chern etc. and many of our contemporaries. Some substantial progress has been made on Finsler geometry, especially global Finsler geometry in recent twenty years. It is not the objective of this book to provide a comprehensive survey. The book is intended to provide basic materials on Finsler geometry for readers who are interested in Finsler geometry, and to bring them into the frontiers of the active research on related topics.

The book is comprised of three parts. The first part consists of Chapters 1–4, in which we introduce Finsler metrics, the Chern connection, geometric invariants, such as Riemannian invariants, non-Riemannian invariants, and projective invariants, and we give some rigidity results for Finsler metrics with some curvature properties. The second part, Chapters 5–6, covers the theory of geodesics via the calculation of variation, using which we establish some comparison theorems, which are fundamental tools to study global Finsler geometry. The last part is made up of Chapters 7–9, in which we present the recent developments in nonlinear analysis on Finsler measure spaces, partly based on the author's recent works on Finsler harmonic functions, the eigenvalue problem, and the heat flow. Chapters 1–5 come out of series of lectures for graduate students at Zhejiang University between 2017 and 2019. Chapters 6–10 are partly based on series of lectures at summer school for graduate students on Finsler geometry held at Chongqing Normal University in July 2019. From then, there has been much progress in nonlinear geometric analysis included in this book. I hope that it keeps going on, and it is my great pleasure if this book provides some help for readers.

I would like to take this opportunity to thank several individuals who have greatly supported my academic journey. I am sincerely grateful to my advisor, Professor An-Min Li from Sichuan University, who first introduced me to the area of differential geometry during my graduate study. I am also deeply thankful to my another advisor, Professor Yibing Shen from Zhejiang University, who brought me into a broader field of research, particularly in Finsler geometry during my graduate study and working at Zhejiang University. I am especially grateful to Professor Zhongmin Shen for his constant helpful discussions and invaluable comments on this book. Lastly, I extend my appreciation to my graduate students for their careful proofreading and assistance in correcting.

This book is written based on the author's research on Finsler geometry supported by National Natural Science Foundation of China (Nos. 12471044, 12071423, 11671352). I would like to thank NSFC for the continuous support. I want to thank Ms Britney Jiang and Ms Angeline Husni for their editorial guidance and much help on this book.

Qiaoling Xia

Hangzhou, P. R. China
February, 2024

Contents

Chapter 1

Minkowski Spaces

In this chapter, we review some basic notions and facts in Minkowski spaces. We refer to [BCS], [ChS], [Sh1], [SS] for more details.

1.1 Minkowski Norm

Let us first recall Euler's Lemma for the homogeneous functions on $\mathbb{R}^n$.

Definition 1.1.1. A real-valued function f on $\mathbb{R}^n$ is called a *positively homogeneous function* of degree r if f satisfies $f(\lambda y) = \lambda^r f(y)$ for all $\lambda > 0$ and $y \in \mathbb{R}^n$.

For any $y \in \mathbb{R}^n$, write $y = (y^i)$ or $y = (y^1, y^2, \ldots, y^n)$. The following lemma is well known.

Lemma 1.1.1 (Euler's Lemma). *A differentiable function $f : \mathbb{R}^n \backslash \{0\} \to \mathbb{R}$ is a positively homogeneous function of degree r if and only if*

$$y^i f_{y^i}(y) = r f(y), \tag{1.1.1}$$

where f_{y^i} are the partial derivatives of f with respect to y^i for each $1 \le i \le n$.

Proof. Suppose that $f(\lambda y) = \lambda^r f(y)$ for $\lambda > 0$. Differentiating this with respect to the parameter λ yields

$$y^i f_{y^i}(\lambda y) = r \lambda^{r-1} f(y).$$

Letting $\lambda = 1$ gives (1.1.1).

1

Conversely, assume that (1.1.1) holds. Then

$$\frac{d}{d\lambda}f(\lambda y) = y^i f_{(\lambda y)^i}(\lambda y) = \frac{1}{\lambda}(\lambda y^i)f_{(\lambda y)^i}(\lambda y) = \frac{r}{\lambda}f(\lambda y).$$

Solving the above equation yields $f(\lambda y) = c\lambda^r$, where c is a constant depending on y. In fact, $c = f(y)$ by taking $\lambda = 1$. Consequently, $f(\lambda y) = \lambda^r f(y)$. $\qquad\qquad\qquad\qquad\qquad\qquad\qquad\qquad\qquad\qquad\square$

From Euler's Lemma, it is easy to check the following properties.

Corollary 1.1.1. *Let f be a positively homogeneous function of degree one on $\mathbb{R}^n$. Then*

(a) $y^i f_{y^i} = f$;
(b) $y^j f_{y^i y^j} = 0$;
(c) $y^k f_{y^i y^j y^k} = -f_{y^i y^j}$;
(d) $y^l f_{y^i y^j y^k y^l} = -2f_{y^i y^j y^k}$;

here all formulae are supposed to be evaluated at y.

Definition 1.1.2. Let V be an n-dimensional vector space. A function $F : V \to \mathbb{R}$ is called a *Minkowski norm* if F satisfies

(1) $F(y) \geq 0$ for any $y \in V$, and $F(y) = 0$ if and only if $y = 0$;
(2) $F(\lambda y) = \lambda F(y)$ for any $y \in V$ and $\lambda > 0$;
(3) F is C^∞ on $V \backslash \{0\}$ such that the $n \times n$ matrix $(g_{ij}(y))$, where $g_{ij}(y) = \left[\frac{1}{2}F^2\right]_{y^i y^j}(y)$, is positive definite at all $y \neq 0$.

The pair (V, F) is called a *Minkowski space*. A Minkowski norm F is said to be *reversible* if $F(y) = F(-y)$. A Minkowski norm F is said to be *absolutely homogeneous* with respect to y if $F(\lambda y) = |\lambda|F(y)$ for any $y \in V$ and $\lambda \in \mathbb{R}$.

Given a basis $\{e_i\}_{i=1}^n$ in V and a fixed nonzero vector $y = y^i e_i \in V$, by identifying $y \in V$ with $y = (y^i) \in \mathbb{R}^n$, we have

$$g_{ij}(y) = (FF_{y^i y^j} + F_{y^i}F_{y^j})(y). \qquad (1.1.2)$$

Then g_{ij} are C^∞ functions on $\mathbb{R}^n \backslash \{0\}$ and homogeneous functions in y of degree zero. For any vectors $u = u^i e_i$, $v = v^i e_i$ in V, the bilinear symmetric form g_y defined by

$$g_y(u, v) := g_{ij}(y)u^i v^j$$

gives an inner product on V, which is called the *fundamental form* on V.

Applying Corollary 1.1.1 to F yields $F_{y^i}y^i = F$. From this and (1.1.2), we have

$$g_{ij}(y)y^j = FF_{y^i}, \quad g_{ij}(y)y^i y^j = g_y(y, y) = F^2(y). \qquad (1.1.3)$$

Let

$$W_y := \{w \in V | g_y(y, w) = 0\}$$

be a subspace in V. Then we have a direct sum decomposition

$$V = \mathrm{span}_{\mathbb{R}}\{y\} \oplus W_y.$$

Let

$$h_y := g_y - \frac{1}{F^2(y)} g_y(y, \cdot) g_y(y, \cdot).$$

Then the components of h_y are

$$h_{ij}(y) = g_{ij}(y) - F^{-2} y_i y_j = g_{ij}(y) - F_{y^i} F_{y^j} = FF_{y^i y^j},$$

where $y_i := g_{ij}y^j$. Obviously,

$$h_y(y, y) = 0, \quad h_y(y, w) = 0, \quad \forall w \in W_y.$$

For any $v = \tau y + w \in V$, where $w \in W_y$, we have

$$h_y(v, v) = g_y(\tau y + w, \tau y + w) - \frac{1}{F^2(y)} g_y(y, \tau y + w)^2 = g_y(w, w) \geq 0.$$

The equality holds if and only if $w = 0$, namely, $v = \tau y$. Thus, h_y is positive semi-definite on V and positive definite on W_y. The rank of the bilinear form h_y is $n - 1$. h_y is called the *angular form* on V.

Lemma 1.1.2. *Let (V, F) be a Minkowski space. Then*

(1) (*Minkowski's inequality*)

$$F(y_1 + y_2) \leq F(y_1) + F(y_2), \quad y_1, y_2 \in V, \qquad (1.1.4)$$

in which equality holds if and only if $y_2 = \tau y_1$ or $y_1 = \tau y_2$ for some $\tau \geq 0$.

(2) (*Cauchy–Schwarz's inequality*) *for any $0 \neq y, v \in V$, we have*

$$g_y(y, v) \leq F(y)F(v), \qquad (1.1.5)$$

in which equality holds if and only if $v = \tau y$ for some $\tau \geq 0$.

Proof. (1) Let $y_1, y_2 \in V$ and $y(t) = (1 - t)y_1 + ty_2$ for $t \in [0, 1]$. Assume that y_1, y_2 are linearly independent. Then $y(t) \neq 0$. Consider the function $\phi(t) = F(y(t))$. Obviously, it is C^∞ on $[0, 1]$. Then

$$\phi''(t) = \frac{1}{F(y(t))} h_y(y_1 - y_2, y_1 - y_2) \geq 0.$$

Thus, $\phi(t)$ is a convex function, which implies that $2\phi(1/2) \leq \phi(0) + \phi(1)$, namely,

$$F(y_1 + y_2) \leq F(y_1) + F(y_2).$$

Next, we consider the case when y_1, y_2 are linearly dependent. Without loss of generality, we assume that $y_1 \neq 0$. Otherwise, the equality holds obviously. Then there exists a constant τ such that $y_2 = \tau y_1$. Consequently,

$$F(y_1 + y_2) = F((1 + \tau)y_1) = \varepsilon(1 + \tau)F(\varepsilon y_1)$$
$$\leq \varepsilon F(\varepsilon y_1) + F(\tau y_1) \leq F(y_1) + F(y_2),$$

where $\varepsilon := \operatorname{sign}(1 + \tau)$. The equality holds if and only if $\tau \geq 0$. Thus, (1) follows.

(2) For any $w \in W_y$, let $\phi(t) = F^2(y + tw)$ for $t \in [0, 1]$. In the same way as (1), we have

$$\phi'(t) = 2g_{y+tw}(y + tw, w),$$

$$\phi''(t) = \frac{2}{F^2(y + tw)} [g_{y+tw}(y + tw, w)]^2 + 2h_{y+tw}(w, w) \geq 0.$$

Note that $\phi'(0) = 0$ and $\phi'(t)$ is monotone increasing. Then $\phi'(t) \geq 0$, which implies that $\phi(1) \geq \phi(0)$, i.e.,

$$F(y) \leq F(y + w), \quad w \in W_y, \tag{1.1.6}$$

in which the equality holds if and only if $w = 0$.

For any $v \in V$, we write $v = \tau y + w$, where $\tau \in \mathbb{R}$ and $w \in W_y$. Observe that

$$g_y(y, v) = \tau g_y(y, y) = \tau F^2(y).$$

If $\tau \leq 0$, then $g_y(y, v) \leq 0$. (1.1.5) is obvious and the equality holds if and only if $\tau = 0$ and $v = 0$. Assume that $\tau > 0$. Then

$$g_y(y, v) = \tau F^2(y) \leq \tau F\left(y + \frac{1}{\tau}w\right) F(y) = F(v)F(y)$$

by (1.1.6). Equality holds if and only if $w = 0$, i.e., $v = \tau y$. This finishes the proof. $\qquad\square$

From the proof of the Cauchy–Schwarz inequality, one obtains the following result.

Corollary 1.1.2. *For any nonzero vector $y \in V$, we have*

$$F(y) \leq F(y + w), \quad \forall w \in W_y,$$

in which equality holds if and only if $w = 0$.

The following corollary follows from an application of the Cauchy–Schwarz inequality.

Corollary 1.1.3. *Let (V, F) be a Minkowski space. Suppose that $y, v \in V \backslash \{0\}$ satisfy $g_y(y, w) = g_v(v, w)$ for any $w \in V$. Then $y = v$.*

Proof. Taking $w = v$ in $g_y(y, w) = g_v(v, w)$ yields

$$F^2(v) = g_y(y, v) \leq F(y)F(v),$$

which implies that $F(v) \leq F(y)$. In the same way, by taking $w = y$, one obtains $F(y) \leq F(v)$. Thus, $F(y) = F(v)$ and hence $g_y(y, v) = F(y)F(v)$. By Lemma 1.1.2 (2), $v = \tau y$ for some $\tau \geq 0$. Since $F(v) = F(y)$, we have $\tau = 1$ and hence $y = v$. $\qquad\square$

When F is reversible, F is the norm on the vector space V in the usual sense. In this case, the Cauchy–Schwarz inequality can be rewritten as

$$|g_y(y, v)| \leq F(y)F(v). \tag{1.1.7}$$

In general, (1.1.7) is not necessarily true for a Minkowski space (V, F).

Example 1.1.1. Let V be a finite dimensional vector space and $\langle , \rangle$ an inner product on V. Choose a basis $\{e_i\}$ in V and define $(a_{ij}) = (\langle e_i, e_j \rangle)$, which is a positive definite and symmetric matrix. Then

$$\alpha(y) = \sqrt{\langle y, y \rangle} = \sqrt{a_{ij} y^i y^j}, \quad y = y^i e_i,$$

is a *Euclidean norm* on V, which is independent of the choice of basis $\{e_i\}$. It is easy to see that it is a Minkowski norm on V. Thus, $(V, \langle , \rangle)$ is a Minkowski space. Note that the Euclidean space with finite dimension n is unique up to a linear isometry. It is isomorphic to the *canonical Euclidean space* $\mathbb{R}^n$ with the *canonical Euclidean norm*

$$F(y) = |y| = \sqrt{(y^1)^2 + \cdots + (y^n)^2}, \quad y = (y^i) \in \mathbb{R}.$$

Example 1.1.2. Let α be the Euclidean norm and β a linear functional on an n-dimensional vector space V. Define

$$F(y) = \alpha(y) + \beta(y).$$

Obviously, F satisfies (2) in Definition 1.1.2. Fix a basis $\{e_i\}$ in V, we express α and β as

$$\alpha(y) = \sqrt{a_{ij}y^i y^j}, \quad \beta(y) = b_i y^i, \quad \forall y = y^i e_i,$$

where (a_{ij}) is a positive definite and symmetric matrix. A direct calculation gives

$$g_{ij}(y) = \frac{F}{\alpha}\left\{ a_{ij} - \frac{y_i}{\alpha}\frac{y_j}{\alpha} + \frac{\alpha}{F}\left(b_i + \frac{y_i}{\alpha} \right)\left(b_j + \frac{y_j}{\alpha} \right) \right\}, \qquad (1.1.8)$$

where $y_i := a_{ik}y^k$. In order to give the formulae for $\det(g_{ij})$ and the inverse matrix $(g^{ij}) := (g_{ij})^{-1}$, we need the following lemma, whose proof is from a linear algebra.

Lemma 1.1.3 ([BCS]). *Let $A = (a_{ij})$, $B = (b_{ij})$ be two symmetric $n \times n$ matrices and $C = (c_i)$ an n-vector in $\mathbb{R}^n$. Assume that B is invertible with the inverse $B^{-1} = (b^{ij})$ and*

$$a_{ij} = b_{ij} + \delta c_i c_j.$$

Then

$$\det(a_{ij}) = (1 + \delta c^2)\det(b_{ij}),$$

where $c = \sqrt{b^{ij}c_i c_j}$. If $1 + \delta c^2 \neq 0$, then A is invertible and the inverse matrix $A^{-1} = (a^{ij})$ is given by

$$a^{ij} = b^{ij} - \frac{\delta c^i c^j}{1 + \delta c^2},$$

where $c^i = b^{ij}c_j$.

By Lemma 1.1.3, one obtains the following formulae:

$$g^{ij} = \frac{\alpha}{F}a^{ij} - \frac{\alpha}{F^2}(b^i y^j + b^j y^i) + \frac{b^2\alpha + \beta}{F^3}y^i y^j, \qquad (1.1.9)$$

$$\det(g_{ij}) = \left(\frac{F}{\alpha} \right)^{n+1} \det(a_{ij}), \qquad (1.1.10)$$

where $b := \|\beta\|_\alpha$. Let

$$\|\beta\|_\alpha := \sup_{y \in V\backslash\{0\}} \frac{\beta(y)}{\alpha(y)} = \sup_{\alpha(y)=1} \beta(y).$$

It is easy to check that $\|\beta\|_\alpha = \sqrt{a^{ij}b_ib_j}$ by Lagrangian multiplier method, where $(a^{ij}) = (a_{ij})^{-1}$. Further, $F = \alpha + \beta$ is a Minkowski norm if and only if $b = \|\beta\|_\alpha < 1$ ([BCS], [CnS]). In fact, the necessity is obvious. Conversely, let $F_t := \alpha + t\beta$ for any $t \in [0,1]$. Then $F_t(y) > 0$ since $F_1(y) = F(y) > 0$. Put $g^t_{ij} := \frac{1}{2}[F^2_t]_{y^iy^j}$. By (1.1.10), we have $\det(g^t_{ij}) > 0$. Note that the eigenvalues $\lambda_1(t), \ldots, \lambda_n(t)$ of the matrix (g^t_{ij}) depend on t continuously and $\lambda_i(0) > 0$. Consequently, $\lambda_i(1) > 0$ for $1 \le i \le n$, which mean that the matrix (g^1_{ij}) is positive definite. In this case, $F = \alpha + \beta$ is called a *Randers norm* on V. $F = \alpha + \beta$ is reversible if and only if $\beta = 0$, which is exactly the Euclidean norm.

1.2   Legendre Transformations

Let (V, F) be an n-dimensional Minkowski space and V^* be the dual space of V. Let $\{e_i\}^n_{i=1}$ be a basis for V and $\{\theta^i\}^n_{i=1}$ be its dual basis. For any $y \in V\backslash\{0\}$, by identifying $y = y^ie_i$ with $y = (y^i) \in \mathbb{R}^n\backslash\{0\}$, one obtains a unique element $\xi = \xi_i\theta^i \in V^*\backslash\{0\}$, where $\xi_i = g_{ij}(y)y^j = y_i$. Thus, one obtains a map $\mathcal{L} : V \to V^*$ given by

$$\mathcal{L}(y) := \begin{cases} g_y(y, \cdot), & y \in V\backslash\{0\}, \\ 0, & y = 0. \end{cases}$$

Equivalently, $\mathcal{L}(y) = \xi = g_{ij}(y)y^j\theta^i \in V^*\backslash\{0\}$ if $y \ne 0$ and $\mathcal{L}(y) = 0$ if $y = 0$. $\mathcal{L}$ is called the *Legendre transformation* from V to V^*. Obviously, it is a nonlinear transformation.

Let

$$S_F = \{y \in V | F(y) = 1\}.$$

It is a closed hypersurface in V, which is diffeomorphic to the standard unit sphere $\mathbb{S}^{n-1}$ in $\mathbb{R}^n$. S_F is called the *indicatrix* of F. For any $\xi \in V^*$, define

$$F^*(\xi) := \sup_{y\in V\backslash\{0\}} \frac{\xi(y)}{F(y)} = \sup_{y\in S_F} \xi(y).$$

It is easy to see that

$$F^*(\lambda\xi) = \lambda F^*(\xi), \quad F^*(\xi + \eta) \le F^*(\xi) + F^*(\eta)$$

for any $\xi, \eta \in V^*$, where $\lambda > 0$. Thus, F^* is a norm on V^*, which is called the *dual norm* of F. We shall show that F^* is a Minkowski norm on V^*.

Lemma 1.2.1. *Let F be the Minkowski norm on V and F^* be the dual norm of F. Then the Legendre transformation $\mathcal{L} : V\backslash\{0\} \to V^*\backslash\{0\}$ is a diffeomorphism with $F(y) = F^*(\mathcal{L}(y))$.*

Proof. Assume that $\mathcal{L}(y) = \mathcal{L}(v)$ for any $y, v \in V \backslash \{0\}$. Then $g_y(y, \cdot) = g_v(v, \cdot)$, which implies $y = v$ from Corollary 1.1.3. This shows that $\mathcal{L}$ is injective. Next, we prove that $\mathcal{L}$ is surjective. Note that S_F is closed. For any $\xi \in V^* \backslash \{0\}$, there exists an element $\bar{y} \in S_F$ such that

$$F^*(\xi) = \xi(\bar{y}) = \sup_{y \in S_F} \xi(y).$$

Let $y = \xi(\bar{y})\bar{y}$. Then $F(y) = \xi(\bar{y})F(\bar{y}) = F^*(\xi)$. For any $v \in V$ and $t \in \mathbb{R}$, consider the function $f(t) = \xi(y + tv) - F(y + tv)F^*(\xi)$. Obviously $f(t) = 0$ if $y + tv = 0$. When $y + tv \neq 0$, it is obvious that

$$\xi\left(\frac{y + tv}{F(y + tv)}\right) \leq F^*(\xi), \quad \text{equivalently,} \quad \xi(y + tv) \leq F(y + tv)F^*(\xi).$$

Thus, $f(t) \leq 0$ for any $t \in \mathbb{R}$. On the other hand, $f(0) = 0$. Hence, $t = 0$ is the maximum point for $f(t)$, which implies that $f'(0) = 0$. Hence, $g_y(y, v) = \xi(v)$ for any $v \in V$, i.e., $\mathcal{L}(y) = \xi$. This shows that $\mathcal{L}$ is surjective. Since $\mathcal{L}$ is C^∞ on $V \backslash \{0\}$, $\mathcal{L}$ is a diffeomorphism. In this case, $F(y) = F^*(\xi) = F^*(\mathcal{L}(y))$. $\qquad\square$

Further, we have the following

Proposition 1.2.1. *For any Minkowski norm F on a vector space V, its dual norm F^* is a Minkowski norm on V^*.*

Proof. It suffices to prove that for any $\xi = \mathcal{L}(y)$, i.e., $\xi_i = g_{ik}(y)y^k$,

$$g^{*ij}(\xi) := \frac{1}{2}[F^{*2}(\xi)]_{\xi_i \xi_j} = g^{ij}(y),$$

where $(g^{ij}(y)) = (g_{ij}(y))^{-1}$. In fact, differentiating $F^2(y) = F^{*2}(\xi)$ with respect to y^i yields

$$[F^2(y)]_{y^i} = [F^{*2}(\xi)]_{\xi_k} g_{ik}(y), \tag{1.2.1}$$

which implies that

$$g^{*kl}(\xi)\xi_l = \frac{1}{2}[F^{*2}(\xi)]_{\xi_k} = \frac{1}{2}g^{ik}(y)[F^2(y)]_{y^i} = y^k.$$

From this and Euler's Lemma, we have $\frac{\partial y^k}{\partial \xi_l} = g^{*kl}(\xi)$. Note that $\left(\frac{\partial y^k}{\partial \xi_l}\right) = \left(\frac{\partial \xi_l}{\partial y^k}\right)^{-1} = (g_{lk}(y))^{-1}$. Consequently, $g^{*ij}(\xi) = g^{ij}(y)$. $\qquad\square$

Let (V, F) be a Minkowski space and (V^*, F^*) be its dual Minkowski space. We also can define the Legendre transformation $\mathcal{L}^* : V^* \to V^{**}$ and the Minkowski norm F^{**} on V^{**}. By identifying V^{**} with V, we have

$$\mathcal{L}^* = \mathcal{L}^{-1}, \quad F^{**} = F.$$

Example 1.2.1 ([HS], [Sh1]). Let $F = \alpha + \beta$ be a Randers norm on a vector space V, where α is a Euclidean norm and β is a linear functional on V with $\|\beta\|_\alpha < 1$. Let $\{e_i\}_{i=1}^n$ be a basis for V and $\{\theta^i\}_{i=1}^n$ be its dual basis. Express α and β as

$$\alpha(y) = \sqrt{a_{ij}y^i y^j}, \quad \beta(y) = b_i y^i, \quad \forall y = y^i e_i \in V.$$

Denote $b := \|\beta\|_\alpha = \sqrt{a^{ij}b_i b_j}$, where $(a^{ij}) = (a_{ij})^{-1}$. It is easy to check that the dual Randers norm $F^* = \alpha^* + \beta^*$ is still of Randers type, where α^* is a Euclidean norm and β^* is a linear functional on V^*, which are expressed by

$$\alpha^*(\xi) = \sqrt{a^{*ij}\xi_i \xi_j}, \quad \beta^*(\xi) = b^{*i}\xi_i, \quad \forall \xi = \xi_i \theta^i,$$

here

$$a^{*ij} = \frac{(1 - b^2)a^{ij} + b^i b^j}{(1 - b^2)^2}, \quad b^{*i} = -\frac{b^i}{1 - b^2}, \quad b^i = a^{ij}b_j.$$

Let $(a_{ij}^*) = (a^{*ij})^{-1}$. Then

$$a_{ij}^* = (1 - b^2)(a_{ij} - b_i b_j).$$

Thus $b^* = \|\beta^*\|_{\alpha^*} = \sup_{\alpha^*(\xi) = 1} \beta^*(\xi)$ given by

$$b^{*2} = a_{ij}^* b^{*i} b^{*j} = \frac{1}{1 - b^2}(a_{ij} - b_i b_j)b^i b^j = b^2.$$

Consequently, by (1.1.8), the Legendre transformation $\mathcal{L} : V \to V^*$ is given by

$$\xi = \mathcal{L}(y) = g_{ij}(y)y^j \theta^i = F(y)\left(\frac{a_{ij}y^j}{\alpha(y)} + b_i\right)\theta^i,$$

and its inverse $\mathcal{L}^* = \mathcal{L}^{-1} : V^* \to V$ is given by

$$y = \mathcal{L}^{-1}(\xi) = g^{*kl}(\xi)\xi_l e_k = F^*(\xi)\left(\frac{a^{*kl}\xi_l}{\alpha^*(\xi)} + b^{*k}\right)e_k.$$

1.3 Cartan Torsion

To characterize how far a Minkowski norm is from a Euclidean norm, E. Cartan introduced a geometric quantity, called *Cartan torsion* for a Minkowski norm([Car]).

Let F be a Minkowski norm on an n-dimensional vector space V. For any nonzero vector $y \in V$, let

$$C_y(u,v,w) := \frac{1}{4} \frac{\partial^3}{\partial s \partial t \partial r} \left[F^2(y + su + tv + rw) \right]_{s=t=r=0},$$

where $u, v, w \in V$. Each C_y is a trilinear symmetric form on V. We call the family $C := \{C_y | y \in V \backslash \{0\}\}$ the *Cartan torsion*.

Let $\{e_i\}_{i=1}^n$ be a basis for V and $\{\theta^i\}_{i=1}^n$ be its dual basis. Define $C_{ijk}(y) := C_y(e_i, e_j, e_k)$. Then

$$C_{ijk} = \frac{1}{4}[F^2]_{y^i y^j y^k} = \frac{1}{2} \frac{\partial g_{ij}}{\partial y^k}.$$

Obviously, C_{ijk} are symmetric with respect to the indices i, j, k. We call $g_y = g_{ij}(y)\theta^i \otimes \theta^j$ and $C_y = C_{ijk}(y)\theta^i \otimes \theta^j \otimes \theta^k$ the *fundamental tensor* and the *Cartan tensor* of F respectively. For any nonzero $y \in V$, we define the mean value I_y of the Cartan torsion by

$$I_y(v) := \text{tr}(C_y(v, \cdot, \cdot)) = g^{ij}(y)C_y(v, e_i, e_j), \quad v \in V.$$

We call the family $I := \{I_y | y \in V \backslash \{0\}\}$ the *mean Cartan torsion*. The tensor

$$I_y = I_i(y)\theta^i = g^{jk}(y)C_{ijk}(y)\theta^i$$

is said to be the *mean Cartan tensor*. Observe that

$$\frac{\partial}{\partial y^i} \left(\det(g_{jk}) \right) = \det(g_{jk}) g^{pq} \frac{\partial g_{pq}}{\partial y^i} = 2\det(g_{jk}) g^{pq} C_{ipq}.$$

We have

$$I_i = g^{jk} C_{ijk} = \frac{\partial}{\partial y^i} \left(\log \sqrt{\det(g_{jk})} \right). \tag{1.3.1}$$

It follows from Euler's Lemma that

$$C_y(y, u, v) = C_y(u, y, v) = C_y(u, v, y) = 0, \quad C_{\lambda y} = \lambda^{-1} C_y$$

and

$$I_y(y) = 0, \quad I_{\lambda y} = \lambda^{-1} I_y,$$

where $\lambda > 0$. In local coordinates, the above identities imply that $C_{ijk}y^i = 0$ and $I_i y^i = 0$.

Obviously, F is Euclidean if and only if $C_y = 0$ for any $y \in V\backslash\{0\}$. In fact, Euclidean norms can be characterized by the mean Cartan torsion. The following result was due to A. Deicke ([Dei]).

Theorem 1.3.1 ([Dei]). *A Minkowski norm on a vector space is Euclidean if and only if $I = 0$.*

Proof. It suffices to prove the sufficiency. For any nonzero $y \in V$, F induces a Riemannian metric $\check{g}_y = g_{ij}(y)dy^i \otimes dy^j$ on V. With respect to the Levi-Civita connection of $\check{g}_y$, the connection coefficients are given by

$$\check{\Gamma}^k_{ij} = \frac{1}{2}g^{kl}\left\{\frac{\partial g_{il}}{\partial y^j} + \frac{\partial g_{jl}}{\partial y^i} - \frac{\partial g_{ij}}{\partial y^l}\right\} = g^{kl}C_{ijl}.$$

Then the Laplacian of g_{ij} is given by

$$\check{\Delta}g_{ij} = g^{kl}\left(\frac{\partial^2 g_{ij}}{\partial y^k \partial y^l} - \check{\Gamma}^p_{kl}\frac{\partial g_{ij}}{\partial y^p}\right) = 2g^{kl}(C_{klj})_{y^i} - 2g^{kl}g^{pq}C_{klq}C_{ijp}$$

$$= 2\left(g^{kl}C_{klj}\right)_{y_i} - 2(g^{kl})_{y^i}C_{klj} - 2I_q g^{pq}C_{ijp} = 4g^{kp}C_{pqi}g^{ql}C_{klj},$$

where we used $I_j = g^{kl}C_{klj} = 0$ by the assumption. Thus, for any $v = v^i \frac{\partial}{\partial y^i} \in T_y(V) \cong V$, we have

$$(\check{\Delta}g_{ij})v^i v^j = 4g^{kp}C_{pqi}g^{ql}C_{klj}v^i v^j = 4\|C_y(v,\cdot,\cdot)\|^2_{\check{g}} \geq 0,$$

which implies that the matrix $(\check{\Delta}g_{ij})$ is positive semi-definite. In particular, $\check{\Delta}g_{ii} \geq 0$ for each i.

Let $B_r := \{y \in V\backslash\{0\}|F(y) < r\}$ for some real number $r > 1$. Note that g_{ii} are positively homogeneous functions in y of degree zero. For each i, g_{ii} attains its maximum in the interior of the domain $\Omega_r := B_r \backslash B_{1/r}$. By the maximum principle, g_{ii} must be constant on Ω_r for each i. Letting $r \to +\infty$, g_{ii} are constants on $V\backslash\{0\}$ and hence $\check{\Delta}g_{ii} = 0$. This means that the trace of the matrix $(\check{\Delta}g_{ij})$ is zero. Since $(\check{\Delta}g_{ij})$ is positive semi-definite, its all eigenvalues are zero. Consequently, $(\check{\Delta}g_{ij})$ is zero identically. By scale-invariance, each g_{ij} is harmonic as a function on $(S_F, \check{g}_y)$. By the maximum principle again, g_{ij} are constants on $(S_F, \check{g}_y)$ for all $1 \leq i, j \leq n$. By scale-invariance again, g_{ij} are constants on $\mathbb{R}\backslash\{0\}$, which means that F is Euclidean. $\square$

To characterize how far a Minkowski norm is from a Randers norm, Matsumoto introduced the following geometric quantity ([Ma2], [Ma4]).

For any $y \in V\backslash\{0\}$, define

$$M_{ijk} := C_{ijk} - \frac{1}{n+1}(I_i h_{jk} + I_j h_{ik} + I_k h_{ij}),$$

where $h_{ij} = FF_{y^i y^j} = g_{ij} - F_{y^i}F_{y^j}$. Let

$$M_y := M_{ijk}(y)\theta^i \otimes \theta^j \otimes \theta^k.$$

Then M_y is a trilinear symmetric tensor on V, which is called a *Matsumoto tensor*. We call the family $\mathbf{M} := \{M_y | y \in V\backslash\{0\}\}$ the *Matsumoto torsion*. It is clear that $\mathbf{M} = \mathbf{0}$ for all two-dimensional Minkowski norms. The Minkowski norm with $\mathbf{M} = 0$ is said to be *C-reducible*. The following example shows that the Randers norm on a vector space V is C-reducible.

Example 1.3.1 ([Ma4]). Let $F = \alpha + \beta$ be a Randers norm on a vector space V as in Example 1.1.2. By (1.3.1) and (1.1.10), we have

$$I_i = \frac{\partial}{\partial y^i}\left(\log\sqrt{\left(\frac{\alpha+\beta}{\alpha}\right)^{n+1}\det(a_{ij})}\right) = \frac{n+1}{2F}\left(b_i - \frac{\beta}{\alpha^2}y_i\right).$$

Differentiating (1.1.8) with respect to y^k yields

$$C_{ijk} = \frac{1}{n+1}\left(I_i h_{jk} + I_j h_{ik} + I_k h_{ij}\right),$$

where $h_{ij} = FF_{y^i y^j}$ are given by

$$h_{ij} = \frac{F}{\alpha}\left(a_{ij} - \frac{y_i y_j}{\alpha^2}\right).$$

This implies that $\mathbf{M} = 0$.

In fact, M. Matsumoto and S. Hojo proved the following stronger result ([Ma4], [MH]).

Theorem 1.3.2. *Let F be a Minkowski norm on a vector space V of dimension $n \geq 3$. Then F is a Randers norm if and only if $\mathbf{M} = 0$.*

The proof is omitted. See [MH] for more details.

Chapter 2

Finsler Manifolds

2.1 Finsler Metrics and Examples

A Finsler metric on a differential manifold M is a family of nonnegative Minkowski norms on TM. Precisely, we have

Definition 2.1.1. Let M be an n-dimensional smooth manifold and TM the tangent bundle over M. A function $F : TM \rightarrow [0, +\infty)$ is called a Finsler metric (or Finsler structure) if it satisfies

(F1) F is C^∞ on the slit tangent bundle $TM_0 := TM \backslash \{0\}$;
(F2) $F_x := F|_{T_x M}$ is a Minkowski norm on $T_x M$ for each $x \in M$.
    The pair (M, F) is called a *Finsler manifold*.

If $F(x, y) = F(x, -y)$, then we say that F is *reversible*. Otherwise F is said to be *nonreversible*. It is easy to see that $\overleftarrow{F}(x, y) := F(x, -y)$ is also a Finsler metric, called the *reverse metric (or reverse structure)* of F.

Definition 2.1.2. On a Finsler manifold (M, F), we define the reversibility function $\Lambda : M \rightarrow \mathbb{R}^+$ by

$$\Lambda(x) = \sup_{y \in T_x M \backslash \{0\}} \frac{\overleftarrow{F}(x, y)}{F(x, y)}.$$

The number $\Lambda := \sup_{x \in M} \Lambda(x)$ is called the reversibility of F.

For any $x \in M$, $\Lambda(x)$ is well defined because of the homogeneity of F and $\Lambda < \infty$ if M is compact. In general, $\Lambda \in [1, \infty]$ and F is reversible if and only if $\Lambda = 1$. If $\Lambda < \infty$, we have

$$\Lambda^{-1} F(x, y) \leq \overleftarrow{F}(x, y) \leq \Lambda F(x, y). \tag{2.1.1}$$

In a local coordinate system $\{(x^i, y^i)\}$, let $g_{ij}(x, y) := \frac{1}{2}\left(F^2(x, y)\right)_{y^i y^j}$ for any $y \in T_x M \backslash \{0\}$. The tensor $g_y = g_{ij}(x, y)dx^i \otimes dx^j$ is a symmetric and positive definite tensor, called the *fundamental tensor* of F. It induces a Riemannian inner product on each tangent space $T_x M$ for any $(x, y) \in TM \backslash \{0\}$, i.e., for any $u = u^i \frac{\partial}{\partial x^i}$, $v = v^i \frac{\partial}{\partial x^i} \in T_x M$,

$$g_y(u, v) = g_{ij}(x, y)u^i v^j$$

In particular, we have $F^2(x, y) = g_y(y, y)$ by the homogeneity of F.

A Finsler metric F is said to be *Riemannian* if F_x is a Euclidean norm on each tangent space $T_x M$. In local coordinates, F is Riemannian if and only if $g_{ij}(x, y) = g_{ij}(x)$ are independent of y. The pair (M, F) is said to be a *Riemannian manifold*. Obviously, every Riemannian metric is reversible.

Given a Finsler metric F on a manifold M and a piecewise C^1 curve $\gamma : [a, b] \to M$ with $\gamma(a) = p$ and $\gamma(b) = q$, we define

$$L_F(\gamma) := \int_a^b F(\gamma(t), \dot{\gamma}(t))dt.$$

It is independent of the choice of the parameter t preserving the orientation. We call $L_F(\gamma)$ the *length* of the curve γ from p to q with respect to F. This length structure L_F induces a *distance* d_F from p to q by

$$d_F(p, q) := \inf_\gamma L_F(\gamma),$$

where the infimum is taken over all piecewise C^1 curves $\gamma : [a, b] \to M$ from $\gamma(a) = p$ to $\gamma(b) = q$. It is easy to see that

$$d_F(p, q) \le d_F(p, r) + d_F(r, q).$$

At any point $x \in M$, there is an open neighborhood U_x of x, a constant $C > 1$ and a diffeomorphism $\varphi : U_x \to \mathbb{B}^n \subset \mathbb{R}^n$ such that

$$C^{-1}|u - v| \le d_F(\varphi^{-1}(u), \varphi^{-1}(v)) \le C|u - v|, \quad u, v \in \mathbb{B}^n.$$

Thus, $d_F(p, q) = 0$ if and only if $p = q$. We conclude that d_F is a metric on M and the manifold topology coincides with the metric topology. It is worth mentioning that the distance d_F may not be symmetric, i.e., $d_F(p, q) \ne d_F(q, p)$, unless F is reversible. If F has finite reversibility Λ, then

$$\Lambda^{-1}d_F(q, p) \le d_F(p, q) \le \Lambda d_F(q, p). \qquad (2.1.2)$$

The *diameter* of M is defined by

$$d := \sup_{p,q \in M} d_F(p, q).$$

The diameter d of M may be infinity. Similarly, for the reverse metric $\overleftarrow{F}$ of F, we can define the length function $\overleftarrow{L}_{\overleftarrow{F}}$ and the distance $\overleftarrow{d}_{\overleftarrow{F}}$, etc. We will simply put arrows $\leftarrow$ on those quantities associated with $\overleftarrow{F}$. We have $\overleftarrow{d}(p,q) = d_F(q,p)$.

Let T^*M be the cotangent bundle over M, which is the dual bundle of TM. Given a Finsler metric F, its *dual metric* $F^* : T^*M \to [0,+\infty)$ defined by

$$F^*(x,\xi) := \sup_{y \in T_x M \backslash \{0\}} \frac{\xi(y)}{F(x,y)} \qquad (2.1.3)$$

is also a Finsler metric by Proposition 1.2.1. From the proof of Proposition 1.2.1, it is easy to see that if the fundamental tensor of F is $g_y = g_{ij}(x,y)dx^i \otimes dx^j$ for any $(x,y) \in TM \backslash \{0\}$, then the fundamental tensor of F^* is given by

$$g^*_\xi = g^{*ij}(x,\xi) \frac{\partial}{\partial x^i} \otimes \frac{\partial}{\partial x^i}$$

for any $(x,\xi) \in T^*(M) \backslash \{0\}$, where $g^{*ij}(x,\xi) = g^{ij}(x,y)$, which are the components of $(g_{ij}(x,y))^{-1}$. Obviously, it induces a Riemannian inner product on each cotangent space $T^*_x(M)$. It follows from Lemma 1.2.1 that $F(x,y) = F^*(x,\mathcal{L}(y))$ for any $(x,y) \in TM_0$, where $\mathcal{L}$ is the Legendre transformation from TM to T^*M. For the sake of simplicity, we write $F(y)$ instead of $F(x,y)$ in the following.

Lemma 2.1.1. *Let (M,F) be a Finsler manifold with reversibility Λ. If the reversibility of F^* is Λ^*, then $\Lambda^* = \Lambda$.*

Proof. Let $\mathcal{L} : TM \to T^*M$ be the Legendre transformation. For any $x \in M$ and $\xi \in T^*_x M \backslash \{0\}$, set $y = \mathcal{L}_x^{-1}(-\xi)$. Then, by Lemma 1.2.1, we have

$$(-\xi)(y) = F^*(-\xi)F(y).$$

On the other hand, by the definition of F^*, we get

$$(-\xi)(y) = \xi(-y) \le F^*(\xi)F(-y).$$

Therefore, we have

$$F^*(-\xi)F(y) \le F^*(\xi)F(-y),$$

that is,

$$\frac{F^*(-\xi)}{F^*(\xi)} \le \frac{F(-y)}{F(y)} \le \Lambda.$$

It implies that $\Lambda^* \leq \Lambda$. In the same way, we have $\Lambda \leq \Lambda^*$. This completes the proof. □

Further, if there are uniform positive constants κ and κ^* such that for any $x \in M$, $v \in T_xM\backslash\{0\}$ and $y \in T_xM$, we have

$$\kappa^* F^2(x,y) \leq g_v(y,y) \leq \kappa F^2(x,y), \tag{2.1.4}$$

then we say that F satisfies *uniform smoothness* and *uniform convexity*, where κ and κ^* are respectively called the *uniform smoothness constant* and *uniform convexity constant*. Obviously, $0 < \kappa^* \leq 1 \leq \kappa$. The constant $\max\{\kappa, 1/\kappa^*\}$ is called the *uniform constant* of F. A Finsler metric F is Riemannian if and only if $\kappa = 1$ if and only if $\kappa^* = 1$ ([Oh2]). The uniform smoothness and the uniform convexity were first introduced in Banach space theory by Ball, Carlen and Lieb in [BCL].

Lemma 2.1.2. *If F satisfies uniform smoothness and uniform convexity with* (2.1.4), *then*

(1) *F has finite reversibility Λ with*

$$\Lambda \leq \min\{\sqrt{\kappa}, \sqrt{1/\kappa^*}\}. \tag{2.1.5}$$

(2) *F^* satisfies uniform smoothness and uniform convexity with*

$$\tilde{\kappa}^* F^{*2}(x,\eta) \leq g_\xi^*(\eta,\eta) \leq \tilde{\kappa} F^{*2}(x,\eta) \tag{2.1.6}$$

for any $\xi \in T_x^*M\backslash\{0\}$ and $\eta \in T_x^*(M)$, where $\tilde{\kappa}^* = 1/\kappa$ and $\tilde{\kappa} = 1/\kappa^*$.

Proof. (1) For any $(x,v) \in TM\backslash\{0\}$, observe that

$$\frac{F^2(-v)}{F^2(v)} = \frac{F^2(-v)}{g_v(v,v)} = \frac{F^2(-v)}{g_v(-v,-v)} \leq \frac{1}{\kappa^*}.$$

Similarly, we have

$$\frac{F^2(-v)}{F^2(v)} = \frac{g_{-v}(-v,-v)}{F^2(v)} = \frac{g_{-v}(v,v)}{F^2(v)} \leq \kappa.$$

Combining these two inequalities with definition of Λ yields (2.1.5).

(2) For any $(x,\xi) \in T^*M\backslash\{0\}$, there is an element $v := \mathcal{L}^{-1}(\xi)$ such that $(x,v) = \left(x, \mathcal{L}^{-1}(\xi)\right) \in TM\backslash\{0\}$ and vice versa by Lemma 1.2.1. Let $\|\cdot\|_\xi = \sqrt{g_\xi^*(\cdot,\cdot)}$ and $\|\cdot\|_v = \sqrt{g_v(\cdot,\cdot)}$ be the norms with respect to the

induced Rimannian inner products g_ξ and g_v by F^* and F, respectively. Note that $\|\cdot\|_\xi$ is the dual norm of $\|\cdot\|_v$. Then, by definition, we have

$$\|\eta\|_\xi = \sup_{y \in T_x M \setminus \{0\}} \frac{\eta(y)}{\|y\|_v} = \sup_{\|y\|_v = 1} \eta(y) \qquad (2.1.7)$$

for any $\eta \in T_x^*(M)$ and $x \in M$. Hence there exists a vector $y \in T_x M$ with $\|y\|_v = 1$ such that $\|\eta\|_\xi = \eta(y)$. By (2.1.3)–(2.1.4), one obtains that

$$\frac{g_\xi^*(\eta, \eta)}{F^{*2}(\eta)} = \frac{\eta(y)^2}{F^{*2}(\eta)} \leq F^2(y) = \frac{F^2(y)}{g_v(y, y)} \leq \frac{1}{\kappa^*},$$

which means that $g_\xi^*(\eta, \eta) \leq \frac{1}{\kappa^*} F^{*2}(\eta)$.

On the other hand, for any $\eta \in T_x^*(M)$, we take $y \in T_x M$ with $\|y\|_v = \sqrt{g_v(y, y)} = 1$ such that $\eta(y) = F(y)F^*(\eta)$. Consequently, we have $F(y)F^*(\eta) = \eta(y) \leq \|\eta\|_\xi \|y\|_v = \|\eta\|_\xi$ by (2.1.7). From this and (2.1.4), we have

$$\frac{g_\xi^*(\eta, \eta)}{F^{*2}(\eta)} = \frac{\|\eta\|_\xi^2}{F^{*2}(\eta)} \geq F^2(y) = \frac{F^2(y)}{g_v(y, y)} \geq \frac{1}{\kappa},$$

which means that $g_\xi^*(\eta, \eta) \geq \frac{1}{\kappa} F^{*2}(\eta)$. Thus (2.1.6) follows. $\qquad \square$

We end this section with the following special class of Finsler metrics, which have been extensively studied ([ChS], [CnS], [GM]).

Example 2.1.1. Let $|\cdot|$ be the standard Euclidean norm induced by the Euclidean inner product $\langle , \rangle$ on $\mathbb{R}^n$ and define

$$\alpha_0 := |y| = \sqrt{(y^1)^2 + \cdots + (y^n)^2}, \quad y \in T_x \mathbb{R}^n \cong \mathbb{R}^n. \qquad (2.1.8)$$

Then α_0 is called the *standard Euclidean metric*, which is a Riemannian metric on $\mathbb{R}^n$.

Further, we consider the unit sphere $\mathbb{S}^n$ in the standard Euclidean space $(\mathbb{R}^{n+1}, |\cdot|)$. For any $p \in \mathbb{S}^n$, we identify $T_p \mathbb{S}^n$ with a hypersurface in R^{n+1} in a natural way. Let $\mathbb{S}_+^n$ and $\mathbb{S}_-^n$ be the upper and lower hemisphere of $\mathbb{S}^n$ and let $\psi_\pm : T_p \mathbb{S}^n \cong \mathbb{R}^n \to \mathbb{S}_\pm^n$ be the stereographic projection from the center, respectively, defined by

$$\psi_\pm(x) := \left(\frac{x}{\sqrt{1 + |x|^2}}, \frac{\pm 1}{\sqrt{1 + |x|^2}} \right).$$

$\psi_\pm$ sends straight lines in $\mathbb{R}^n$ to great circles on $\mathbb{S}^n_\pm$. The pull-back metric on $\mathbb{R}^n$ to $\mathbb{S}^n_+$ by ψ_+ is given by

$$\alpha_{+1} := \frac{\sqrt{(1+|x|^2)|y|^2 - \langle x,y\rangle^2}}{1+|x|^2}, \quad y \in T_x\mathbb{R}^n \cong \mathbb{R}^n. \qquad (2.1.9)$$

The pair $(\mathbb{R}^n, \alpha_{+1})$ is called the *projective spherical model*.

Similarly, let $\mathbb{B}^n_1$ be a unit ball in $(\mathbb{R}^n, |\cdot|)$ and let

$$\alpha_{-1} := \frac{\sqrt{(1-|x|^2)|y|^2 + \langle x,y\rangle^2}}{1-|x|^2}, \quad y \in T_x\mathbb{B}^n_1 \cong \mathbb{R}^n. \qquad (2.1.10)$$

Then α_{-1} is a Riemannian metric on $\mathbb{B}^n_1$, which is called the *Klein metric*. The pair $(\mathbb{B}^n_1, \alpha_{-1})$ is called the *Klein model*.

The Riemannian metrics given by (2.1.8)–(2.1.10) can be expressed in a unified form

$$\alpha_\mu := \frac{\sqrt{(1+\mu|x|^2)|y|^2 - \mu\langle x,y\rangle^2}}{1+\mu|x|^2}, \quad y \in T_x\mathbb{B}^n_{r_\mu} \cong \mathbb{R}^n, \qquad (2.1.11)$$

where $r_\mu := 1/\sqrt{-\mu}$ if $\mu < 0$ and $r_\mu := +\infty$ if $\mu \geq 0$. It is easy to see that the Riemannian metric α_μ defined on an open set $\mathbb{B}^n_{r_\mu} \subset \mathbb{R}^n$ has constant sectional curvature μ.

Besides the Riemannian metrics, there are many non-Riemannian Finsler metrics as follows.

Example 2.1.2. Let $\alpha = \sqrt{a_{ij}(x)y^iy^j}$ be a Riemannian metric and $\beta = b_i(x)y^i$ a 1-form on an n-dimensional manifold M. Let

$$b := \|\beta\|_\alpha = \sup_{y \in T_xM\backslash\{0\}} \frac{\beta(x,y)}{\alpha(x,y)}, \quad x \in M.$$

Then $b = \sqrt{a^{ij}(x)b_ib_j}$ from Example 1.1.2. Further, when $b < 1$, $F(x,y) := \alpha(x,y) + \beta(x,y)$ defines a Finsler metric on M, which is called a *Randers metric* on M. (M, F) is called a *Randers manifold* or a *Randers space*. Randers metrics were first introduced by a physicist G. Randers when he studied general relativity ([Ran]). Randers spaces are a class of very important Finsler spaces, which have many plentiful geometric properties ([CnS], [ChS]).

Example 2.1.3. Let Ω be a bounded open domain with boundary $\partial\Omega$ in $\mathbb{R}^n$. Suppose that Ω is *convex*, i.e., any line segment joining two points in

Ω is strictly contained in Ω. Then, for any point $x_0 \in \Omega$, there is a unique functional $\varphi : \mathbb{R}^n \to [0, \infty)$ such that

$$\varphi(\lambda y) = \lambda \varphi(y), \quad \lambda > 0, \quad y \in \mathbb{R}^n$$

and

$$\varphi(z - x_0) = 1, \quad z \in \partial \Omega.$$

Ω is said to be *strongly convex* if φ is a Minkowski norm on $\mathbb{R}^n$ for some $x_0 \in \Omega$. One can show that the strong convexity of Ω is independent of the choice of a particular point $x_0 \in \Omega$ (cf. Exercise 1.2.7 in [BCS]).

For any bounded open and stongly convex domain $\Omega \subset \mathbb{R}^n$ and $0 \neq y \in T_x\Omega \cong \mathbb{R}^n$, define $F : T_x\Omega \to (0, \infty)$ such that

$$z := x + \frac{y}{F(x, y)} \in \partial \Omega. \tag{2.1.12}$$

It is easy to check that $F = F(x, y)$ is a Finsler metric on Ω, called the *Funk metric*. With this metric F, we define

$$\bar{F}(x, y) := \frac{1}{2} \left(F(x, y) + F(x, -y) \right),$$

called the *Hilbert metric* on Ω. It is a reversible Finsler metric.

In particular, when $\Omega = \mathbb{B}_1^n$ is the standard unit ball in $(\mathbb{R}^n, |\cdot|)$ centered at the origin of $\mathbb{R}^n$ and $\varphi(y) = |y|$, the Funk metric F is determined by $|x + F^{-1}(x, y)y| = 1$. From this, one obtains

$$F = \frac{\sqrt{(1 - |x|^2)|y|^2 + \langle x, y \rangle^2} + \langle x, y \rangle}{1 - |x|^2}, \tag{2.1.13}$$

which is a Randers metric with $\alpha = \alpha_{-1}$ defined by (2.1.10) and $\beta := \frac{\langle x, y \rangle}{1 - |x|^2}$ for $y \in T_x\mathbb{B}_1^n \cong \mathbb{R}^n$. From this, the Hilbert metric is given by

$$F = \frac{\sqrt{(1 - |x|^2)|y|^2 + \langle x, y \rangle^2}}{1 - |x|^2}, \tag{2.1.14}$$

which is the *Klein metric* $\alpha = \alpha_{-1}$ (see (2.1.10)).

For the Funk metric F on Ω, we have the following important property.

Proposition 2.1.1 ([Ok]). *The Funk metric F defined on a strongly convex domain $\Omega \subset \mathbb{R}^n$ satisfies $F_{x^k} = F F_{y^k}$.*

Proof. Since F is a Funk metric on Ω, by (2.1.12), we have

$$\varphi\left(x + \frac{y}{F(x,y)} - x_0\right) = 1$$

for any nonzero $y \in T_x\Omega \cong \mathbb{R}^n$. Differentiating this with respect to x^j and y^j respectively yields

$$\left(\delta_j^i - F^{-2}F_{x^j}y^i\right)\varphi_{w^i}(w) = 0, \quad \left(\delta_j^i - F^{-1}F_{y^j}y^i\right)\varphi_{w^i}(w) = 0,$$

where $w = x + F^{-1}y - x_0$. From these two equations, one obtains

$$\left(F_{x^j} - FF_{y^j}\right)\varphi_{w^i}(w)y^i = 0.$$

Observe that $v = (v^i) \in \mathbb{R}^n$ is tangent to $\partial\Omega$ if and only if $\varphi_{w^i}(w)v^i = \varphi^{-1}(w)g_w(w,v) = 0$, where g_w is the fundamental form of φ. Since y is not in $T_w\Omega$, $\varphi_{w^i}(w)y^i \neq 0$. Consequently, $F_{x^j} = FF_{y^j}$. $\qquad\square$

Example 2.1.4. Let α and β be given as in Example 2.1.2. Define

$$F := \alpha\phi(s), \quad s := \frac{\beta}{\alpha}, \tag{2.1.15}$$

where $\phi(s)$ is a positive C^∞ function on $(-b_0, b_0)$ satisfying

$$\phi - s\phi' + (b^2 - s^2)\phi'' > 0, \quad |s| \le b < b_0. \tag{2.1.16}$$

By Lemma 2.1.3 below, $F = \alpha\phi(s)$ is a Finsler metrics on M, called an (α, β)-*metric*. (α, β)-metrics were first studied by M. Matsumoto ([Ma1]) in 1972. They have many applications in physics and biology (ecology) ([AM], [BS], [RSS]).

By a direct calculation or a maple program ([ChS]), the fundamental tensor is given by

$$g_{ij} = \rho a_{ij} + \rho_0 b_i b_j + \rho_1(b_i\alpha_j + b_j\alpha_i) - s\rho_1\alpha_i\alpha_j, \tag{2.1.17}$$

where $\alpha_i = \alpha_{y^i}$ and

$$\rho = \phi^2 - s\phi\phi', \quad \rho_0 = \phi\phi'' + (\phi')^2, \quad \rho_1 = -s(\phi\phi'' + \phi'\phi') + \phi\phi'.$$

By Lemma 1.1.3, one obtains

$$\det(g_{ij}) = \phi^{n+1}(\phi - s\phi')^{n-2}\left(\phi - s\phi' + (b^2 - s^2)\phi''\right)\det(a_{ij}). \tag{2.1.18}$$

Lemma 2.1.3. *When $n \ge 2$, $F = \alpha\phi(\beta/\alpha)$ is a Finsler metric on M if and only if ϕ satisfies (2.1.16).*

Proof. Suppose that (2.1.16) is true. By taking $b = s$ in (2.1.16), we get $\phi(s) - s\phi'(s) > 0$ for $|s| < b_0$. Consider a family of functions

$$\phi_t(s) := 1 - t + t\phi(s), \quad t \in [0, 1].$$

Let $F_t = \alpha\phi_t(s)$ and $g_{ij}^t = \frac{1}{2}[F_t^2]_{y^i y^j}$. Note that

$$\phi_t - s\phi_t' = 1 - t + t(\phi - s\phi') > 0,$$
$$\phi_t - s\phi_t' + (b^2 - s^2)\phi_t'' = 1 - t + t\left(\phi - s\phi' + (b^2 - s^2)\phi''\right) > 0.$$

Thus $\det(g_{ij}^t) > 0$ by (2.1.18) for all $t \in [0, 1]$. Since (g_{ij}^0) is positive definite, (g_{ij}^t) is positive definite for any $t \in [0, 1]$. In particular, (g_{ij}^1) is positive definite. Thus, $F_1 = F$ is a Finsler metric.

Conversely, assume that $F = \alpha\phi(s)$ is a Finsler metric. Then $\phi(s) > 0$ for $|s| < b_0$. If n is even, then (2.1.18) and $\det(g_{ij}) > 0$ imply (2.1.16) for any $|s| \leq b$. If $n(\geq 2)$ is odd, then $\det(g_{ij}) > 0$ implies that $\phi(s) - s\phi'(s) \neq 0$ for $|s| \leq b < b_0$. Since $\phi(0) > 0$, we have $\phi(s) - s\phi'(s) > 0$ for $|s| \leq b$. Thus, $\phi(s) - s\phi'(s) > 0$ for $|s| < b_0$ by the arbitrariness of b. The conclusion follows from (2.1.18). $\square$

Randers metrics are special (α, β)-metrics defined by $\phi = 1 + s$. If $\phi = (1 + s)^2$, then

$$F = \frac{(\alpha + \beta)^2}{\alpha},$$

which is called a *square metric*. It is a Finsler metric if and only if $\|\beta\|_\alpha < 1$ by (2.1.16). In particular, for $\alpha = \alpha_{-1}$ given by (2.1.10) and $\beta = \frac{\langle x, y \rangle}{1 - |x|^2}$, the square metric

$$F = \frac{\left(\sqrt{(1 - |x|^2)|y|^2 + \langle x, y \rangle^2} + \langle x, y \rangle\right)^2}{(1 - |x|^2)^2 \sqrt{(1 - |x|^2)|y|^2 + \langle x, y \rangle^2}} \quad (2.1.19)$$

defined on a unit ball $\mathbb{B}^n$ in $(\mathbb{R}^n, |\cdot|)$. It is the well known *Berwald metric* ([Ber3]).

Example 2.1.5. Let α and β be given as in Example 2.1.2. Define

$$F := \alpha\phi(b^2, s), \quad b := \|\beta\|_\alpha, \quad s := \frac{\beta}{\alpha}, \quad (2.1.20)$$

where $\phi(b^2, s)$ is a positive C^∞ function satisfying

$$\phi - s\phi_2 > 0, \quad \phi - s\phi_2 + (b^2 - s^2)\phi_{22} > 0, \quad |s| \leq b < b_0 \quad (2.1.21)$$

when $n \geq 3$ or

$$\phi - s\phi_2 + (b^2 - s^2)\phi_{22} > 0, \quad |s| \leq b < b_0 \tag{2.1.22}$$

when $n = 2$, where ϕ_1 and ϕ_2 stand for the usual partial derivatives of $\phi(b^2, s)$ with respect to the first variable b^2 and the second variable s respectively. The notations ϕ_{11}, ϕ_{12}, ϕ_{22}, etc. stand for similar meaning. By similar arguments as in Example 2.1.4, $F = \alpha\phi(b^2, s)$ is a Finsler metrics on M if and only if ϕ satisfies (2.1.21). This class of Finsler metrics are called *general* (α, β)-*metrics* ([YZ]). When $\phi(b^2, s)$ is independent of b^2, general (α, β)-metrics are exactly (α, β)-metrics.

For $F = \alpha\phi(b^2, s)$, the fundamental tensors are given by

$$g_{ij} = \rho a_{ij} + \rho_0 b_i b_j + \rho_1 (b_i \alpha_j + b_j \alpha_i) - s\rho_1 \alpha_i \alpha_j, \tag{2.1.23}$$

where $\alpha_i = \alpha_{y^i}$ and

$$\rho = \phi(\phi - s\phi_2), \quad \rho_0 = \phi\phi_{22} + (\phi_2)^2, \quad \rho_1 = (\phi - s\phi_2)\phi_2 - s\phi\phi_{22}.$$

By Lemma 1.1.3, one obtains

$$\det(g_{ij}) = \phi^{n+1}(\phi - s\phi_2)^{n-2}\left(\phi - s\phi_2 + (b^2 - s^2)\phi_{22}\right)\det(a_{ij}).$$

In particular, taking $\alpha(y) = |y|$ as a standard Euclidean metric on $\mathbb{R}^n$ and $\beta = \langle x, y \rangle$ as an inner product in $\mathbb{R}^n$ for any $x \in \mathbb{R}^n$ and $y \in T_x\mathbb{R}^n \cong \mathbb{R}^n$. Then $b = \|\beta\|_\alpha = |x|$ and the general (α, β)-metric is given by

$$F = |y|\phi(|x|^2, s), \quad s = \frac{\langle x, y \rangle}{|y|},$$

which is called *spherically symmetric metrics* ([Zh], [GM]). Geometrically, spherically symmetric metrics F on a domain $\Omega \subset \mathbb{R}^n$ are invariant under any rotations in $\mathbb{R}^n$, that is, $F(\mathcal{A}x, \mathcal{A}y) = F(x, y)$, where $\mathcal{A}$ is an orthogonal transformation on $\mathbb{R}^n$. It is easy to see that both Funk metrics and Berwald metrics are spherically symmetric metrics.

Example 2.1.6 ([Br1], [Br2]). Let V be a three dimensional vector space and $V \otimes \mathbb{C}$ be its complex vector space. Take a basis $\{e_1, e_2, e_3\}$ for V and define a quadratic form on $V \otimes \mathbb{C}$ by

$$Q(u, v) = e^{i\alpha}u^1v^1 + e^{i\beta}u^2v^2 + e^{-i\alpha}u^3v^3,$$

where $u = u^i e_i$, $v = v^i e_i$ are vectors in V and $\alpha, \beta \in \mathbb{R}$. For $X \in V \backslash \{0\}$, let $[X] := \{tX | t > 0\}$. Then $S^2 := \{[X] | X \in V \backslash \{0\}\}$ is diffeomorphic to the standard sphere $\mathbb{S}^2$ in $\mathbb{R}^3$. For a vector $Y \in V$, denote by $[X, Y] \in T_{[X]}S^2$ the tangent vector to the curve $c(t) := [X + tY]$ at $t = 0$. Define $F : TS^2 \to \mathbb{R}$ by

$$F([X,Y]) := \mathrm{Re}\left\{\sqrt{\frac{Q(X,X)Q(Y,Y) - Q(X,Y)^2}{Q(X,X)^2}} - i\frac{Q(X,Y)}{Q(X,X)}\right\}, \quad (2.1.24)$$

where $\mathrm{Re}(\cdot)$ denotes the real part of a complex number. Clearly, F is well defined. Assume that $|\beta| \leq \alpha < \frac{\pi}{2}$. Then F is a family of Finsler metrics on S^2 involving the parameters α, β, which are called the *Bryant's metrics*.

Now we give an explicit expression for a special class of Bryant metrics on S^2 with $\beta = \alpha$ ([Sh1]). Take an arbitrary vector $y \in T_x\mathbb{R}^2$, let

$$X := (x, 1) \in \mathbb{R}^3, \quad Y := (y, 0) \in \mathbb{R}^3.$$

We have $Q(X, X) = e^{i\alpha}|x|^2 + e^{-i\alpha}$, $Q(X, Y) = e^{i\alpha}\langle x, y\rangle$ and $Q(Y, Y) = e^{i\alpha}|y|^2$. Hence

$$Z = \frac{Q(X,X)Q(Y,Y) - Q(X,Y)^2}{Q(X,X)^2} = \frac{e^{2\alpha i}(|x|^2|y|^2 - \langle x, y\rangle^2) + |y|^2}{e^{2\alpha i}|x|^4 + 2|x|^2 + e^{-2\alpha i}},$$

and

$$\mathrm{Re}\left\{-i\frac{Q(X,Y)}{Q(X,X)}\right\} = \frac{\sin(2\alpha)\langle x, y\rangle}{|x|^4 + 2|x|^2\cos(2\alpha) + 1}.$$

For a complex number z, the real part of $\sqrt{z}$ is given by $\mathrm{Re}(\sqrt{z}) = \sqrt{\frac{\mathrm{Re}(z)+|z|}{2}}$. Note that

$$|Z| = \frac{\sqrt{(|x|^2|y|^2 - \langle x, y\rangle^2)^2 + 2\cos(2\alpha)(|x|^2|y|^2 - \langle x, y\rangle^2)|y|^2 + |y|^4}}{|x|^4 + 2|x|^2\cos(2\alpha) + 1}$$

and

$$\mathrm{Re}(Z) = \frac{|x|^2|y|^2 - \langle x, y\rangle^2 + \cos(2\alpha)|y|^2}{|x|^4 + 2|x|^2\cos(2\alpha) + 1} + 2\left(\frac{\sin(2\alpha)\langle x, y\rangle}{|x|^4 + 2|x|^2\cos(2\alpha) + 1}\right)^2.$$

Consequently,

$$F([X,Y]) = \sqrt{\frac{\mathrm{Re}(Z) + |Z|}{2}} + \frac{\sin(2\alpha)\langle x, y\rangle}{|x|^4 + 2|x|^2\cos(2\alpha) + 1}.$$

Later on, Shen extended Bryant's metric to $\mathbb{S}^n$, which is locally expressed by

$$F = \sqrt{\frac{\sqrt{\mathcal{A}} + \mathcal{B}}{2\mathcal{D}} + \left(\frac{\mathcal{C}}{\mathcal{D}}\right)^2} + \frac{\mathcal{C}}{\mathcal{D}}, \quad (2.1.25)$$

where

$$\mathcal{A} := \left(|y|^2 \cos(2\theta) + |x|^2|y|^2 - \langle x, y \rangle^2\right)^2 + \left(|y|^2 \sin(2\theta)\right)^2;$$

$$\mathcal{B} := |y|^2 \cos(2\theta) + |x|^2|y|^2 - \langle x, y \rangle^2;$$

$$\mathcal{C} := \langle x, y \rangle \sin(2\theta); \quad \mathcal{D} := |x|^4 + 2|x|^2 \cos(2\theta) + 1,$$

where $0 < \theta < \pi/2$ ([Sh4]). Bryant's metrics are spherically symmetric metrics.

Example 2.1.7. Let m be a positive even number and define

$$F = \sqrt[m]{a_{i_1 \cdots i_m}(x) y^{i_1} \cdots y^{i_m}},$$

which is called the *m-th root metric*. It was first studied by Shimada in 1979 ([Shi]). When $m = 4$, it is called the *fourth root Finsler metric*. In the case of four dimension, the fourth root metric in the form $F = \sqrt[4]{y^1 y^2 y^3 y^4}$ is called the *Berwald–Moore metric*. This metric is singular in y and not positive definite. The Berwald–Moore metric was considered by physicists as an important subject for a possible model of space-time ([Pa]).

2.2 Cartan Tensor and Hilbert Form

2.2.1 *Fundamental Tensor and Cartan Tensor*

Let (M, F) be an n-dimensional Finsler manifold and π be a projection from the tangent bundle TM to M. We define an equivalent relation on the slit tangent bundle TM_0: $(x, y_1) \sim (x, y_2)$ if and only if $y_1 = \lambda y_2$ for some positive number λ, where $(x, y_1), (x, y_2) \in TM_0$. We denote by $(x, [y])$ the equivalent class of $(x, y) \in TM_0$. The quotient space $TM_0/\sim$ is called a *projective tangent bundle* over M, denoted by PTM. In particular, $SM := \{(x, y) \in TM | F(x, y) = 1\}$ is said to be a *projective sphere bundle* over M. Since SM is isomorphic to PTM, we have $\dim(PTM) = \dim(SM) = 2n - 1$. The fiber $S_x M$ of SM at any point $x \in M$ is given by

$$S_x M = \{y \in T_x M | F_x(y) = F(x, y) = 1\},$$

which is called the *indicatrix* of F at x, also denoted by I_x. It is a hypersurface in the Minkowski space $(T_x M, F_x)$. Note that the shapes of $S_x M$ might be different at different points $x \in M$. If F is Riemannian, then $S_x M$ is a Euclidean sphere for each $x \in M$.

Let $\pi : PTM \to M$ be a projection and π^*TM be the pull-back tangent bundle over PTM, namely, a vector bundle whose fiber over a typical point

$(x, [y]) \in PTM$ is a copy of T_xM, that is,

$$\pi^*TM|_{(x,[y])} := \{(x, [y]; v)|v \in T_xM\} \cong T_xM.$$

We have the following commutative diagram:

$$
\begin{array}{ccc}
\pi^*TM & \xrightarrow{\ \pi\ } & TM \\
\tilde{\pi} \downarrow & & \tilde{\pi} \downarrow \\
PTM \cong SM & \xrightarrow{\ \pi\ } & M
\end{array}
$$

Similarly, let T^*M be the cotangent bundle over M. It is the dual bundle of TM, whose fiber at $x \in M$ is the cotangent space T_x^*M. Consequently, the dual bundle of π^*TM is π^*T^*M, whose fiber is defined by

$$\pi^*T^*M|_{(x,[y])} = \{\omega = (x, [y]; \theta)|\theta \in T_x^*M\} \cong T_x^*M.$$

In a local coordinate system (x^i), the local frame field $\{\partial_i|_x := \frac{\partial}{\partial x^i}\}$ for TM determines a local frame field $\{\partial_i\}$ for π^*TM, where

$$\partial_i|_{(x,[y])} := (x, [y]; \partial_i|_x), \quad \partial_i|_x = \frac{\partial}{\partial x^i}.$$

This gives a linear isomorphism between $\pi^*TM|_{(x,[y])}$ and T_xM for every $y \in T_xM$. Also, we may define the pull-back tangent bundle π^*TM or cotangent bundle π^*T^*M over TM_0 via the projection $\pi : TM_0 \to M$ in the same way. In fact, we just need to use (x, y) instead of $(x, [y])$ in previous arguments.

Given a Finsler metric F on M, $F(x, y) = F(x, y^i \frac{\partial}{\partial x^i}|_x)$ is regarded as a function of $(y^i) \in \mathbb{R}^n$ at each point $x \in M$. Since $F(x, y)$ is a positively homogeneous function of degree one with respect to y, F may be regarded as a function defined on PTM or SM. The *fundamental tensor* and the *Cartan tensor* of F are given by

$$g = g_{ij}(x, y)dx^i \otimes dx^j, \quad C = C_{ijk}(x, y)dx^i \otimes dx^j \otimes dx^k,$$

where

$$g_{ij}(x, y) = \frac{1}{2}[F^2(x, y)]_{y^i y^j} = F F_{y^i y^j} + F_{y^i} F_{y^j},$$

$$C_{ijk} := \frac{1}{4}[F^2(x, y)]_{y^i y^j y^k} = \frac{1}{2}\frac{\partial g_{ij}}{\partial y^k}.$$

They are respectively $(0, 2)$-type and $(0, 3)$-type symmetric tensors defined on π^*TM, From Euler's Lemma, one obtains

$$F = F_{y^i} y^i, \quad F_{y^i y^j} y^i = 0, \quad F F_{y^i} = g_{ij}(x, y) y^j, \tag{2.2.1}$$

$$F^2(x, y) = g_{ij}(x, y) y^i y^j, \quad C_{ijk}(x, y) y^i = 0. \tag{2.2.2}$$

Obviously, F is a Riemannian metric if and only if $C = 0$, equivalently, $C_{ijk} = 0$ for all $1 \le i, j, k \le n$.

Define $I(y) = \mathrm{tr}_g C(y, \cdot, \cdot)$ for any $y \in T_x M$. In local coordinates,

$$I = I_i(x, y)dx^i, \quad I_i = g^{jk} C_{ijk}. \tag{2.2.3}$$

Obviously, $I_i y^i = 0$ and

$$I_i = \frac{1}{2} g^{jk} \frac{\partial g_{ij}}{\partial y^k} - \frac{1}{2} g^{jk} \frac{\partial g_{jk}}{\partial y^i} = \frac{\partial}{\partial y^i} \left(\log \sqrt{\det(g_{ij})} \right). \tag{2.2.4}$$

By Deicke's theorem, we have the following equivalent characterization.

Theorem 2.2.1 ([Dei]). *A Finsler metric F is Riemannian if and only if $I = 0$.*

To characterize Randers metrics, Matsumoto introduces the following $(0, 3)$-type tensor called *Matsumoto tensor* ([Ma4], [MH]).

$$\mathbf{M} = M_{ijk}(x, y)dx^i \otimes dx^j \otimes dx^k, \ M_{ijk} := C_{ijk} - \frac{1}{n+1}(I_i h_{jk} + I_j h_{ki} + I_k h_{ij}),$$

where h_{ij} are components of the angular metric $h := g - \frac{1}{F^2(x,y)} g(\cdot, y) g(\cdot, y)$. It follows from Theorem 1.3.2 that

Theorem 2.2.2 ([Ma4], [MH]). *Let (M, F) be an $n(\ge 3)$-dimensional Finsler manifold. Then F is a Randers metric if and only if the Matsumoto tensor vanishes.*

2.2.2 Hilbert Form

Let (M, F) be an n-dimensional Finsler manifold and $(x^i, y^i)_{i=1}^n$ be the local coordinates on TM. There is a *distinguished* section ℓ on $\pi^* TM$ defined by

$$\ell = \ell^i \frac{\partial}{\partial x^i} := \frac{y^i}{F} \frac{\partial}{\partial x^i}.$$

Its dual is called the *Hilbert form ω*, which is the section of $\pi^* T^* M$, that is,

$$\omega := F_{y^i} dx^i.$$

Both ℓ and ω are globally defined on the manifold TM_0. Hilbert form was first introduced by Hilbert when he formulated 23 unsolved problems in his Paris address in 1900. With the Hilbert form, the length of a piecewise C^1 curve $\gamma : [a, b] \to M$ can be formulated by

$$L_F(\gamma) = \int_\gamma \omega.$$

The right-hand side is called *Hilbert's invariant integral*. Further, one can check that

$$\omega \wedge (d\omega)^{n-1} \neq 0.$$

Hence, the Hilbert form ω induces a *contact structure* on PTM.

It is worth mentioning that $\{\frac{\partial}{\partial x^i}\}$ can not form a local frame on TM (resp., PTM) although it is a local frame on M. In fact, let $x^i = x^i$ $(\tilde{x}^1, \ldots, \tilde{x}^n)$ be a local coordinate transformation on M. Corresponding, the chain rule gives

$$y^i = \frac{\partial x^i}{\partial \tilde{x}^k} \tilde{y}^k, \quad \frac{\partial y^i}{\partial \tilde{y}^k} = \frac{\partial x^i}{\partial \tilde{x}^k}.$$

Thus, we have

$$\frac{\partial}{\partial x^i} = \frac{\partial \tilde{x}^j}{\partial x^i} \frac{\partial}{\partial \tilde{x}^j} + \frac{\partial \tilde{y}^j}{\partial x^i} \frac{\partial}{\partial \tilde{y}^j} = \frac{\partial \tilde{x}^j}{\partial x^i} \frac{\partial}{\partial \tilde{x}^j} + \frac{\partial^2 \tilde{x}^j}{\partial x^k \partial x^i} y^k \frac{\partial}{\partial \tilde{y}^j},$$

$$\frac{\partial}{\partial y^i} = \frac{\partial \tilde{x}^j}{\partial y^i} \frac{\partial}{\partial \tilde{x}^j} + \frac{\partial \tilde{y}^j}{\partial y^i} \frac{\partial}{\partial \tilde{y}^j} = \frac{\partial \tilde{x}^j}{\partial x^i} \frac{\partial}{\partial \tilde{y}^j}.$$

From this, we can see that $\{\frac{\partial}{\partial x^i}\}$ does not satisfy the transformation law of tensors, but $\left\{\frac{\partial}{\partial y^i}\right\}$ does. To remedy this case, we define the quantities

$$N^i{}_j := \gamma^i_{jk} y^k - C^i_{jk} \gamma^k_{pq} y^p y^q, \tag{2.2.5}$$

where $C^i_{jk} := g^{ip} C_{pjk}$ and

$$\gamma^i_{jk} := \frac{1}{2} g^{il} \left(\frac{\partial g_{lk}}{\partial x^j} + \frac{\partial g_{jl}}{\partial x^k} - \frac{\partial g_{jk}}{\partial x^l} \right), \tag{2.2.6}$$

which are called the *formal Christoffel symbols of the second kind* of F. $N^i{}_j$ are positively homogeneous functions of degree one, which are said to be the *nonlinear connection* on TM_0. It can be shown that

$$N^j{}_i = \frac{\partial x^j}{\partial \tilde{x}^q} \frac{\partial \tilde{x}^p}{\partial x^i} \tilde{N}^q{}_p - \frac{\partial \tilde{x}^p}{\partial x^i} \frac{\partial^2 x^j}{\partial \tilde{x}^p \partial \tilde{x}^t} \tilde{y}^t.$$

Since $\frac{\partial \tilde{x}^p}{\partial x^i} \frac{\partial x^j}{\partial \tilde{x}^p} = \delta^i_j$, we have

$$\frac{\partial}{\partial \tilde{x}^t} \left(\frac{\partial x^j}{\partial \tilde{x}^p} \right) \frac{\partial \tilde{x}^p}{\partial x^i} = -\frac{\partial x^j}{\partial \tilde{x}^p} \frac{\partial}{\partial \tilde{x}^t} \left(\frac{\partial \tilde{x}^p}{\partial x^i} \right) = -\frac{\partial x^j}{\partial \tilde{x}^p} \frac{\partial x^k}{\partial \tilde{x}^t} \frac{\partial}{\partial x^k} \left(\frac{\partial \tilde{x}^p}{\partial x^i} \right).$$

Hence,

$$N^j{}_i = \frac{\partial x^j}{\partial \tilde{x}^q} \frac{\partial \tilde{x}^p}{\partial x^i} \tilde{N}^q{}_p + \frac{\partial x^j}{\partial \tilde{x}^p} \frac{\partial x^k}{\partial \tilde{x}^t} \frac{\partial^2 \tilde{x}^p}{\partial x^i \partial x^k} \tilde{y}^t = \frac{\partial x^j}{\partial \tilde{x}^q} \frac{\partial \tilde{x}^p}{\partial x^i} \tilde{N}^q{}_p + \frac{\partial x^j}{\partial \tilde{x}^p} \frac{\partial^2 \tilde{x}^p}{\partial x^i \partial x^k} y^k.$$

From this, it is easy to check that

$$\frac{\delta}{\delta x^i} := \frac{\partial}{\partial x^i} - N^j{}_i \frac{\partial}{\partial y^j}$$

satisfy $\frac{\delta}{\delta x^i} = \frac{\partial \tilde{x}^j}{\partial x^i} \frac{\delta}{\delta \tilde{x}^j}$. Thus, $\{\frac{\delta}{\delta x^i}, \frac{\partial}{\partial y^i}\}$ forms a local frame on TM_0. Its dual frame is $\{dx^i, \delta y^i\}$, where

$$\delta y^i := dy^i + N^i{}_j dx^j.$$

On the other hand, it turns out that the manifold TM_0 has a natural Riemannian metric

$$\tilde{g} = g_{ij} dx^i \otimes dx^j + g_{ij} \frac{\delta y^i}{F} \otimes \frac{\delta y^j}{F}, \tag{2.2.7}$$

known as the *Sasaki metric* on TM_0. With respect to this metric, the *horizontal subspace* spanned by $\{\frac{\delta}{\delta x^i}\}$ is orthogonal to the *vertical subspace* spanned by $\{\frac{\partial}{\partial y^i}\}$. Consequently, the tangent bundle of the slit tangent bundle TM_0 has the following direct sum decomposition:

$$T(TM_0) = HTM \oplus VTM = \operatorname{span}\left\{\frac{\delta}{\delta x^i}\right\} \oplus \operatorname{span}\left\{\frac{\partial}{\partial y^i}\right\}. \tag{2.2.8}$$

We still use π to denote the projection from TM_0 to M. Then $d\pi : T(TM_0) \to TM$ and $\ker(d\pi) = VTM$. HTM and VTM are, respectively, called the *horizontal subbundle* and *vertical subbundle* of $T(TM_0)$, whose fibers are respectively spanned by $\{\frac{\delta}{\delta x^i}\}$ and $\{\frac{\partial}{\partial y^i}\}$. Similarly, the cotangent bundle of TM_0 has the following direct sum decomposition:

$$T^*(TM_0) = H^*TM \oplus V^*TM = \operatorname{span}\left\{dx^i\right\} \oplus \operatorname{span}\left\{\delta y^i\right\}. \tag{2.2.9}$$

Observe that

$$d(F^2(y)) = d\left(g_{ij} y^i y^j\right) = \frac{\partial g_{ij}}{\partial x^k} y^i y^j dx^k + \frac{\partial g_{ij}}{\partial y^k} y^i y^j dy^k + 2g_{ij} y^i dy^j$$

$$= \frac{\partial g_{ij}}{\partial x^k} y^i y^j dx^k + 2g_{ij} y^i dy^j. \tag{2.2.10}$$

On the other hand, by (2.2.6), we have

$$\frac{\partial g_{lk}}{\partial x^j} + \frac{\partial g_{jl}}{\partial x^k} - \frac{\partial g_{jk}}{\partial x^l} = 2g_{il} \gamma^i_{jk}. \tag{2.2.11}$$

Taking turns the indices l, k, j in (2.2.11) twice and then adding these to (2.2.11) yields

$$\frac{\partial g_{lk}}{\partial x^j} = g_{li} \gamma^i_{kj} + g_{ik} \gamma^i_{lj}. \tag{2.2.12}$$

From this and (2.2.5), we have

$$\frac{\partial g_{ij}}{\partial x^k} y^i y^j = \left(g_{il} \gamma^l_{jk} + g_{jl} \gamma^l_{ik} \right) y^i y^j = 2 g_{il} \gamma^l_{jk} y^i y^j$$

$$= 2 g_{il} \left(N^l_{k} + C^l_{kt} \gamma^t_{pq} y^p y^q \right) y^i = 2 g_{il} N^l_{k} y^i. \qquad (2.2.13)$$

Putting (2.2.13) into (2.2.10) gives $d(F^2) = 2 g_{ij} y^i \delta y^j$, that is, $dF = F_{y^j} \delta y^j \in \Gamma(V^*TM)$ (the set of sections over V^*TM). This proves the following result.

Proposition 2.2.1. *F is horizontally constant, i.e., for any horizontal vector field $X = X^i \frac{\delta}{\delta x^i} \in \Gamma(HTM)$ (the set of sections over HTM), we have $X(F) = dF(X) = 0$.*

Note that the Sasaki metric $\tilde{g}$ on TM_0 is well defined on SM. We still use $\tilde{g}$ to denote $\tilde{g}|_{SM}$. Recall that the Hilbert form $\omega = F_{y^i} dx^i$ is a 1-form on $\pi^* T^* M$. Define a 1-form on TM_0

$$\tilde{\omega} := d(\log F) = F^{-1} F_{y^i} \delta y^i.$$

It is easy to see that $\tilde{\omega}$ is a global 1-form on PTM, which is dual to the "radial vector" $y = y^i \frac{\partial}{\partial y^i}$ so that $\omega^{2n} := \tilde{\omega}$ vanishes on SM. Now, we use the following convention of index ranges:

$$1 \le i, j, \cdots \le n, \quad 1 \le a, b, \cdots \le n-1, \quad 1 \le A, B, \cdots \le 2n-1.$$

Take a positively oriented orthonormal coframe $\{\omega^i\}$ with $\omega^n = \omega$ on $\pi^* T^* M$ with respect to the Sasaki metric $\tilde{g}$ and set

$$\omega^i = v^i_j dx^j, \quad \omega^{n+i} = F^{-1} v^i_j \delta y^j, \qquad (2.2.14)$$

Then $\det(v^i_j) = \sqrt{\det(g_{ij})}$ and $v^n_j = F_{y^j}$. Thus, the collection $\{\omega^i, \omega^{n+i}\}$ forms a local orthonormal frame for $T^*(TM_0)$ and $\{\omega^i, \omega^{n+a}\}$ forms a local orthonormal frame for $T^* SM$, with respect to the Sasaki metric $\tilde{g}$. Then

$$\tilde{g} = \delta_{ij} \omega^i \otimes \omega^j + \delta_{ij} \omega^{n+i} \otimes \omega^{n+j}, \qquad (2.2.15)$$

and the restriction of $\tilde{g}$ to SM is a Riemannian metric

$$\hat{g} := \tilde{g}|_{SM} = \delta_{ij} \omega^i \otimes \omega^j + \delta_{ab} \omega^{n+a} \otimes \omega^{n+b} = \delta_{AB} \omega^A \otimes \omega^B. \qquad (2.2.16)$$

Under the local frame $\left\{\frac{\delta}{\delta x^i}, \frac{\partial}{\partial y^i}\right\}$, (2.2.16) can be rewritten as

$$\hat{g} = g_{ij}dx^i \otimes dx^j + F^{-1}F_{y^iy^j}\delta y^i \otimes \delta y^j$$
$$= g_{ij}dx^i \otimes dx^j + F_{y^iy^j}\delta y^i \otimes \delta y^j, \qquad (2.2.17)$$

where we used that $F(x,y) = 1$.

2.3 Volume Measure

2.3.1 *Volume Form*

Definition 2.3.1. Let M be an n-dimensional C^∞ orientable manifold. A *volume form* $d\mu$ on M is a collection of nondegenerate n-forms $\{d\mu_i = \sigma_i(x)dx^1 \wedge dx^2 \wedge \cdots \wedge dx^n\}$ preserving the orientation on coordinate neighborhoods $\{U_i, \varphi_i\}$ such that

(1) $M = \cup_i U_i$;
(2) when $U_i \cap U_j \neq \emptyset$, $d\mu_i = d\mu_j$ on $U_i \cap U_j$, i.e., if $d\mu_i = \sigma_i(x)dx^1 \wedge dx^2 \wedge \cdots \wedge dx^n$ and $d\mu_j = \sigma_j(\tilde{x})d\tilde{x}^1 \wedge d\tilde{x}^2 \wedge \cdots \wedge d\tilde{x}^n$, then

$$\sigma_i(x) = \det\left(\frac{\partial \tilde{x}^i}{\partial x^j}\right)\sigma_j(\tilde{x}), \quad \det\left(\frac{\partial \tilde{x}^i}{\partial x^j}\right) > 0.$$

If each $\sigma_i(x)$ is smooth, then we call $d\mu$ a smooth volume form.

By (2), we have

$$\int_{U_i \cap U_j} f d\mu_i = \int_{U_i \cap U_j} f d\mu_j, \quad f \in C_0^\infty(U_i \cap U_j),$$

where $C_0^\infty(U)$ means the set of smooth functions on an open set $U \subset M$ with a compact support. Let $\{\psi_i\}$ be a partition of unity subordinated to the covering $\{U_i\}$. Then for any function $f \in C_0^\infty(M)$, the integral

$$\int_M f d\mu = \sum_i \int_{U_i} \psi_i f d\mu_i$$

is well defined. Thus each volume form $d\mu$ defines a *volume measure* m on M, simply called the *measure* on M. A Finsler manifold (M, F, m) equipped with a smooth measure m is called a *Finsler measure space*. In particular, for a bounded open domain Ω, the volume of Ω with respect to the measure m is given by

$$m(\Omega) = \int_\Omega d\mu.$$

Consider an oriented manifold M equipped with a smooth volume form $d\mu$. We can view $d\mu$ as an n-form on M and hence for any two volume forms $d\mu_1$ and $d\mu_2$ on M, there is a positive C^∞ function φ such that $d\mu_1 = \varphi d\mu_2$. In local coordinates $(x^i)_{i=1}^n$, we can write

$$d\mu = \sigma(x)dx^1 \wedge \cdots \wedge dx^n.$$

The simplest volume form on $\mathbb{R}^n$ is the Euclidean volume form $d\mu := dx^1 \wedge \cdots \wedge dx^n$. For a Riemannian manifold (M, g), the Riemannian volume form is given by

$$d\mu_g := \sqrt{\det(g_{ij}(x))}dx^1 \wedge \cdots \wedge dx^n,$$

which is uniquely determined by the Riemannian metric g. On a Finsler manifold (M, F), the volume form can not be uniquely determined by F and it may not be formulated in an explicit expression. In Finsler geometry, there are two frequently used volume forms: *Busemann–Hausdorff volume form* and *Holmes–Thompson volume form*, which can be explicitly formulated in some special cases. These two volume forms are reduced to the Riemannian volume form when F is Riemannian.

2.3.2   *Busemann–Hausdorff Volume Form*

Let (M, F) be an n-dimensional Finsler manifold. For any $x \in M$, choose a local coordinate system $\{U, x^i\}$ such that $\{\frac{\partial}{\partial x^i}\}$ forms a basis in $T_x M$ and $\{dx^i\}$ is its dual basis. Denote

$$B_1^n(x) = \left\{(y^i) \in \mathbb{R}^n \,\middle|\, F\left(x, y^i \frac{\partial}{\partial x^i}\right) < 1\right\}.$$

Then $B_1^n(x)$ is an open and strongly convex bounded subset in $\mathbb{R}^n$. Define

$$d\mu_{BH} := \sigma_{BH}dx^1 \wedge \cdots \wedge dx^n, \quad \sigma_{BH} = \frac{\text{Vol}(\mathbb{B}_1^n)}{\text{Vol}(B_1^n(x))}, \qquad (2.3.1)$$

where $\mathbb{B}_1^n$ is the unit ball in $\mathbb{R}^n$ and Vol means the Euclidean volume. It is easy to see that $d\mu_{BH}$ is well defined and it determines a regular measure, denoted by m_{BH}. Busemann proved that m_{BH} is the Hausdorff measure of the induced metric d_F if F is reversible. $d\mu_{BH}$ is called the *Busemann-Hausdorff volume form* ([BK], [Bu]). In some references, the volume form determined by m_{BH} is also denoted by dm_{BH} instead of $d\mu_{BH}$. In general, the Euclidean volume of $B_1^n(x)$ can not be expressed by F in an explicit form. However we can explicitly formulate $d\mu_{BH}$ for some special metrics.

Example 2.3.1. Let (V, F) be an n-dimensional Minkowski space and $\{e_i\}_{i=1}^n$ be a basis of V. Then

$$B_1^n = \{(y^i) \in \mathbb{R}^n | F(y^i e_i) < 1\}$$

has constant Euclidean volume. Thus,

$$\sigma_{BH}(x) = \frac{\text{Vol}(\mathbb{B}_1^n)}{\text{Vol}(B_1^n)} = \text{constant}.$$

Consequently, the volume of B_1^n with respect to $d\mu_{BH}$ is given by

$$\text{Vol}_{BH}(B_1^n) = \int_{B_1^n} d\mu_{BH} = \int_{B_1^n} \sigma_{BH}(x) dx^1 \wedge \cdots \wedge dx^n = \text{Vol}(\mathbb{B}_1^n).$$

Example 2.3.2. Let $F = \alpha + \beta$ be a Randers metric on an n-dimensional manifold M as in Example 2.1.2 and $d\mu_\alpha$ be the Riemanninn volume form for α. To compute $d\mu_{BH}$, we choose a local orthonormal basis $\{e_i\}_{i=1}^n$ for (T_xM, α_x) such that $\beta_x(y) = by^1$, where $b := \|\beta\|_\alpha < 1$ and $y = y^i e_i \in T_xM$ for any $x \in M$. Thus, the open subset

$$B_1^n(x) = \{(y^i) \in \mathbb{R}^n | F(x, y^i e_i) < 1\} = \{(y^i) \in \mathbb{R}^n | \alpha_x(y) + \beta_x(y) < 1\}$$

$$= \left\{ (y^i) \in \mathbb{R}^n | (1 - b^2) \left(y^1 + \frac{b}{1 - b^2} \right)^2 \right.$$

$$\left. + (y^2)^2 + \cdots + (y^n)^2 < \frac{1}{1 - b^2} \right\}.$$

Let $\psi : \mathbb{R}^n \to \mathbb{R}^n$, $(y^i) \to (\tilde{y}^i)$ be a change of coordinates in $\mathbb{R}^n$ such that

$$\tilde{y}^1 = (1 - b^2) \left(y^1 + \frac{b}{1 - b^2} \right), \quad \tilde{y}^a = \sqrt{1 - b^2} y^a, \quad 2 \le a \le n.$$

Then the determinant of the Jacobi matrix is equal to $(1 - b^2)^{\frac{n+1}{2}}$ and $\psi(B_1^n(x)) = \mathbb{B}_1^n$. Consequently,

$$\text{Vol}(\mathbb{B}_1^n) = \int_{\mathbb{B}_1^n} d\tilde{y}^1 \wedge \cdots \wedge d\tilde{y}^n$$

$$= (1 - b^2)^{\frac{n+1}{2}} \int_{B_1^n(x)} dy^1 \wedge \cdots \wedge dy^n = (1 - b^2)^{\frac{n+1}{2}} \text{Vol}(B_1^n(x)),$$

which means that $\sigma_{BH}(x) = (1 - b^2)^{\frac{n+1}{2}}$ and hence

$$d\mu_{BH} = (1 - b^2)^{\frac{n+1}{2}} \omega^1 \wedge \omega^2 \wedge \cdots \wedge \omega^n = (1 - b^2)^{\frac{n+1}{2}} d\mu_\alpha, \quad (2.3.2)$$

where $\{\omega^i\}_{i=1}^n$ is the dual basis of $\{e_i\}_{i=1}^n$ in a neighborhood $U \subset M$ of x and $d\mu_\alpha$ is the volume form of α. Note that $d\mu_{BH}$ and $d\mu_\alpha$ are global n-form on M. In a local coordinate system (U, x^i), (2.3.2) may be rewritten as

$$d\mu_{BH} = (1 - b^2)^{\frac{n+1}{2}} \sqrt{\det(a_{ij}(x))} \, dx^1 \wedge dx^2 \wedge \cdots \wedge dx^n.$$

Assume that M is closed. Then

$$\mathrm{Vol}_{BH}(M) = \int_M (1 - b^2)^{\frac{n+1}{2}} d\mu_\alpha \leq \mathrm{Vol}_\alpha(M)$$

and equality holds if and only if $\beta = 0$.

2.3.3 *Holmes–Thompson Volume Form*

A Finsler metric F on a manifold M induces the Sasaki metric $\tilde{g}$ given by (2.2.7) on TM_0. Then the Riemannian volume form of $\tilde{g}$ is

$$d\mu_{TM_0} = \frac{\det(g_{ij}(x, y))}{F^n} dx \wedge \delta y = \frac{\det(g_{ij}(x, y))}{F^n} dx \wedge dy,$$

where $dx = dx^1 \wedge \cdots \wedge dx^n$ and $dy = dy^1 \wedge \cdots \wedge dy^n$. Note that $y = y^i \frac{\partial}{\partial y^i}$ is the "radial vector" of $S_x M = \{y \in T_x M | F(y) = 1\}$. The induced volume form on SM by $d\mu_{TM_0}$ is given by

$$d\mu_{SM} = y \rfloor d\mu_{TM_0} = \Omega dx \wedge d\tau,$$

where $y \rfloor d\mu_{TM_0}$ means the contraction of $d\mu_{TM_0}$ with y, that is,

$$y \rfloor d\mu_{TM_0}(\hat{X}_1, \ldots, \hat{X}_{2n-1}) := d\mu_{TM_0}(y, \hat{X}_1, \ldots, \hat{X}_{2n-1})$$

for any vector fields $\hat{X}_1, \ldots, \hat{X}_{2n-1}$ on TM_0, and

$$\Omega = \det\left(\frac{g_{ij}}{F}\right), \quad d\tau = \sum_{i=1}^n (-1)^{i-1} y^i dy^1 \wedge \cdots \wedge \hat{dy^i} \wedge \cdots \wedge dy^n,$$

where $\hat{dy^i}$ means the missing of dy^i. Actually, $d\mu_{SM}$ is just the Riemannian volume form on $(SM, \hat{g})$. Let c_{n-1} be the Euclidean area of the unit sphere $\mathbb{S}^{n-1}$ in $\mathbb{R}^n$. Define σ_{HT} by

$$\sigma_{HT}(x) := \frac{1}{c_{n-1}} \int_{S_x M} \Omega d\tau. \quad (2.3.3)$$

Then $dm_{HT} := \sigma_{HT}(x)dx$ is a well defined n-form on (M, F), which is called the *Holmes-Thompson volume form* ([Th]).

Example 2.3.3. Consider the Randers metric $F = \alpha + \beta$ in Example 2.3.2. For any $x \in M$, we identify $T_x M$ with $\mathbb{R}^n$. Let

$$S_x M = \{y = (y^i) \in \mathbb{R}^n | F(x, y) = 1\}$$

be an indicatrix in $T_x M$. Then, by (1.1.10), the volume form $d\mu_{S_x M}$ on $S_x M$ is given by

$$d\mu_{S_x M} = \Omega d\tau = \frac{F}{\alpha^{n+1}} \det(a_{ij}) d\tau.$$

Note that the right-hand side of the above equality is a homogeneous form of degree zero in y. Without loss of generality, we assume that $\alpha(y) = 1$. Then

$$d\mu_{S_x M} = F \det(a_{ij}) d\tau.$$

Denote by $A = (a_{ij})$ the metric matrix of α. Since A is positive definite, there is an orthogonal matrix P such that $A = P^T \Lambda^2 P$, where $\Lambda^2 = \mathrm{diag}(\lambda_1, \ldots, \lambda_n)$ is a matrix consisting of positive eigenvalues of A. Set $z = \Lambda P y$. Then $y \in S_x := \{y = (y^i) \in \mathbb{R}^n | \alpha(y) = 1\}$ if and only if $z \in \mathbb{S}^{n-1}$ (Euclidean unit sphere in $\mathbb{R}^n$). Moreover,

$$d\tau = \sum_{i=1}^{n} (-1)^{i-1} y^i dy^1 \wedge \cdots \wedge \hat{dy^i} \wedge \cdots \wedge dy^n$$

$$= \frac{1}{\sqrt{\det(a_{ij})}} \sum_{i=1}^{n} (-1)^{i-1} z^i dz^1 \wedge \cdots \wedge \hat{dz^i} \wedge \cdots \wedge dz^n = \frac{d\mu_{\mathbb{S}^{n-1}}}{\sqrt{\det(a_{ij})}}.$$

Hence, by (2.3.3), we have

$$\sigma_{HT}(x) = \frac{1}{c_{n-1}} \int_{S_x M} \frac{F}{\alpha^{n+1}} \det(a_{ij}) d\tau = \frac{1}{c_{n-1}} \int_{S_x} (1 + \beta) \det(a_{ij}) d\tau$$

$$= \frac{\sqrt{\det(a_{ij})}}{c_{n-1}} \int_{\mathbb{S}^{n-1}} d\mu_{\mathbb{S}^{n-1}} = \sqrt{\det(a_{ij})},$$

which implies that $dm_{HT} = d\mu_\alpha$, where we used the fact that the integral of the odd function $\beta(y) = b_i y^i$ on $\mathbb{S}^{n-1}$ is zero. If M is closed, then $\mathrm{Vol}_{HT}(M) = \mathrm{Vol}_\alpha(M)$.

2.4 Divergence Lemma

Assume that (M, F) is an n-dimensional oriented Finsler manifold with a smooth boundary ∂M. Let $i : \partial M \to M$ be an inclusion map and $F|_{\partial M} := i^* F$. It is easy to check that $(\partial M, F|_{\partial M})$ is an $(n-1)$-dimensional Finsler submanifold (hypersurface) in M.

Let $\xi \in T_x^* M$ be a unit covector such that $T_x(\partial M) = \ker(\xi)$. Obviously ξ is unique up to a sign. There is a unique unit vector $\nu_+ = \mathcal{L}^{-1}(\xi) \in T_x M$ such that $\xi(\nu_+) = 1$. Also, there is a unique vector $\nu_- = \mathcal{L}^{-1}(-\xi) \in T_x M$ such that $-\xi(\nu_-) = 1$. In general, $\nu_+ \neq -\nu_-$ unless F is reversible. ν_+ (resp., ν_-) is called the *normal vector* to ∂M. Thus, for any $x \in \partial M$, there exist exactly two normal vectors ν with $g_\nu(\nu, \nu) = 1$ such that

$$T_x(\partial M) = \{X \in T_x M | g_\nu(\nu, X) = 0\}.$$

By the definition of Legendre transformation, $\pm\xi(X) = g_{\nu_\pm}(\nu_\pm, X)$ for any $X \in T_x M$. We choose ξ so that the normal vector ν is outward pointing. Take a local oriented basis $\{e_i\}_{i=1}^n$ for TM along ∂M such that $e_n|_{\partial M} = \nu$ and $\{e_i\}_{i=1}^{n-1}$ is a local basis for $i_*(T(\partial M))$. Its dual basis is $\{\omega^i\}_{i=1}^n$. Given a smooth measure m on M, its volume form $dm = \sigma(x)\omega^1 \wedge \cdots \wedge \omega^n$ on M induces a volume form dm_ν on ∂M by

$$dm_\nu := \sigma(x)i^*(\omega^1 \wedge \cdots \wedge \omega^{n-1}).$$

Recall that for a smooth vector field X on M, the $(n-1)$-form $X \lrcorner dm$ is defined by

$$X \lrcorner dm(X_1, \ldots, X_{n-1}) := dm(X, X_1, \ldots, X_{n-1}).$$

for any vector fields $X_1, \ldots, X_{n-1}$ on M. The *divergence* of X with respect to m is defined by

$$d(X \lrcorner dm) := \mathrm{div}_F(X)dm. \tag{2.4.1}$$

It is easy to verify that

$$i^*(X \lrcorner dm) = \xi(X)dm_\nu = g_\nu(\nu, X)dm_\nu.$$

Let $dm = \sigma(x)dx$ in a local coordinate system. The divergence $\mathrm{div}_F(X)$ of a vector field $X = X^i \frac{\partial}{\partial x^i}$ is locally given by

$$\mathrm{div}_F(X) = \frac{1}{\sigma(x)} \frac{\partial}{\partial x^i} \left(\sigma(x)X^i\right). \tag{2.4.2}$$

Applying the Stokes theorem to (2.4.1), we obtain the following divergence lemma.

Lemma 2.4.1 (Divergence Lemma). *Let (M, F, m) be an oriented Finsler measure space with a smooth boundary. Let ν denote the outward pointing normal vector. Then for any smooth vector field X on M,*

$$\int_M \mathrm{div}_F(X)dm = \int_{\partial M} g_\nu(\nu, X)dm_\nu. \tag{2.4.3}$$

In fact, there is another definition of the normal vector of ∂M. Explicitly, for every $X \in \Gamma(TM)$ (the set of sections on TM), there is a unique normal vector field ν_X such that

$$g_X(\nu_X, Y) = 0, \quad g_X(\nu_X, \nu_X) = 1, \quad g_\nu(\nu, \nu_X) > 0, \tag{2.4.4}$$

where $Y \in \Gamma(T(\partial M))$. It is easy to check that $g_X(\nu, \nu_X) > 0$. Let

$$T_+^\nu M := \{Y \in \Gamma(TM) \mid g_\nu(\nu, Y) > 0\},$$

$$T_-^\nu M := \{Y \in \Gamma(TM) \mid g_\nu(\nu, Y) < 0\},$$

$$T_+^{\nu_X} M := \{Y \in \Gamma(TM) \mid g_X(\nu_X, Y) > 0\},$$

$$T_-^{\nu_X} M := \{Y \in \Gamma(TM) \mid g_X(\nu_X, Y) < 0\}.$$

Proposition 2.4.1 ([WX]). *For any $X, Y \in \Gamma(TM)$, we have*

$$g_\nu(\nu, Y) = 0 \Leftrightarrow Y \in \Gamma(T(\partial M)) \Leftrightarrow g_X(\nu_X, Y) = 0$$

and $T_\pm^\nu M = T_\pm^{\nu_X} M$.

Proof. The first claim follows from the definitions of ν and ν_X. For the second claim, it suffices to prove that $T_+^\nu M = T_+^{\nu_X} M$. Another follows in a similar way. In fact, we have either $T_+^\nu M \subset T_+^{\nu_X} M$ or $T_+^\nu M \subset T_-^{\nu_X} M$. Otherwise, there are two vector fields $Y_1, Y_2 \in T_+^\nu M$ such that $g_X(\nu_X, Y_1) > 0$ and $g_X(\nu_X, Y_2) < 0$. Then by the continuity of $g_X(\nu_X, \cdot)$ in $T_+^\nu M$, there exists $Y_0 \in T_+^\nu M$ with $g_X(\nu_X, Y_0) = 0$, which means that $g_\nu(\nu, Y_0) = 0$. This is impossible. Taking into consideration that $\nu \in T_+^{\nu_X} M$, we have $T_+^\nu M \subset T_+^{\nu_X} M$. A similar argument implies that $T_+^{\nu_X} M \subset T_+^\nu M$. The proof is finished. $\square$

Chapter 3

Connections and Structure Equations

In this chapter, we shall introduce the connections and the structure equations on Finsler manifolds. Further, we shall derive the first and second Bianchi identities from the structure equations.

3.1 Chern Connection

The connection on a Finsler manifold (M, F) means an affine connection on the pull-back bundle $\pi : \pi^*TM \to TM_0$. So far as we know, there are several ways to define the connections on a Finsler manifold (M, F), such as, the Cartan connection, the Berwald connection, the Rund connection and the Chern connection, etc. We will focus on the Chern connection introduced by Chern ([Ch1]), which is a natural generalization of the Levi-Civita connection in Riemannian case and seems to be an appropriate representative of this subject. The relationships among the Cartan connection, the Berwald connection, the Rund connection and the Chern connection will be given at the end of this section.

Let (M, F) be an n-dimensional Finsler manifold and TM_0 the slit tangent bundle with a natural projection $\pi : TM_0 \to M$. Denote by π^*TM the pull-back tangent bundle and π^*T^*M the pull-back cotangent bundle over TM_0 (see Section 2.2.1).

Theorem 3.1.1 (Chern). *Let (M, F) be an n-dimensional Finsler manifold. Then the pull-back bundle π^*TM admits a unique linear connection, called the Chern connection. For an arbitrary local frame $\{e_i\}_{i=1}^n$ for π^*TM and its dual frame $\{\omega^i\}_{i=1}^n$ for π^*T^*M, the connection 1-forms $\{\omega_i^j\}$ defined on TM_0 are characterized by the following equations:*

$$d\omega^i = \omega^j \wedge \omega_j^i, \tag{3.1.1}$$

$$dg_{ij} = g_{ik}\omega_j^k + g_{kj}\omega_i^k + 2C_{ijk}\omega^{n+k}, \tag{3.1.2}$$

where $g_{ij} = g(e_i, e_j)$, $C_{ijk} = C(e_i, e_j, e_k)$, and

$$\omega^{n+i} := dy^i + y^j \omega_j^i, \quad y = y^i e_i.$$

Proof. We prove this in a local coordinate system (x^i, y^i) in TM_0. Let $e_i = \frac{\partial}{\partial x^i}$ and $\omega^i = dx^i$. The local 1-form ω_j^i can be expressed by

$$\omega_j^i = \Gamma_{jk}^i dx^k + \Pi_{jk}^i dy^k.$$

Then (3.1.1) is equivalent to

$$0 = d^2 x^i = dx^j \wedge (\Gamma_{jk}^i dx^k + \Pi_{jk}^i dy^k),$$

which implies that $\Pi_{jk}^i = 0$ and $\Gamma_{jk}^i = \Gamma_{kj}^i$. Thus, we have $\omega_j^i = \Gamma_{jk}^i dx^k$. Plugging this into (3.1.2) yields

$$dg_{ij} = g_{ik}\Gamma_{jl}^k dx^l + g_{kj}\Gamma_{il}^k dx^l + 2C_{ijk}(dy^k + y^t\Gamma_{tl}^k dx^l).$$

Denote $\tilde{N}_j^i := y^t \Gamma_{jt}^i$. Then the above equation is equivalent to

$$\frac{\partial g_{ij}}{\partial x^l} = g_{ik}\Gamma_{jl}^k + g_{kj}\Gamma_{il}^k + 2C_{ijk}\tilde{N}_l^k. \tag{3.1.3}$$

By permutating the indices, we have

$$\frac{\partial g_{jl}}{\partial x^i} = g_{jk}\Gamma_{li}^k + g_{kl}\Gamma_{ij}^k + 2C_{jlk}\tilde{N}_i^k, \tag{3.1.4}$$

$$\frac{\partial g_{li}}{\partial x^j} = g_{lk}\Gamma_{ij}^k + g_{ki}\Gamma_{lj}^k + 2C_{lik}\tilde{N}_j^k. \tag{3.1.5}$$

Adding (3.1.4) and (3.1.5) and then subtracting (3.1.3) yields

$$\Gamma_{ij}^k = \gamma_{ij}^k - g^{kl}\left\{C_{jtl}\tilde{N}_i^t + C_{itl}\tilde{N}_j^t - C_{ijt}\tilde{N}_l^t\right\}, \tag{3.1.6}$$

where γ_{ij}^k is defined by (2.2.6). By contracting (3.1.6) with y^i, and using $C_{ijk}y^i = 0$ and (2.2.5), one obtains

$$\tilde{N}_j^k = \gamma_{ij}^k y^i - g^{kl}C_{jtl}\tilde{N}_i^t y^i = \gamma_{ij}^k y^i - g^{kl}C_{jtl}\Gamma_{ip}^t y^p y^i$$

$$= \gamma_{ij}^k y^i - g^{kl}C_{jtl}\gamma_{ip}^t y^p y^i = N_j^k.$$

Hence, $N_j^i = y^t \Gamma_{jt}^i$ and

$$\Gamma_{ij}^k = \gamma_{ij}^k - g^{kl}\left\{C_{jtl}N_i^t + C_{itl}N_j^t - C_{ijt}N_l^t\right\}. \tag{3.1.7}$$

It follows from (2.2.5)–(2.2.6) and (3.1.7) that Γ^k_{ij} (hence ω^k_j) are uniquely determined by F. In this case,

$$\omega^{n+i} = dy^i + N^i{}_j dx^j = \delta y^i.$$

This finishes the proof. $\qquad\square$

From the proof of Theorem 3.1.1, we have the following useful formulae. Note that $\tilde{N}^k{}_j = N^k{}_j$ and $C_{ijk} = \frac{1}{2}\frac{\partial g_{ij}}{\partial y^k}$. Then (3.1.3) is rewritten as

$$\frac{\partial g_{ij}}{\partial x^l} = g_{ik}\Gamma^k_{jl} + g_{kj}\Gamma^k_{il} + 2C_{ijk}N^k{}_l. \qquad (3.1.8)$$

and (3.1.7) is reformulated by

$$\Gamma^k_{ij} = \frac{1}{2}g^{kl}\left\{\frac{\delta g_{jl}}{\delta x^i} + \frac{\delta g_{il}}{\delta x^j} - \frac{\delta g_{ij}}{\delta x^l}\right\}. \qquad (3.1.9)$$

(3.1.7) also implies that

$$N^i{}_j(y) = y^k\Gamma^i_{jk} = \frac{1}{2}g^{il}\left\{\frac{\partial g_{jl}}{\partial x^k} + \frac{\partial g_{kl}}{\partial x^j} - \frac{\partial g_{jk}}{\partial x^l}\right\}y^k - g^{il}C_{jtl}N^t{}_k y^k.$$

$$(3.1.10)$$

Let

$$G^i := \frac{1}{2}N^i{}_j y^j = \frac{1}{2}\Gamma^i_{jk}y^j y^k. \qquad (3.1.11)$$

From this and (3.1.6), we have

$$G^i = \frac{1}{2}\gamma^i_{jk}y^j y^k = \frac{1}{4}g^{il}\left\{2\frac{\partial g_{jl}}{\partial x^k} - \frac{\partial g_{jk}}{\partial x^l}\right\}y^j y^k. \qquad (3.1.12)$$

Differentiating this with respect to y^j yields

$$\frac{\partial G^i}{\partial y^j} = \gamma^i_{jk}y^k - C^i_{jk}\gamma^k_{pq}y^p y^q = N^i{}_j. \qquad (3.1.13)$$

Moreover, observe that

$$[F^2]_{x^l} = \frac{\partial g_{jk}}{\partial x^l}y^j y^k, \quad [F^2]_{x^k y^l}y^k = 2\frac{\partial g_{jl}}{\partial x^k}y^j y^k.$$

Then (3.1.12) can be rewritten as

$$G^i = \frac{1}{4}g^{il}\left\{[F^2]_{x^k y^l}y^k - [F^2]_{x^l}\right\}. \qquad (3.1.14)$$

Clearly, G^i are uniquely determined by F and positively homogeneous functions of degree two with respect to y. We call G^i the *geodesic coefficients* of F, which will be appeared in the geodesic equation (see (3.1.21)). The geodesic coefficients G^i give rise to a globally defined vector field G on TM_0

$$G(x,y) := y^i \frac{\partial}{\partial x^i} - 2G^i(x,y)\frac{\partial}{\partial y^i}. \tag{3.1.15}$$

It is of C^∞ on TM_0 and C^1 at the zero section in TM. The vector field G is said to be the *spray* induced by F and G^i are also said to be the *spray coefficients* of F. In particular, for (α,β)-metrics and general (α,β)-metrics, the geodesic coefficients G^i are respectively given by Lemma 3.1.1 in [ChS] and Proposition 3.4 in [YZ] with the help of (3.1.12) or (3.1.14) and a Maple program (cf. Section A.3 in [ChS]).

Now we give an invariant description of the Chern connection. Given a non-vanishing smooth vector field Y on M, one can introduce the *weighted Riemannian metric* g_Y on M given by

$$g_Y(u,w) = g_{ij}(x,Y_x)u^i w^j, \quad \text{for} \quad u,w \in T_x M \quad \text{and} \quad x \in M. \tag{3.1.16}$$

Then $F^2(Y) = g_Y(Y,Y)$. Define a linear connection D^Y by

$$D^Y_{\frac{\partial}{\partial x^i}}\frac{\partial}{\partial x^j} := \Gamma^k_{ij}(x,Y_x)\frac{\partial}{\partial x^k}. \tag{3.1.17}$$

Then $\omega^k_i(x,Y_x) = \Gamma^k_{ij}(x,Y_x)dx^j$ define the connection 1-forms of D^Y and (3.1.1)–(3.1.2) are equivalent to the following two equations:

(1) **Torsion free:**

$$D^Y_U V - D^Y_V U = [U,V], \tag{3.1.18}$$

(2) **Almost compatibility with the metric g_Y:**

$$W(g_Y(U,V)) = g_Y(D^Y_W U, V) + g_Y(U, D^Y_W V) + 2C_Y(D^Y_W Y, U, V) \tag{3.1.19}$$

for any $U, V, W \in \Gamma(TM)$, where C_Y is the Cartan tensor of F, i.e.,

$$C_Y(U,V,W) := C_{ijk}(x,Y)U^i V^j W^k = \frac{1}{4}\frac{\partial^3 F^2(x,Y)}{\partial Y^i \partial Y^j \partial Y^k}U^i V^j W^k.$$

We also call the linear connection D^Y the *Chern connection* of F and $D^Y_U V$ the *covariant derivative* of V in the direction U with respect to the Chern connection D and the reference vector Y.

Let $\gamma : [a,b] \to M$ be a C^2 curve on (M,F) and $Y = Y(t)$ be a C^1 vector field on M along γ. We extend Y to be a C^1 vector field on M such that $Y|_\gamma = Y(t)$. We say that Y is *parallel* along γ if it satisfies $D_{\dot\gamma}^{\dot\gamma} Y = 0$. In local coordinates, $Y = Y^i(t)\frac{\partial}{\partial x^i}$ is parallel along γ if and only if

$$\dot{Y}^i(t) + Y^j(t)N^i{}_j(\gamma(t), \dot\gamma(t)) = 0, \quad 1 \le i \le n, \qquad (3.1.20)$$

which are a system of ordinary differential equations in Y^i. For any point $x \in M$ and an initial vector $y \in T_x M$, there exists a unique local parallel vector field Y on M such that $Y(x) = y$. Thus, we obtain a map

$$P_\gamma : T_{\gamma(a)}M \to T_{\gamma(b)}M, \quad P_\gamma(y) = Y_b, \quad y \in T_{\gamma(a)}M$$

when $\gamma(b)$ is sufficiently close to $\gamma(a)$. We call P_γ the *parallel translation* along γ. In particular, if the tangent vector $\dot\gamma$ of a smooth curve γ is parallel along itself, i.e., $D_{\dot\gamma}^{\dot\gamma}\dot\gamma = 0$, then γ is said to be a *geodesic* on M. Locally, the geodesic equation can be written as

$$\ddot\gamma(t) + 2G^i(\gamma(t), \dot\gamma(t)) = 0. \qquad (3.1.21)$$

By the theory of ODE, there is a unique geodesic $\gamma : (-\varepsilon, \varepsilon) \to M$ with $\gamma(0) = x$ and $\dot\gamma(0) = y$ for some sufficiently small $\varepsilon > 0$. Geodesics are uniquely determined by the Finsler metric and independent of the choice of connections.

Let $U = U^i(t)\frac{\partial}{\partial x^i}$ and $V = V^i(t)\frac{\partial}{\partial x^i}$ be two parallel vector fields along the geodesic γ with $U(a) = u, V(a) = v \in T_{\gamma(a)}M$. Then

$$\dot{U}^i(t) + U^j(t)N^i{}_j(\gamma(t), \dot\gamma(t)) = 0, \quad \dot{V}^i(t) + V^j(t)N^i{}_j(\gamma(t), \dot\gamma(t)) = 0.$$
$$(3.1.22)$$

From these and (3.1.8), a direct calculation shows that

$$\frac{d}{dt}\left(g_{\dot\gamma}(U(t), V(t))\right) = \frac{d}{dt}\left(g_{ij}(\dot\gamma(t))U^i(t)V^j(t)\right) = 0. \qquad (3.1.23)$$

Consequently, $g_{\dot\gamma(b)}(P_\gamma(u), P_\gamma(v)) = g_{\dot\gamma(a)}(u, v)$. This proves the following result.

Proposition 3.1.1. *Let $\gamma : [a, b] \to M$ be a geodesic on a Finsler space (M, F). Then the parallel translation P_γ preserves the inner product $g_{\dot\gamma}$ along γ.*

Proposition 3.1.1 shows that the parallel translation P_γ is a C^∞ diffeomorphism from $T_{\gamma(a)}M \backslash \{0\}$ to $T_{\gamma(b)}M \backslash \{0\}$. In general, it is not linear and does not preserve the Finsler metric F. However, for a Berwald space

(M, F), it is a linear isometry from $(T_{\gamma(a)}M, F_{\gamma(a)})$ to $(T_{\gamma(b)}M, F_{\gamma(b)})$ (cf. Proposition 4.3.2 in [ChS]).

In some references, the Chern connection is also called the *Rund connection* because it was found by Rund independently ([Run]). Besides this, there are other connections, for example,

(1) **Cartan connection**: Its connection forms are given by $^{c}\omega^i_j = \omega^i_j + C^i_{jk}\delta y^k$, where ω^i_j are the Chern connection forms and $C^i_{jk} = g^{il}C_{jkl}$. This connection has nonzero torsion but is compatible with the metric g.

(2) **Berwald connection**: Its connection forms are given by $^{b}\omega^i_j = \omega^i_j + \dot{A}^i_{\ jk}dx^k$, where ω^i_j are the Chern connection forms and

$$A^i_{jk} := FC^i_{jk}, \quad \dot{A}^i_{jk} := A^i_{jk|m}\ell^m,$$

$$A^i_{jk|m} := \frac{\delta A^i_{jk}}{\delta x^m} + A^p_{jk}\Gamma^i_{pm} - A^i_{pk}\Gamma^p_{jm} - A^i_{jp}\Gamma^p_{km}.$$

The Berwald connection is torsion free but is not compatible with the metric g. Its connection coefficients are given by

$$B^i_{jk} = \Gamma^i_{jk} + \dot{A}^i_{jk},$$

where Γ^i_{jk} are the Chern connection coefficients. It is easy to see that $B^i_{jk} = \frac{1}{2}[G^i]_{y^j y^k}$.

When F is Riemannian, all these connections are reduced to the Riemannian connection (i.e., Levi-Civita connection).

3.2 Structure Equations

Let (M, F) be an n-dimensional Finsler manifold and $\{e_i\}_{i=1}^n$ be a local frame for π^*TM with dual coframe $\{\omega^i\}_{i=1}^n$ for π^*T^*M. According to Theorem 3.1.1, the Chern connection forms $\{\omega^i_j\}$ with respect to $\{e_i\}$ are uniquely determined by (3.1.1)–(3.1.2). It is clear that $\{\omega^i, \omega^{n+i}\}$ is a local coframe for $T^*(TM_0)$. The *curvature form* Ω^i_j are defined by

$$\Omega^i_j := d\omega^i_j - \omega^k_j \wedge \omega^i_k. \tag{3.2.1}$$

Let $\Omega(U, V)W := \Omega^i_j(U, V)W^j e_i$, where $U = U^i e_i, V = V^i e_i$, and $W = W^i e_i$ are sections on π^*TM. For any non-vanishing smooth vector field Y on M with $Y_x = y \in T_xM$, (3.2.1) is equivalent to

$$\Omega(U, V)W = D^Y_U D^Y_V W - D^Y_V D^Y_U W - D^Y_{[U,V]}W. \tag{3.2.2}$$

Differentiating (3.1.1) yields

$$d^2\omega^i = d\omega^j \wedge \omega^i_j - \omega^j \wedge d\omega^i_j$$
$$= \omega^k \wedge \omega^j_k \wedge \omega^i_j - \omega^j \wedge (\Omega^i_j + \omega^k_j \wedge \omega^i_k) = -\omega^j \wedge \Omega^i_j.$$

Since $d^2\omega^i = 0$, we have

$$\omega^j \wedge \Omega^i_j = 0, \qquad (3.2.3)$$

which is called the *first Bianchi identity*. Since the Ω^i_j are 2-forms, they can be expressed in terms of $\omega^i \wedge \omega^j$, $\omega^i \wedge \omega^{n+j}$ and $\omega^{n+i} \wedge \omega^{n+j}$. However, by (3.2.3) and Cartan's Lemma, Ω^i_j are expressed by

$$\Omega^i_j = \frac{1}{2} R_j{}^i{}_{kl} \omega^k \wedge \omega^l + P_j{}^i{}_{kl} \omega^k \wedge \omega^{n+l}, \qquad (3.2.4)$$

where $R_j{}^i{}_{kl} = -R_j{}^i{}_{lk}$. This implies that $R_Y(U,V)W = \Omega(U,V)W$, where $R_Y(U,V)W := R_j{}^i{}_{kl}(Y)U^k V^l W^j e_i$. Consequently, by (3.2.2)

$$R_Y(U,V)W = D^Y_U D^Y_V W - D^Y_V D^Y_U W - D^Y_{[U,V]} W. \qquad (3.2.5)$$

We call R the *Riemannian curvature tensor* of F. Plugging (3.2.4) into (3.2.3) yields

$$R_j{}^i{}_{kl} + R_k{}^i{}_{lj} + R_l{}^i{}_{jk} = 0, \qquad P_j{}^i{}_{kl} = P_k{}^i{}_{jl}. \qquad (3.2.6)$$

We also call (3.2.6) the *first Bianchi identity*. Let

$$\Omega^i := d\omega^{n+i} - \omega^{n+j} \wedge \omega^i_j.$$

Differentiating $\omega^{n+i} = dy^i + y^j \omega^i_j$ gives

$$d\omega^{n+i} = dy^j \wedge \omega^i_j + y^j d\omega^i_j$$
$$= (\omega^{n+j} - y^k \omega^j_k) \wedge \omega^i_j + y^j \omega^k_j \wedge \omega^i_k + y^j \Omega^i_j$$
$$= \omega^{n+j} \wedge \omega^i_j + y^j \Omega^i_j.$$

Thus, one obtains

$$\Omega^i = y^j \Omega^i_j. \qquad (3.2.7)$$

From this and (3.2.4), we can express Ω^i in the form

$$\Omega^i = \frac{1}{2} R^i{}_{kl} \omega^k \wedge \omega^l - L^i{}_{kl} \omega^k \wedge \omega^{n+l}, \qquad (3.2.8)$$

where $R^i_{\ kl} := y^j R_j{}^i{}_{kl}$ with $R^i_{\ kl} = -R^i_{\ lk}$ and $L^i_{\ kl} := -y^j P_j{}^i{}_{kl}$. Thus, one obtains the *structure equations*:

$$\begin{cases} d\omega^i = \omega^j \wedge \omega^i_j, \\ d\omega^i_j = \omega^k_j \wedge \omega^i_k + \Omega^i_j, \\ dw^{n+i} = \omega^{n+j} \wedge \omega^i_j + \Omega^i, \end{cases} \qquad (3.2.9)$$

where Ω^i_j and Ω^i are defined by (3.2.4) and (3.2.8), respectively.

In a local coordinate system (x^i, y^i) in TM_0, the associated local coframe $\{\omega^i, \omega^i_j\}$ on $T^*(TM_0)$ are given by $\omega^i = dx^i$ and $\omega^{n+i} = \delta y^i = dy^i + N^i_{\ j}dx^j$. Since $\omega^i_j = \Gamma^i_{jk}dx^k$, $(3.2.9)_1$ implies that $\Gamma^i_{jk} = \Gamma^i_{kj}$. From this and $(3.2.9)_2$, we have

$$\Omega^i_j = \left(\frac{\delta \Gamma^i_{jl}}{\delta x^k} + \Gamma^t_{jl}\Gamma^i_{tk} \right) dx^k \wedge dx^l + \frac{\partial \Gamma^i_{jk}}{\partial y^l} \delta y^l \wedge dx^k.$$

Comparing this with (3.2.4) yields

$$R_j{}^i{}_{kl} = \frac{\delta \Gamma^i_{jl}}{\delta x^k} - \frac{\delta \Gamma^i_{jk}}{\delta x^l} + \Gamma^t_{jl}\Gamma^i_{tk} - \Gamma^t_{jk}\Gamma^i_{lt}, \qquad (3.2.10)$$

$$P_j{}^i{}_{kl} = -\frac{\partial \Gamma^i_{jk}}{\partial y^l}. \qquad (3.2.11)$$

Consequently, by (3.2.11) and (3.1.13),

$$L^i_{\ kl} = -y^j P_j{}^i{}_{kl} = \frac{\partial N^i_{\ k}}{\partial y^l} - \Gamma^i_{kl} = [G^i]_{y^k y^l} - \Gamma^i_{kl}, \qquad (3.2.12)$$

which means that $L^i_{\ kl} = L^i_{\ lk}$ and $y^l L^i_{\ kl} = 0$. Let $L_{ijk} := g_{it} L^t_{\ jk} = -y^t P_{tijk}$, where $P_{tijk} = g_{il} P_t{}^l{}_{jk}$. Then, by (3.1.7) and (2.2.6), we have

$$g_{il}\Gamma^l_{tj} = \frac{1}{2} \left(\frac{\partial g_{ti}}{\partial x^j} + \frac{\partial g_{ij}}{\partial x^t} - \frac{\partial g_{tj}}{\partial x^i} \right) - C_{jpi}N^p_{\ t} - C_{tpi}N^p_{\ j} + C_{tjp}N^p_{\ i}.$$

From this, one obtains

$$L_{ijk} = y^t g_{il} \frac{\partial \Gamma^l_{tj}}{\partial y^k} = y^t \left(\frac{\partial \left(g_{il}\Gamma^l_{tj} \right)}{\partial y^k} - 2C_{ilk}\Gamma^l_{tj} \right)$$

$$= \left(\frac{\delta C_{ijk}}{\delta x^t} - C_{pjk}\Gamma^p_{it} - C_{ipk}\Gamma^p_{jt} - C_{ijp}\Gamma^p_{kt} \right) y^t = C_{ijk|t}y^t. \qquad (3.2.13)$$

Obviously, L_{ijk} are symmetric with respect to the indices i, j, k and

$$L_{ijk}y^i = L_{ijk}y^j = L_{ijk}y^k = 0. \qquad (3.2.14)$$

We call $P := P^i_{j\ kl} \frac{\partial}{\partial x^i} \otimes dx^j \otimes dx^k \otimes dx^l$ and $L := L_{ijk} dx^i \otimes dx^j \otimes dx^k$ the *Chern curvature tensor* and the *Landsberg curvature tensor* of F, respectively. Note that $\frac{\delta y^j}{\delta x^k} = -N^j_{\ k}$ and $N^i_{\ j} y^j = 2G^i$ by (3.1.11). It follows from (3.2.10) that

$$
\begin{aligned}
R^i_{\ kl} = y^j R_j{}^i{}_{kl} &= \frac{\delta N^i_{\ l}}{\delta x^k} - \frac{\delta N^i_{\ k}}{\delta x^l} \\
&= \frac{\partial N^i_{\ l}}{\partial x^k} - \frac{\partial N^i_{\ k}}{\partial x^l} + N^t_{\ l} \frac{\partial N^i_{\ k}}{\partial y^t} - N^t_{\ k} \frac{\partial N^i_{\ l}}{\partial y^t}.
\end{aligned}
\tag{3.2.15}
$$

Set

$$
R^i_{\ k} := R^i_{\ kl} y^l = y^j R_j{}^i{}_{kl} y^l. \tag{3.2.16}
$$

Then

$$
\begin{aligned}
R^i_{\ k} &= \left\{ \frac{\partial N^i_{\ l}}{\partial x^k} - N^t_{\ k} \frac{\partial N^i_{\ l}}{\partial y^t} - \frac{\partial N^i_{\ k}}{\partial x^l} + N^t_{\ l} \frac{\partial N^i_{\ k}}{\partial y^t} \right\} y^l \\
&= \frac{\partial(y^l N^i_{\ l})}{\partial x^k} - N^t_{\ k} \frac{\partial(y^l N^i_{\ l})}{\partial y^t} + N^i_{\ l} N^l_{\ k} - y^l \frac{\partial N^i_{\ k}}{\partial x^l} + 2G^t \frac{\partial N^i_{\ k}}{\partial y^t} \\
&= 2\frac{\partial G^i}{\partial x^k} - y^l \frac{\partial^2 G^i}{\partial x^l \partial y^k} + 2G^j \frac{\partial^2 G^i}{\partial y^k \partial y^j} - \frac{\partial G^i}{\partial y^j} \frac{\partial G^j}{\partial y^k}.
\end{aligned}
\tag{3.2.17}
$$

By (3.2.15) and (3.2.17), one can verify that

$$
R^i_{\ kl} = \frac{1}{3} \left(\frac{\partial R^i_{\ k}}{\partial y^l} - \frac{\partial R^i_{\ l}}{\partial y^k} \right). \tag{3.2.18}
$$

Thus, $R^i_{\ k}$ and $R^i_{\ kl}$ can be determined each other by vertical differentiation. The tensors $R^i_{\ k}$, $R^i_{\ kl}$ and $R_j{}^i{}_{kl}$ only depend on the metric F but independent of the connections. They are all called *Riemann curvature tensors* if there is no confusion.

3.3 Bianchi Identities and Ricci Identities

In this section, we shall employ the covariant differentiation to derive the relations among curvatures and then give some Ricci identities for the covariant differentiations of tensors on the slit tangent bundle TM_0.

Let $\{e_i\}_{i=1}^n$ be a local frame for $\pi^* TM$ and $\{\omega^i\}_{i=1}^n$ be the dual coframe for $\pi^* T^* M$. With respect to $\{e_i\}$, the Chern connection forms are denoted by $\{\omega^i_j\}$ and $\{\omega^i, \omega^{n+i}\}$ form a local coframe for $T^*(TM_0)$. For any tensor field on TM_0, we can define its covariant differentiations with respect to the Chern connection in a standard way. Because of the direct sum decomposition of $T^*(TM_0)$ (see (2.2.8)–(2.2.9)), the covariant differentiation of a tensor should be decomposed into the horizontal part and the vertical part. For example, for a scalar function f on TM_0, we define the covariant differentiation for f by

$$\nabla f := df = f_{|k}\omega^k + f_{\cdot k}\omega^{n+k}, \tag{3.3.1}$$

where $f_{|k}$ (resp., $f_{\cdot k}$) stand for the horizontal (resp., vertical) covariant derivatives of f. In a similar way, given a $(1,1)$-type tensor field $T = T^i_j e_i \otimes \omega^j$, define its covariant differentiation ∇T by

$$\nabla T = \nabla T^i_j e_i \otimes \omega^j,$$

where ∇T^i_j are defined by

$$\nabla T^i_j := dT^i_j + T^k_j \omega^i_k - T^i_k \omega^k_j = T^i_{j|k}\omega^k + T^i_{j\cdot k}\omega^{n+k}, \tag{3.3.2}$$

where $T^i_{j|k}$ (resp., $T^i_{j\cdot k}$) stand for the horizontal (resp., vertical) covariant derivatives of T^i_j. In a local coordinate system (x^i, y^i), for a scalar function f on TM_0, we have

$$f_{|k} = \frac{\delta f}{\delta x^k}, \quad f_{\cdot k} = \frac{\partial f}{\partial y^k}. \tag{3.3.3}$$

For the $(1,1)$-type tensor $T = T^i_j e_i \otimes \omega^j$, $T^i_{j|k}$ and $T^i_{j\cdot k}$ are given by

$$T^i_{j|k} = \frac{\delta T^i_j}{\delta x^k} + T^t_j \Gamma^i_{tk} - T^i_t \Gamma^t_{jk}, \quad T^i_{j\cdot k} = \frac{\partial T^i_j}{\partial y^k}. \tag{3.3.4}$$

The covariant differentiation satisfies the product rule. For example, for $T = T^i_j e_i \otimes \omega^j$ and $\Phi = \Phi_{ij}\omega^i \otimes \omega^j$, we have

$$(T^i_j \Phi_{kl})_{|p} = T^i_{j|p}\Phi_{kl} + T^i_j \Phi_{kl|p}.$$

For the fundamental tensor $g = g_{ij}\omega^i \otimes \omega^j$, the horizontal covariant derivatives of g_{ij} with respect to the Chern connection are given by

$$g_{ij|k} = \frac{\delta g_{ij}}{\delta x^k} - g_{il}\Gamma^l_{jk} - g_{lj}\Gamma^l_{ik}. \tag{3.3.5}$$

By the definition of $\frac{\delta}{\delta x^k}$, we have $\frac{\delta g_{ij}}{\delta x^k} = \frac{\partial g_{ij}}{\partial x^k} - 2C_{ijl}N^l_k$. Hence, (3.1.1)–(3.1.2) are equivalent to

$$\Gamma^k_{ij} = \Gamma^k_{ji}, \quad g_{ij|k} = 0, \quad g_{ij\cdot k} = 2C_{ijk}, \tag{3.3.6}$$

which show that the Chern connection is torsion free and almost compatible with g. From (3.3.6) and $y^i_{|k} = 0$ for any $y = y^i e_i$, it is easy to check that F is a horizontal constant, i.e., $F_{|k} = 0$. This gives another proof of Proposition 2.2.1.

Recall that the *first Bianchi identity*

$$\omega^j \wedge \Omega_j^i = 0,$$

which is derived by differentiating (3.1.1). This is equivalent to (3.2.6). Now differentiating (3.1.2), and using it again and (3.2.9), we get

$$g_{ik}\Omega_j^k + g_{kj}\Omega_i^k + 2C_{ijk}\Omega^k + 2\left(C_{ijk|l}\omega^l + C_{ijk\cdot l}\omega^{n+l}\right) \wedge \omega^{n+k} = 0.$$

Plugging (3.2.4) and (3.2.8) into the above equation and using $C_{ijk\cdot l} = C_{ijl\cdot k}$ yield

$$\frac{1}{2}\left(R_{ijkl} + R_{jikl} + 2C_{ijt}R^t{}_{kl}\right)\omega^k \wedge \omega^l$$

$$+ \left(P_{ijkl} + P_{jikl} + 2C_{ijl|k} - 2C_{ijt}L^t{}_{kl}\right)\omega^k \wedge \omega^{n+l} = 0,$$

where $R_{ijkl} := g_{jt}R_i{}^t{}_{kl}$ and $P_{ijkl} := g_{jt}P_i{}^t{}_{kl}$. Consequently, one obtains

$$R_{ijkl} + R_{jikl} = -2C_{ijt}R^t{}_{kl}, \quad P_{ijkl} + P_{jikl} = -2C_{ijl|k} + 2C_{ijt}L^t{}_{kl}.$$

$$(3.3.7)$$

Contracting the first equation of (3.3.7) with y^i, y^j yields $y_t R^t{}_{kl} = 0$. On the other hand, $R_{ijkl} = -R_{ijlk}$ and (3.2.6) is equivalent to

$$R_{ijkl} + R_{kjli} + R_{ljik} = 0, \quad P_{ijkl} = P_{kjil}. \tag{3.3.8}$$

Hence, we have

$$2(R_{klij} - R_{ijkl})$$

$$= (R_{klij} + R_{lkij}) - (R_{ijkl} + R_{jikl}) + R_{klij} - R_{ijkl} - R_{lkij} + R_{jikl}$$

$$= (R_{klij} + R_{lkij}) - (R_{ijkl} + R_{jikl}) + R_{klij} + R_{kjli} + R_{ljik} + R_{ikjl}$$

$$\quad + R_{jkli} + R_{jikl}$$

$$= (R_{klij} + R_{lkij}) - (R_{ijkl} + R_{jikl}) + (R_{kjli} + R_{jkli}) + (R_{ljik} + R_{jlik})$$

$$\quad + R_{jlki} + R_{klij} + R_{ikjl} + R_{jikl}$$

$$= (R_{klij} + R_{lkij}) - (R_{ijkl} + R_{jikl}) + (R_{kjli} + R_{jkli}) + (R_{ljik} + R_{jlik})$$

$$\quad + (R_{likj} + R_{ilkj}) + (R_{ikjl} + R_{kijl}).$$

Plugging the first equation of (3.3.7) into the above identity leads to

$$R_{klij} - R_{ijkl} = -C_{klt}R^t{}_{ij} + C_{ijt}R^t{}_{kl} - C_{kjt}R^t{}_{li} - C_{lit}R^t{}_{kj}$$
$$-C_{jlt}R^t{}_{ik} - C_{ikt}R^t{}_{jl}. \tag{3.3.9}$$

Let $R_{ij} := g_{ik}R^k{}_j$. Contracting (3.3.9) with y^i and y^k respectively yields $R_{ij} = R_{ji}$.

Further, differentiating $\Omega^i_j = d\omega^i_j - \omega^k_j \wedge \omega^i_k$ and using it again yield

$$d\Omega^i_j = -\Omega^k_j \wedge \omega^i_k + \omega^k_j \wedge \Omega^i_k, \tag{3.3.10}$$

called the *second Bianchi identity*. By inserting (3.2.4) into (3.3.10), one obtains

$$\frac{1}{2}\left(R_j{}^i{}_{kl|t} - P_j{}^i{}_{ks}R^s{}_{lt}\right)\omega^k \wedge \omega^l \wedge \omega^t$$
$$+\frac{1}{2}\left(R_j{}^i{}_{kl\cdot t} - 2P_j{}^i{}_{kt|l} + 2P_j{}^i{}_{ks}L^s{}_{lt}\right)\omega^k \wedge \omega^l \wedge \omega^{n+t}$$
$$+P_j{}^i{}_{kl\cdot t}\omega^k \wedge \omega^{n+l} \wedge \omega^{n+t} = 0,$$

which implies that

$$R_j{}^i{}_{kl|t} + R_j{}^i{}_{lt|k} + R_j{}^i{}_{tk|l} = P_j{}^i{}_{ks}R^s{}_{lt} + P_j{}^i{}_{ls}R^s{}_{tk} + P_j{}^i{}_{ts}R^s{}_{kl}, \tag{3.3.11}$$
$$R_j{}^i{}_{kl\cdot t} = P_j{}^i{}_{kt|l} - P_j{}^i{}_{lt|k} - P_j{}^i{}_{ks}L^s{}_{lt} + P_j{}^i{}_{ls}L^s{}_{kt}, \quad P_j{}^i{}_{kl\cdot t} = P_j{}^i{}_{kt\cdot l}. \tag{3.3.12}$$

Contracting (3.3.11) and (3.3.12) with y^j, respectively, leads to

$$R^i{}_{kl|t} + R^i{}_{lt|k} + R^i{}_{tk|l} = -L^i{}_{ks}R^s{}_{lt} - L^i{}_{ls}R^s{}_{tk} - L^i{}_{ts}R^s{}_{kl}, \tag{3.3.13}$$
$$R_t{}^i{}_{kl} = R^i{}_{kl\cdot t} + L^i{}_{kt|l} - L^i{}_{lt|k} - L^i{}_{ks}L^s{}_{lt} + L^i{}_{ls}L^s{}_{kt}. \tag{3.3.14}$$

(3.3.11)–(3.3.12) or (3.3.13)–(3.3.14) are also called the *second Bianchi identity*. Contracting (3.3.13) with y^l yields

$$R^i{}_{k|t} - R^i{}_{t|k} + R^i{}_{tk|l}y^l = R^s{}_tL^i{}_{ks} - R^s{}_kL^i{}_{ts}. \tag{3.3.15}$$

Finally, we give some Ricci identities for the covariant differentiations of tensors on TM_0. We still take the scalar function f and the $(1,1)$-type tensor T on TM_0 as examples given in the beginning of this section. For a scalar function f on TM_0, by differentiation (3.3.1) and using the structure equations, one obtains

$$\left(df_{|k} - f_{|j}\omega_k^j + \frac{1}{2}f_{\cdot j}R^j{}_{lk}\omega^l\right) \wedge \omega^k + \left(df_{\cdot k} - f_{\cdot j}\omega_k^j - f_{\cdot j}L^j{}_{lk}\omega^l\right) \wedge \omega^{n+k} = 0.$$

$$(3.3.16)$$

Define the covariant differentiations of $f_{|k}$ and $f_{\cdot k}$ by

$$\nabla f_{|k} := df_{|k} - f_{|j}\omega_k^j = f_{|k|l}\omega^l + f_{|k\cdot l}\omega^{n+l},$$

$$\nabla f_{\cdot k} := d(f_{\cdot k}) - f_{\cdot j}\omega_k^j = f_{\cdot k|l}\omega^l + f_{\cdot k\cdot l}\omega^{n+l}.$$

Inserting these into (3.3.16) yields the Ricci identities for f:

$$f_{|k|l} - f_{|l|k} = f_{\cdot j}R^j{}_{kl}, \quad f_{\cdot k|l} - f_{|l\cdot k} = f_{\cdot j}L^j{}_{lk}, \quad f_{\cdot k\cdot l} = f_{\cdot l\cdot k}. \quad (3.3.17)$$

Obviously, the last identity is trivial. Similarly, for the $(1,1)$-type tensor T, differentiating (3.3.2), and using the structure equations and (3.3.2) again give

$$\left(dT^i_{j|k} - T^i_{j|l}\omega_k^l - T^i_{l|k}\omega_j^l + T^l_{j|k}\omega_l^i\right) \wedge \omega^k$$

$$+ \left(dT^i_{j\cdot k} - T^i_{j\cdot l}\omega_k^l - T^i_{l\cdot k}\omega_j^l + T^l_{j\cdot k}\omega_l^i\right) \wedge \omega^{n+k}$$

$$= \frac{1}{2}\left(T^k_j R_k{}^i{}_{st} - T^i_k R_j{}^k{}_{st} - T^i_{j\cdot k}R^k{}_{st}\right)\omega^s \wedge \omega^t$$

$$+ \left(T^k_j P_k{}^i{}_{st} - T^i_k P_j{}^k{}_{st} + T^i_{j\cdot k}L^k{}_{st}\right)\omega^s \wedge \omega^{n+t}. \quad (3.3.18)$$

Define the covariant differentiations of $T^i{}_{j|k}$ and $T^i{}_{j\cdot k}$ by

$$\nabla T^i_{j|k} := dT^i_{j|k} - T^i_{j|l}\omega_k^l - T^i_{l|k}\omega_j^l + T^l_{j|k}\omega_l^i = T^i_{j|k|l}\omega^l + T^i_{j|k\cdot l}\omega^{n+l},$$

$$\nabla T^i_{j\cdot k} := dT^i_{j\cdot k} - T^i_{j\cdot l}\omega_k^l - T^i_{l\cdot k}\omega_j^l + T^l_{j\cdot k}\omega_l^i = T^i_{j\cdot k|l}\omega^l + T^i_{j\cdot k\cdot l}\omega^{n+l}.$$

By substituting these in (3.3.18), one obtains the following Ricci identities:

$$T^i_{j|k|l} - T^i_{j|l|k} = T^i_s R_j{}^s{}_{kl} - T^s_j R_s{}^i{}_{kl} + T^i_{j\cdot s}R^s{}_{kl},$$

$$T^i_{j|k\cdot l} - T^i_{j\cdot l|k} = T^i_s P_j{}^s{}_{kl} - T^s_j P_s{}^i{}_{kl} - T^i_{j\cdot s}L^s{}_{kl}.$$

In the same way, we also have the Ricci identities for the covariant derivatives of any (r,s)-type tensor on TM_0 with respect to the Chern connection.

Chapter 4

Curvature Invariant Quantities

In Riemannian geometry, the Riemann curvature tensor is the most important geometric invariant quantity that measures the flexibility of the space at a point. It is firstly extended by Berwald to Finsler geometry ([Ber1]–[Ber2]). With this tensor, we shall define the flag curvature and the Ricci curvature. These curvatures are called the *Riemannian geometric (invariant) quantities*. Besides this, we shall introduce some non-Riemannian geometric (invariant) quantities, such as, the Berwald curvature, the S-curvature, the Landsberg curvature etc., which vanish on Riemannian manifolds. The Riemannian geometric quantities characterize the shapes of Finsler manifolds. While the non-Riemannian geometric quantities describe the color of Finsler manifolds. This is because that the indicatrix of $F_x := F|_{T_x M}$ on the tangent space $T_x M$ can be viewed as the infinitesimal color pattern at $x \in M$. In this sense, Finsler geometry is more "colorful" than Riemannian geometry.

4.1 Riemannian Geometric Quantities

In this section, we shall introduce the Riemannian geometric quantities, such as the flag curvature and the Ricci curvature, etc., which are generalizations of the sectional curvature and the Ricci curvature in Riemannian geometry.

4.1.1 *Flag Curvature and Ricci Curvature*

Let (M, F) be an n-dimensional Finsler manifold. Let $\{e_i\}_{i=1}^n$ be a local frame for TM and $\{\theta^i\}_{i=1}^n$ be a local dual coframe for T^*M. Then $\{e_i := (x, y; e_i|_x)\}_{i=1}^n$ is a local frame for π^*TM and $\{\omega^i := \pi^*\theta^i\}$ is its local dual coframe for $\pi^*(T^*M)$. The *Riemann curvature tensor* $R = R^i{}_k e_i \otimes \omega^k$ can

51

be viewed as a family of linear maps

$$R = \left\{ R_y = (R^i{}_k) : T_xM \to T_xM \text{ is a linear map} | (x, y) \in TM_0 \right\},$$

where $R^i{}_k$ are defined by (3.2.16) and

$$R_y(u) = R^i{}_k(x, y) u^k e_i, \quad u = u^i e_i \in T_xM.$$

It follows from (3.2.16) and $R_{ij} = R_{ji}$ that $R_{\lambda y} = \lambda^2 R_y$ for some $\lambda > 0$ and

$$R_y(y) = 0, \quad g_y(R_y(u), v) = g_y(u, R_y(v)).$$

For any nonzero vector $y \in T_xM$, consider a tangent plane $\Pi \subset T_xM$ containing y and let

$$\mathbf{K}(\Pi, y) := \frac{g_y(R_y(u), u)}{g_y(y, y) g_y(u, u) - g_y(u, y)^2}, \tag{4.1.1}$$

where $u \in \Pi$ such that $\Pi = \text{span}\{y, u\}$. One can easily verify that $\mathbf{K}(\Pi, y)$ is independent of the choice of $u \in \Pi$ such that $\Pi = \text{span}\{y, u\}$. $\mathbf{K}(\Pi, y)$ is called the *flag curvature* of F with flagpole y, where Π is called a *flag* with flagpole y. In dimension two, $\Pi = T_xM$ is a unique tangent plane. Thus, $\mathbf{K}(\Pi, y)$ is a scalar function on TM_0, which is exactly the Gauss curvature. In general, $\mathbf{K}(\Pi, y)$ is a function of flag Π and flagpole y.

Definition 4.1.1. Let (M, F) be a Finsler manifold. Then

(1) F is said to be of *scalar flag curvature* if $\mathbf{K}(\Pi, y) = \mathbf{K}(x, y)$ is a scalar function on TM_0;
(2) F is said to be of *isotropic flag curvature* if $\mathbf{K}(\Pi, y) = \mathbf{K}(x)$ is a scalar function on M;
(3) F is said to be of *constant flag curvature* if $\mathbf{K}(\Pi, y)$ is a constant.

For the first case in Definition 4.1.1, if $\mathbf{K}(\Pi, y) = \frac{3\theta}{F} + \sigma$ for some 1-form θ and scalar function σ on M, we say that F is of *weakly isotropic flag curvature*. In particular, when $\theta = df$ for some function f on M, we say that F is of *almost isotropic flag curvature*.

From (4.1.1), F is of scalar flag curvature $\mathbf{K} = \mathbf{K}(x, y)$ if and only if for any $y, u \in T_xM \backslash \{0\}$,

$$R_y(u) = \mathbf{K}(g_y(y, y)u - g_y(y, u)y). \tag{4.1.2}$$

In local coordinates, the above equality is equivalent to

$$R^i{}_k(y) = \mathbf{K}\left(F^2\delta^i_k - g_{kl}y^l y^i\right). \tag{4.1.3}$$

Define $\mathrm{Ric}(y) := \mathrm{trace}_{g_y} R_y$, namely,

$$\mathrm{Ric}(y) := g^{ij}g_y(R_y(e_i), e_j) = g^{ij}g_y(R^k{}_i e_k, e_j) = R^k{}_k,$$

where $g_{ij} = g_y(e_i, e_j)$ and $(g^{ij}) = (g_{ij})^{-1}$. Ric is a well defined scalar function on TM_0. We call Ric the *Ricci curvature* (or *Ricci scalar*) of F. F is called an *Einstein metric* if there is a scalar function $\sigma = \sigma(x)$ on M such that

$$\mathrm{Ric}(y) = (n-1)\sigma F^2(y). \tag{4.1.4}$$

Remark 4.1.1. For the reverse metric $\overleftarrow{F}$, since $\overleftarrow{g}_y = g_{-y}$ and $\overleftarrow{G}^i(y) = G^i(-y)$, the associated flag curvature and the Ricci curvature are given by

$$\overleftarrow{\mathbf{K}}(\Pi, y) = \mathbf{K}(\Pi, -y), \quad \overleftarrow{\mathrm{Ric}}(y) = \mathrm{Ric}(-y). \tag{4.1.5}$$

We write $y = y^i e_i$ and $u = u^i e_i$. For the flag $\Pi = \mathrm{span}\{y, u\}$, we have

$$\mathbf{K}(\Pi, y) = \frac{R_{jl}(y)u^j u^l}{(g_{ik}g_{jl} - g_{il}g_{jk})y^i u^j y^k u^l} = \frac{R_{ijlk}(y)y^i u^j y^k u^l}{(g_{ik}g_{jl} - g_{il}g_{jk})y^i u^j y^k u^l}, \tag{4.1.6}$$

where $R_{ijkl} := g_{tj}R_i{}^t{}_{kl}$, and

$$\mathrm{Ric} = g^{ij}R_{ij} = R^i{}_i, \tag{4.1.7}$$

where

$$R_{ik} = g_{ij}R^j{}_k = g_{ij}y^s R_s{}^j{}_{kl}y^l = R_{jikl}y^j y^l = -R_{ijkl}y^j y^l, \tag{4.1.8}$$

in which we used the first equality in (3.3.7). In particular, if we choose a local orthonormal frame $\{e_i\}_{i=1}^n$ with respect to g_y such that $e_n = y/F(y)$, then $g_{ij} = g_y(e_i, e_j) = \delta_{ij}$. In this case, for the flag $\Pi_a = \mathrm{span}\{e_a, e_n\}$ $(1 \leq a \leq n-1)$, the flag curvature with flagpole e_n is given by

$$\mathbf{K}(\Pi_a, e_n) = R_{aa} \text{ (no sum)}.$$

Consequently, the Ricci curvature in the direction e_n is

$$\mathrm{Ric}(e_n) = \sum_{a=1}^{n-1} \mathbf{K}(\Pi_a, e_n) = \sum_{a=1}^{n-1} R_{aa}.$$

Proposition 4.1.1. *Let F be a Riemannian metric on a manifold M. For any $x \in M$ and any tangent plane $\Pi \subset T_x M$, the flag curvature*

$\mathbf{K}(\Pi, y) = \mathbf{K}(\Pi)$ *is independent of* $y \in \Pi \backslash \{0\}$ *and Ricci scalar Ric is the usual Ricci curvature of* F.

Proof. Since F is Riemannian, we have $C_{ijk} = 0$ and $g_{ij} = g_{ij}(x)$, $R_i{}^j{}_{kl} = R_i{}^j{}_{kl}(x)$ are functions of $x \in M$. Thus, $R_{ijkl} = g_{jt} R_i{}^t{}_{kl}$ are also functions of x. It follows from (3.3.7) and (3.3.9) that

$$R_{ijkl} = -R_{jikl} = -R_{ijlk}.$$

From this and (4.1.8), we get

$$R_{jl}(y) u^j u^l = R_{ijlk}(x) y^i y^k u^j u^l = R_{jikl}(x) y^j y^l u^i u^k = R_{ijlk}(x) y^j y^l u^i u^k$$
$$= R_{jl}(u) y^j y^l,$$

where $u = u^i e_i \in \Pi \backslash \{0\}$ such that $\Pi = \text{span}\{y, u\}$. Thus, by (4.1.6), we have

$$\mathbf{K}(\Pi, y) = \mathbf{K}(\Pi, u),$$

which means that $\mathbf{K}(\Pi, y) = \mathbf{K}(\Pi)$ is independent of y.

Note that $\text{Ric}(y) = y^i R_i{}^k{}_{kj}(x) y^j$ is a quadratic form in y. Let

$$\text{Ric}_{ij}(x) := \frac{1}{2}[\text{Ric}]_{y^i y^j}, \quad \text{equivalently,} \quad \text{Ric}(y) = \text{Ric}_{ij}(x) y^i y^j.$$

Then $\text{Ric}_{ij} = R_i{}^k{}_{kj}$ are the components of the Ricci tensor for the Riemannian metric F. Denote $S(X, Y) := \text{Ric}_{ij}(x) X^i Y^j$ for any $X, Y \in \Gamma(TM)$, which is the Ricci tensor of F. By the homogeneity of Ric and S, the Ricci scalar $\text{Ric}(y) = S(y, y)$ is exactly the Ricci curvature in the unit vector y. $\quad\square$

Finally we prove the following well known lemma ([Ber4]).

Lemma 4.1.1 (Schur's Lemma). *Let* (M, F) *be an* $n(\geq 3)$-*dimensional connected Finsler manifold. If* (M, F) *is of isotropic flag curvature, then it is of constant flag curvature.*

Proof. As we know, (M, F) is of isotropic flag curvature $\mathbf{K} = \mathbf{K}(x)$ if and only if

$$R^i{}_k = \mathbf{K}\left(F^2 \delta^i_k - g_{kl} y^l y^i\right). \tag{4.1.9}$$

By (3.2.18), we have

$$R^i{}_{kl} = \mathbf{K}(F F_{y^l} \delta^i_k - F F_{y^k} \delta^i_l). \tag{4.1.10}$$

Note that $F_{|k} = y^i{}_{|k} = 0$. Plugging (4.1.9) and (4.1.10) into (3.3.15) yields

$$\mathbf{K}_{|t}(F^2\delta_k^i - FF_{y^k}y^i) - \mathbf{K}_{|k}(F^2\delta_t^i - FF_{y^t}y^i) + \mathbf{K}_{|l}(FF_{y^k}\delta_t^i - FF_{y^t}\delta_k^i)y^l = 0.$$

Letting $i = k$ and taking sum over i on both sides of the above equality give

$$(n-2)\left\{F^2\mathbf{K}_{|t} - FF_{y^t}(\mathbf{K}_{|l}y^l)\right\} = 0.$$

Since $n \geq 3$, we have

$$F^2\mathbf{K}_{|t} = FF_{y^t}(\mathbf{K}_{|l}y^l). \tag{4.1.11}$$

Differentiating this with respect to y^k and then contracting with g^{kt} yields $\mathbf{K}_{|l}y^l = 0$. Consequently, $\mathbf{K}_{|t} = 0$ by (4.1.11), that is, $\mathbf{K}$ is locally a constant. Since M is connected, $\mathbf{K}$ is a constant on M. $\qquad\square$

Example 4.1.1. Minkowski spaces are Finsler spaces with zero flag curvature. This is a trivial example. There are many nontrivial Finsler metrics with zero flag curvature. For example, let F be a Berwald metric given by (2.1.19), which is defined on a unit ball in $(\mathbb{R}^n, |\cdot|)$. It is easy to check that it is of the flag curvature $\mathbf{K} = 0$ by (3.1.14) and (3.2.17).

Example 4.1.2. Let F be the Funk metric on a strongly convex domain $\Omega \subset \mathbb{R}^n$ (see Example 2.1.3). By Proposition 2.1.1, we have

$$F_{x^k} = FF_{y^k}. \tag{4.1.12}$$

From this, one obtains that $F_{x^k}y^k = F^2$ and $F_{x^k y^i}y^k = F_{x^i}$. Thus, by (3.1.14), the geodesic coefficients G^i of F are given by $G^i = \frac{1}{2}Fy^i$. Plugging this into (3.2.17) and using (4.1.12) yield

$$R^i{}_k = -\frac{1}{4}\left(F^2\delta_k^i - g_{kl}y^l y^i\right),$$

which means that the flag curvature $\mathbf{K} = -\frac{1}{4}$. In particular, the Funk metric F on $\mathbb{B}_1^n$ defined by (2.1.13) is of constant negative flag curvature $-\frac{1}{4}$.

Similarly, for the Klein metric $\bar{F}(x, y) = \frac{1}{2}\left(F(x, y) + F(x, -y)\right)$, we have

$$\bar{F}_{x^k}y^k = \bar{F}(x, y)\left(F(x, y) - F(x, -y)\right), \quad \bar{F}_{x^k y^i}y^k = \bar{F}_{x^i}.$$

The geodesic coefficients $\bar{G}^i$ of $\bar{F}$ are given by $\bar{G}^i = \frac{1}{2}\left(F(x, y) - F(x, -y)\right)y^i$. Plugging this into (3.2.17) yields

$$\bar{R}^i{}_k = -\left(\bar{F}^2(y)\delta_k^i - \bar{g}_{kl}(y)y^l y^i\right),$$

which means that the flag curvature $\mathbf{K} = -1$. In particular, the Klein metric F on $\mathbb{B}^n_1$ defined by (2.1.14) has constant negative flag curvature -1.

Example 4.1.3. Bryant's metric (2.1.24) defined as in Example 2.1.6 has constant curvature $\mathbf{K} = 1$ ([Br1], [Br2]). Further, this metric can be extended to $\mathbb{S}^n$ and locally expressed by (2.1.25). It is of constant curvature $\mathbf{K} = 1$.

     Examples 4.1.1–4.1.3 show that there are non-Riemannian Finsler metrics with constant flag curvature. It is a natural problem to classify Finsler metrics of constant flag curvature. Bao–Shen–Robles classified Randers metrics of constant flag curvature via navigation technique ([BSR]). Using the same technique, Yoshikawa–Okubo classified Kropina metrics of constant flag curvature ([YO]). For general Finsler metrics of constant flag curvature, the classification is open up to now. More generally, it is a question to study the classifications of Finsler metrics of scalar (resp., weakly or almost isotropic) flag curvature. Anyway, much progress has been made on these problems in recent decades (cf. [ShY], [CS], [Xia2]–[Xia3] etc. and references therein).

4.1.2   *Geodesic Fields*

In this section, we will give alternative formulae for the Riemannian geometric quantities. These formulae are not suitable for computations but very useful in comparison geometry and global analysis on manifolds.

     A nonzero smooth vector field Y on a Finsler manifold (M, F) is said to be a *geodesic field* if $D^Y_Y Y = 0$, i.e., all integral curves of Y are geodesic. In this case, $Y = Y^i(x)\frac{\partial}{\partial x^i}$ satisfies

$$Y^i \frac{\partial Y^k}{\partial x^i} + 2G^k(Y) = 0. \qquad (4.1.13)$$

For any $y \in T_x M$ linearly independent of $u \in T_x M$, we extend y, u to vector fields Y, U around x such that Y is a geodesic field. Let $\hat{D}$ be the Levi-Civita connection of weighted Riemannian metric $\hat{g} := g_Y$ and $\hat{R}_Y$ the Riemann curvature tensor of $\hat{g}$. Then we have

Lemma 4.1.2 ([Sh1]). $D^Y_Y U = \hat{D}_Y U$ and $R_Y = \hat{R}_Y$. In particular, Y is also a geodesic field of g_Y and $R_y = \hat{R}_y$ for any $y = Y_x \in T_x M$.

4.2 Non-Riemannian Geometric Quantities

Finsler spaces are colorful spaces as well as curved spaces. Roughly speaking, the (mean) Cartan torsion describes the "color" of the space at a point. A Finsler space (M, F) is a Riemannian space if and only if the (mean) Cartan torsion of F vanishes. From this view, Finsler spaces are much more "colorful" than Riemannian spaces. Besides the Cartan torsion, there are other geometric quantities which vanish on Riemannian spaces, such as, the Chern curvature, the Landsberg curvature, the S-curvature, the Weyl curvature and the Douglas curvature, etc. The Chern curvature and the Landsberg curvature describe the rates of change of the Cartan torsion over the space, while the S-curvature characterizes the rate of change of the distorsion over the space.

4.2.1 Chern Curvature and Landsberg Curvature

Let (M, F) be a Finsler manifold. In a local coordinate system (x^i, y^i) in TM, the functions $g_{ij}(y) = \frac{1}{2}[F^2]_{y^i y^j}(y)$ and $C_{ijk}(y) := \frac{1}{4}[F^2]_{y^i y^j y^k}(y)$ on TM define the inner product g_y and the Cartan torsion C_y on $T_x M$, respectively.

Let $P_j{}^i{}_{kl}$ and L_{ijk} be the Chern curvature tensor and the Landsberg curvature tensor of (M, F) given by (3.2.11) and (3.2.13), respectively. For any $y \in T_x M \backslash \{0\}$ and $u = u^i \frac{\partial}{\partial x^i}$, $v = v^i \frac{\partial}{\partial x^i}$, $w = w^i \frac{\partial}{\partial x^i} \in T_x M$, define

$$L_y(u, v, w) = L_{ijk}(y) u^i v^j w^k, \quad P_y(u, v, w) = P_j{}^i{}_{kl}(y) u^j v^k w^l \frac{\partial}{\partial x^i}.$$

We call $P = \{P_y | (x, y) \in TM_0\}$ and $L = \{L_y | (x, y) \in TM_0\}$ the *Chern curvature* and the *Landsberg curvature* of (M, F), respectively. They are related to the Chern curvature tensor P and the Landsberg curvature tensor L ([Ch1])–[Ch2], [La1]–[La2]).

Let $\gamma(t)$ be a geodesic on (M, F) and $U(t), V(t), W(t)$ be parallel vector fields along γ. Then (3.2.13) implies that

$$L_{\dot\gamma}(U(t), V(t), W(t))\big|_{t=0} = \frac{d}{dt}\big|_{t=0}(C_{\dot\gamma}(U(t), V(t), W(t))), \quad (4.2.1)$$

which shows that the Landsberg curvature measures the rate of change of the Cartan torsion along γ.

Definition 4.2.1. A Finsler metric F is called a *Berwald metric* if $P = 0$ and a *Landsberg metric* if $L = 0$. A differential manifold equipped with the Berwald metric (resp., Landsberg metric) is called a *Berwald space* (resp., *Landsberg space*) or a *Berwald manifold* (resp., *Landsberg manifold*) respectively.

Obviously, every Berwald space is a Landsberg space. It is unknown if there is a Landsberg metric which is not a Berwald metric until now. By (3.2.11), the geodesic coefficients G^i are quadratic in y, namely,

$$G^i(x,y) = \frac{1}{2}\Gamma^i_{jk}(x)y^j y^k, \quad \text{equivalently,} \quad N^i{}_k(x,y) = \Gamma^i_{jk}(x)y^j \quad (4.2.2)$$

on a Berwald space. The Cartan torsion is constant along any geodesic on a Landsberg space from (4.2.1). Moreover, from $L_{ijk} = g_{it}L^t{}_{jk} = -y^t P_{tijk}$, it is easy to see that there is a relationship between the Berwald curvature and the Landsberg curvature as follows.

Proposition 4.2.1. *The Landsberg curvature L is related to the Chern curvature P by*

$$L_y(u,v,w) = -g_y(u, P_y(y,v,w)) = g_y(P_y(u,v,w), y).$$

Next, we introduce some geometric properties of Berwald spaces and Landsberg spaces. As an observation in Section 3.1, the parallel translation P_γ along any geodesic γ preserves the inner product g_γ but does not preserve the Finsler norm in general. However, P_γ preserves the Finsler norm F for a Berwald space (M, F). This is due to a good property of Berwald spaces, i.e., $\Gamma^i_{jk}(x,y) = \Gamma^i_{jk}(x)$ independent of the tangent vector $y \in T_xM$. Berwald spaces can be viewed as Finsler spaces modeled on a single Minkowski space.

Proposition 4.2.2 ([Ic1]). *Let $\gamma : [a,b] \to M$ be a geodesic on a Berwald space (M, F). Then $F(\gamma(t), P_{\gamma(t)}(y)) = F(\gamma(a), y)$ for any $y \in T_{\gamma(a)}M$.*

Proof. Let $V(t) = V^i(t)\frac{\partial}{\partial x^i}$ be a parallel vector field along γ. Then

$$\dot{V}^i(t) = -V^j(t)N^i{}_j(\gamma, \dot{\gamma}).$$

Note that F is Berwaldian. Then we have

$$N^k{}_i(x,y)v^i = \Gamma^k_{ij}(x)y^j v^i = N^k{}_j(x,v)y^j$$

for any $y, v \in T_xM$. Thus,

$$\frac{d}{dt}(F(\gamma(t), V(t))) = dF\left(\dot{\gamma}^i\frac{\partial}{\partial x^i} + \dot{V}^i(t)\frac{\partial}{\partial y^i}\right)$$

$$= dF\left(\dot{\gamma}^i\frac{\partial}{\partial x^i} - V^j(t)N^i{}_j(\gamma(t), \dot{\gamma}(t))\frac{\partial}{\partial y^i}\right)$$

$$= dF\left(\dot{\gamma}^i \frac{\partial}{\partial x^i} - \dot{\gamma}^j(t) N^i{}_j(\gamma(t), V(t)) \frac{\partial}{\partial y^i}\right)$$

$$= dF\left(\dot{\gamma}^i \frac{\delta}{\delta x^i}\right) = 0,$$

in which the last equality follows from Proposition 2.2.1. The proof is finished. □

Corollary 4.2.1 ([Ic1]). *Let (M, F) be a Berwald space. For any piecewise smooth curve $\gamma(t)$ from p to q in M, the parallel translation P_γ is a linear isometry between $(T_p M, F_p)$ and $(T_q M, F_q)$.*

At each point $x \in M$, $F_x = F|_{T_x M}$ induces a Riemannian metric $\breve{g}_x = g_{ij}(y) dy^i \otimes dy^j$ on the slit tangent space $T_x M_0 = T_x M \backslash \{0\}$. Then we have the following conclusion.

Proposition 4.2.3 ([Ic2]). *Let (M, F) be a Landsberg manifold. Then for any piecewise C^∞ curve γ from p to q in M, the parallel translation P_γ along γ preserves the induced Riemannian metrics $\breve{g}_x$ on $T_x M_0$, i.e., $P_\gamma : (T_p M_0, \breve{g}_p) \to (T_q M_0, \breve{g}_q)$ is an isometry.*

For its proof, we refer to that of Proposition 4.4.1 in [ChS]. We end this section with introducing the mean Landsberg curvature. Let $\{e_i\}_{i=1}^n$ be a basis for $T_x M$ with dual basis $\{\omega^i\}_{i=1}^n$. For any $y \in T_x M \backslash \{0\}$, recall that the *mean Cartan torsion* is the family $I = \{I_y | y \in T_x M \backslash \{0\}, x \in M\}$, where I_y is the mean of C_y, which is given by (2.2.3). Similarly, we can define the mean J_y of L_y by

$$J_y(u) = g^{ij} L_y(u, e_i, e_j), \quad \text{equivalently,} \quad J_y = J_k \omega^k, \quad J_k = g^{ij} L_{ijk},$$

where $(g^{ij}) = (g_{ij})^{-1}$, here $g_{ij} = g_y(e_i, e_j)$. The family $J = \{J_y | y \in T_x M \backslash \{0\}, x \in M\}$ is called the *mean Landsberg curvature*. A Finsler metric is said to be a *weakly Landsberg metric* if $J = 0$. In dimension two, I and J determine C and L, respectively.

Let $\gamma(t)$ be a geodesic on (M, F) and $V(t)$ be a parallel vector field along γ. It follows from (4.2.1) that

$$J_{\dot{\gamma}}(V(t))|_{t=0} = \frac{d}{dt}\Big|_{t=0} \big(I_{\dot{\gamma}}(V(t))\big). \tag{4.2.3}$$

In local coordinates, $J_y = J_i(x, y) dx^i \otimes dx^j$ with

$$J_i = I_{i|k} y^k. \tag{4.2.4}$$

Thus, the mean Cartan torsion I is constant along any geodesic on a weakly Landsberg space.

4.2.2   S-Curvature and the Related Curvatures

Let (M, F, m) be a Finsler measure space. Take an arbitrary basis $\{e_i\}_{i=1}^n$ for $T_x M$ and its dual basis $\{\omega^i\}_{i=1}^n$ for $T_x^* M$. Write $dm = \sigma(x)\omega^1 \wedge \cdots \wedge \omega^n$. For any $y \in T_x M \backslash \{0\}$, define

$$\tau(x, y) = \log\left(\frac{\sqrt{\det(g_{ij}(x, y))}}{\sigma(x)}\right), \qquad (4.2.5)$$

where $g_{ij}(x, y) = g_y(e_i, e_j)$. τ is called the *distortion* of (M, F, m). It is a positively homogeneous function of degree one.

Observe that

$$\tau_{y^i} = \frac{\partial}{\partial y^i}\left(\log\sqrt{\det(g_{ij})}\right) = g^{jk}C_{ijk} = I_i. \qquad (4.2.6)$$

Thus, by Deicke's Theorem, we obtain the following result.

Proposition 4.2.4. *On a Finsler measure space (M, F, m), the following statements are equivalent:*

(1) *F is Riemannian;*
(2) *$C = 0$;*
(3) *$I = 0$;*
(4) *$\tau(x, y)$ is independent of y.*

To measure the rate of change of the distortion τ along a geodesic, we need introduce S-curvature.

Definition 4.2.2. Let (M, F, m) be a Finsler measure space and $\gamma(t)$ be a geodesic with $\gamma(0) = x$ and $\dot\gamma(0) = y$. Define

$$S(x, y) := \frac{d}{dt}\Big(\tau(\gamma(t), \dot\gamma(t))\Big)|_{t=0}. \qquad (4.2.7)$$

We call S the *S-curvature* of F with respect to the measure m.

S-curvature was first introduced by Shen ([Sh3]). It is a positively homogeneous function of degree one in y and measures the rate of change of the distortion τ along the geodesic γ. Locally, (4.2.7) can be expressed by

$$S(x, y) = \tau_{|k}y^k. \qquad (4.2.8)$$

Letting $i = j$ in the first equality of (3.1.13) yields

$$\frac{\partial G^i}{\partial y^i} = \frac{1}{2}g^{jl}\frac{\partial g_{jl}}{\partial x^k}y^k - 2I_k G^k,$$

Thus, we have by (4.2.5)

$$y^i \frac{\partial \tau}{\partial x^i} = \frac{1}{2} g^{jl} \frac{\partial g_{jl}}{\partial x^i} y^i - y^i \frac{\partial}{\partial x^i} (\log \sigma) = \frac{\partial G^i}{\partial y^i} + 2I_k G^k - y^i \frac{\partial}{\partial x^i} (\log \sigma),$$

which means that

$$\tau_{|i} y^i = \frac{\partial G^i}{\partial y^i} - y^i \frac{\partial}{\partial x^i} (\log \sigma), \tag{4.2.9}$$

where we used (4.2.6). From (4.2.8)–(4.2.9), one obtains

$$S(x, y) = N^i{}_i(x, y) - y^i \frac{\partial}{\partial x^i} \big(\log \sigma(x) \big). \tag{4.2.10}$$

Proposition 4.2.5 ([Sh1]). *Let (M, F, m_{BH}) be a Berwald measure space equipped with the Busemannn-Hausdorff measure m_{BH}. Then $S = 0$.*

Proof. Let $\gamma(t)$ be a geodesic on M with $\gamma(0) = x$ and $\dot\gamma(0) = y \in T_x M$. Take an arbitrary parallel frame field $\{e_i(t)\}_{i=1}^n$ along γ with dual coframe field $\{\omega^i(t)\}_{i=1}^n$. Along the geodesic γ, the Busemann-Hausdorff volume form dm_{BH} can be expressed by $dm_{BH}|_\gamma = \sigma_{BH}(\gamma(t))\omega^1 \wedge \cdots \wedge \omega^n$ from (2.3.1). By Propositions 3.1.1 and 4.2.2, we have

$$g_{\dot\gamma}(e_i(t), e_j(t)) = g_{\dot\gamma(0)}(e_i(0), e_j(0)) = constant,$$

$$F(y^i e_i(t)) = F(y^i e_i(0)) = constant.$$

These imply that σ_{BH} is constant along γ and hence the distorsion τ is constant along γ. Thus $S(\gamma, \dot\gamma) = 0$. Since γ is arbitrary, we have $S = 0$. $\square$

In general, S-curvature does not vanish. In fact, there are many examples whose S-curvatures are not zero (see Example 4.2.1 below).

Definition 4.2.3. Let (M, F, m) be an n-dimensional Finsler manifold. Then

(1) F is said to be of *weakly isotropic S-curvature* if there is a scalar function $c = c(x)$ and a 1-form η on M such that

$$S(x, y) = (n + 1)(cF + \eta); \tag{4.2.11}$$

(2) F is said to be of *almost isotropic S-curvature* if there is a scalar function $c = c(x)$ and a closed 1-form η on M such that (4.2.11) holds;

(3) F is said to be of *isotropic S-curvature* if $S(x, y) = (n + 1)cF$ for some scalar function c;

(4) F is said to be of *constant S-curvature* if $S(x,y) = (n+1)cF$ for some constant c.

Next we introduce three non-Riemannian geometric quantities related to S-curvature. Let

$$E_{ij} := \frac{1}{2}S_{\cdot i \cdot j}, \quad H_{ij} := \frac{1}{2}S_{\cdot i \cdot j|k}y^k, \quad \Xi_i := S_{\cdot i|j}y^j - S_{|i},$$

where $S_{\cdot i}$ means the vertical derivative of S with respect to the Chern connection, that is, $S_{\cdot i} = S_{y^i}$. Similarly, $S_{\cdot i \cdot j} = S_{y^i y^j}$. By (4.2.10), we have

$$E_{ij} = \frac{1}{2}[G^k]_{y^i y^j y^k}.$$

The E-curvature is independent of the measure m.

Definition 4.2.4. We call the symmetric tensors $E_y := E_{ij}(x,y)dx^i \otimes dx^j$, $H_y := H_{ij}(x,y)dx^i \otimes dx^j$ and the 1-form $\Xi_y := \Xi_i(x,y)dx^i$ the E-curvature tensor, the H-curvature tensor and the Ξ-curvature tensor of F, respectively. The families $\{E_y | y \in T_x M \backslash \{0\}, \ x \in M\}$, $\{H_y | y \in T_x M \backslash \{0\}, \ x \in M\}$ and $\{\Xi_y | y \in T_x M \backslash \{0\}, \ x \in M\}$ are called the *E-curvature*, the *H-curvature* and the Ξ-*curvature* of F, respectively. The E-curvature is also said to be the *mean Berwald curvature*.

Since $S(x,y)$ is a positively homogeneous function of degree one, we have

$$E_y(y,v) = H_y(y,v) = \Xi_y(y) = 0.$$

We say that F is of *isotropic E-curvature* if there is a scalar function $c = c(x)$ on M such that

$$E = \frac{1}{2}(n+1)cF^{-1}h,$$

where $h = h_{ij}dx^i \otimes dx^j$ is the angular metric, here $h_{ij} = FF_{y^i y^j}$. F is said to be of *almost vanishing H-curvature* or *almost vanishing Ξ-curvature* if there is a 1-form $\theta = \theta_i(x)y^i$ on M such that

$$H = \frac{1}{2}(n+1)\theta F^{-1}h,$$

or

$$\Xi_i = -(n+1)F^2 \left(\frac{\theta}{F}\right)_{y^i}.$$

Obviously, if F is of weakly isotropic S-curvature, then F is of isotropic E-curvature (resp., almost vanishing H-curvature, almost vanishing Ξ-curvature). The converse is not true in general. There are Finsler metrics with $\Xi = 0$ and $H = 0$, while the S-curvature is not almost isotropic (see Example 1.1, [Sh7]). However, for a Randers metric $F = \alpha + \beta$, F is of almost isotropic S-curvature if and only if it is of almost vanishing H-curvature if and only if it is of almost vanishing Ξ-curvature. In particular, F is of constant S-curvature if and only if it is of vanishing H-curvature if and only if it is of vanishing Ξ-curvature ([Sh7], [Xia1]).

Example 4.2.1. Let F be the Funk metric on a strongly convex domain Ω in $\mathbb{R}^n$. The geodesic coefficients are given by $G^i = \frac{1}{2}Fy^i$ (see Example 4.1.2). From this, we have

$$N^i_{\ i} = \frac{1}{2}(Fy^i)_{y^i} = \frac{1}{2}(n+1)F.$$

On the other hand, since Ω is strongly convex, there is a Finsler metric F such that $\Omega\backslash\{x\} = B^n_1 = \{(y^i) \in \mathbb{R}^n | F(y) < 1\}$ for any $x \in \Omega$, which implies that $\text{Vol}(B^n_1) = \text{Vol}(\Omega)$. Consequently, the Busemann Hausdorff volume form $dm_{BH} = \sigma_{BH}(x)dx$ has a constant density, i.e., $\sigma_{BH} = $ constant. By (4.2.10), we have

$$S(y) = \frac{1}{2}(n+1)F(y),$$

namely, F is of constant S-curvature $\frac{1}{2}$. Further,

$$E_{ij} = \frac{1}{2}S_{y^iy^j} = \frac{1}{2}(n+1)F_{y^iy^j} = \frac{n+1}{4}F^{-1}h_{ij},$$

which means that F is of constant E-curvature $\frac{1}{2}$.

4.3 Projectively Geometric Quantities

We shall introduce two important projectively invariant quantities: Weyl projective curvature and Douglas curvature. They are invariant under the projective transformation of Finsler manifolds. For this, we need understand the projective equivalence between two Finsler metrics.

4.3.1 *Projective Equivalence*

Definition 4.3.1. Two Finsler metrics $\bar{F}$ and F on a manifold M are said to be *projectively equivalent* if they have the same positively oriented geodesics as point sets of oriented curves. More precisely, for any geodesic $\bar{\gamma}(\bar{t})$ of $\bar{F}$, after an appropriate oriented reparametrization, $\bar{t} = \bar{t}(t)$ such that $d\bar{t}/dt > 0$, the new curve $\gamma(t) := \bar{\gamma}(\bar{t}(t))$ is a geodesic of F, and vice versa.

If a map $\psi : (M, F) \to (M, F)$ is a diffeomorphism such that $\psi^* F$ is projectively equivalent to F, then ψ is said to be a *projective transformation* of (M, F).

Proposition 4.3.1. *Two Finsler metrics $\bar{F}$ and F on a manifold M are projectively equivalent if and only if the geodesic coefficients $\bar{G}^i$ and G^i of $\bar{F}$ and F are related by*

$$\bar{G}^i = G^i + Py^i, \tag{4.3.1}$$

where $P = P(x, y)$ is a positively homogeneous function on TM_0 of degree one in y.

Proof. Assume that the two Finsler metrics F and $\bar{F}$ are projectively equivalent. For any $y \in T_x M \backslash \{0\}$, let $\gamma(t)$ be a geodesic with $\gamma(0) = x$ and $\dot{\gamma}(0) = y$. Then there is a transformation $\bar{t} = \bar{t}(t)$ with $\bar{t}(0) = 0$ and $\dot{\bar{t}}(0) = 1$ such that $\dot{\bar{t}} := d\bar{t}/dt > 0$ around $t = 0$ and $\bar{\gamma}(\bar{t}) = \gamma(t)$ is a geodesic of $\bar{F}$ with $\bar{\gamma}(0) = x$ and $\dot{\bar{\gamma}}(0) = y$. By (3.1.21), one obtains

$$2G^i(x, y) = -\ddot{\gamma}^i(0) = -\ddot{\bar{\gamma}}^i(0) - \ddot{\bar{t}}(0)\dot{\bar{\gamma}}^i(0) = 2\bar{G}^i(x, y) - \ddot{\bar{t}}(0)y^i.$$

From the above equation, one can see that $P := \frac{1}{2}\ddot{\bar{t}}(0)$ depends only on $(x, y) \in TM_0$, Hence, $P = P(x, y)$ is a function of (x, y). Moreover, it is a positively homogeneous function on TM_0 of degree one in y. Thus, one obtains (4.3.1).

Conversely, if the geodesic coefficients $\bar{G}^i$ and G^i of Finsler metrics $\bar{F}$ and F are related by (4.3.1), one can easily show that $\bar{F}$ and F are projectively equivalent. $\square$

There is another equivalent way to characterize two projectively equivalent Finsler metrics, in which we can express $P(x, y)$ in terms of the Finsler metric.

Proposition 4.3.2 ([Rap]). *Let $\bar{F}$ and F be Finsler metrics on a manifold M. $\bar{F}$ is projectively equivalent to F if and only if $\bar{F}$ satisfies the following differential equation:*

$$\bar{F}_{|k \cdot l}y^k - \bar{F}_{|l} = 0, \tag{4.3.2}$$

where "$|$" and "$\cdot$", respectively, denote the horizontal and vertical covariant derivatives with respect to the Chern connection of F. In this case, the geodesic coefficients are related by $\bar{G}^i = G^i + Py^i$, here $P = \frac{\bar{F}_{|k}y^k}{2\bar{F}}$.

Proof. Let $\{dx^i, \delta y^i\}_{i=1}^n$ be a local cotangent frame on TM_0. We view $\bar{F}$ as a function on TM_0. With respect to the Chern connection of F, we have

$$\bar{F}_{|i} = \bar{F}_{x^i} - N_i^k \bar{F}_{y^k}, \quad \bar{F}_{\cdot i} = \bar{F}_{y^i}, \quad \bar{F}_{|i \cdot j} = (\bar{F}_{|i})_{y^j}.$$

Hence,

$$\bar{F}_{x^k} y^k = \bar{F}_{|k} y^k + 2G^l \bar{F}_{y^l},$$

$$\bar{F}_{x^k y^l} y^k - \bar{F}_{x^l} = \bar{F}_{|k \cdot l} y^k - \bar{F}_{|l} + 2G^j \bar{F}_{y^j y^l}.$$

By (3.1.14) and using the above equations, one obtains that $\bar{G}^i = G^i + P y^i + Q^i$, where

$$P = \frac{\bar{F}_{|k} y^k}{2\bar{F}}, \quad Q^i = \frac{1}{2} \bar{F} \bar{g}^{il} \left\{ \bar{F}_{|k \cdot l} y^k - \bar{F}_{|l} \right\}.$$

Thus, $\bar{F}$ is projectively equivalent to F if and only if $Q^i = 0$, that is, (4.3.2) holds. $\qquad\square$

In particular, if F is the standard Euclidean metric in $\mathbb{R}^n$, then $G^i = 0$. Thus, for any open subset $\mathcal{U} \subset \mathbb{R}^n$, $\bar{F}$ is projectively equivalent to F on $\mathcal{U}$ if and only if the geodesics of $\bar{F}$ are straight lines. Straight lines in $\mathcal{U}$ can be parametrized by $\gamma(t) = a + f(t)b$, where $a, b \in \mathbb{R}^n$ are constant vectors and $f(t)$ is a positive C^∞ function with $f(0) = 0$ and $f'(0) = 1$. We make the following definition.

Definition 4.3.2. A Finsler metric $F = F(x,y)$ on an open subset $\mathcal{U} \subset \mathbb{R}^n$ is said to be *projectively flat* if all geodesics of F are straight lines in $\mathcal{U}$. A Finsler metric F on a manifold M is said to be *locally projectively flat* if at any point, there is a local coordinate system (x^i) in which F is projectively flat.

It follows from Proposition 4.3.2 and Definition 4.3.2 that

Corollary 4.3.1 ([Ham]). *A Finsler metric $F = F(x,y)$ on an open subset $\mathcal{U}$ in $\mathbb{R}^n$ is projectively flat if and only if*

$$F_{x^k y^l} y^k - F_{x^l} = 0. \tag{4.3.3}$$

In this case, $G^i = P y^i$, where $P = \frac{F_{x^k} y^k}{2F}$, which is called the projective factor of F.

Further, (4.3.3) is equivalent to

$$F_{x^k y^l} = F_{x^l y^k}. \tag{4.3.4}$$

Indeed, contracting (4.3.4) with y^k yields (4.3.3) by the homogeneity of F. Conversely, taking the derivatives on both sides of (4.3.3) in y^j gives

$$F_{x^k y^l y^j} y^k + F_{x^j y^l} = F_{x^l y^j}. \tag{4.3.5}$$

On the other hand, by (4.3.3) and the homogeneity of F, we have

$$F_{x^k y^l y^j} y^k = \left(F_{x^i y^k} y^i \right)_{y^l y^j} y^k = F_{x^i y^k y^l y^j} y^i y^k + F_{x^j y^k y^l} y^k + F_{x^l y^k y^j} y^k$$

$$= -F_{x^i y^l y^j} y^i,$$

which implies that $F_{x^k y^l y^j} y^k = 0$. From this and (4.3.5), one obtains (4.3.4).

Note that the equation (4.3.3) is linear with respect to F. If F_1 and F_2 are projectively flat Finsler metrics on $\mathcal{U}$, then $F := c_1 F_1 + c_2 F_2$ is projectively flat on $\mathcal{U}$ only if F is a Finsler metric, where c_1, c_2 are constants. If $F = F(x, y)$ is projectively flat on $\mathcal{U}$, then its reverse metric $\overleftarrow{F}(y) = F(x, -y)$ is projectively flat on $\mathcal{U}$. Thus the symmetrization $\bar{F}(x, y) := \frac{1}{2}(F(x, y) + F(x, -y))$ is also projectively flat on $\mathcal{U}$.

Let $F = \alpha + \beta$ be a Randers metric, where $\alpha = \sqrt{a_{ij}(x) y^i y^j}$ is a Riemannian metric and $\beta = b_i(x) y^i$ is a 1-form on a manifold M. Let

$$r_{ij} := \frac{1}{2}(b_{i;j} + b_{j;i}), \quad s_{ij} = \frac{1}{2}(b_{i;j} - b_{j;i}),$$

$$e_{ij} = r_{ij} + b_i s_j + b_j s_i, \quad s^i{}_j = a^{ik} s_{kj}, \quad s_j = b_i s^i{}_j = b^i s_{ij},$$

where $b_{i;j}$ are the covariant derivatives of b_i with respect to α and we use a^{ij} to raise or lower the indices of b_i, r_{ij}, s_{ij}, etc.

Proposition 4.3.3 ([BM]). *A Randers metric $F = \alpha + \beta$ is locally projectively flat if and only if α is locally projectively flat and β is closed.*

Proof. Assume that $F = \alpha + \beta$ is locally projectively flat. There is a local coordinate system (x^i) in which $G^i = \tilde{P} y^i$, where $\tilde{P}$ is a projective factor. On the other hand, by (3.1.14), the geodesic coefficients G^i of F are given by

$$G^i = G^i_\alpha + P y^i + Q^i,$$

where $P = \frac{1}{2F} e_{00} - s_0$ and $Q^i = \alpha s^i{}_0$, here $e_{00} = e_{ij} y^i y^j$, $s^i{}_0 = s^i{}_j y^j$ and $s_0 = s_i y^i$. Thus we have

$$G^i_\alpha + P y^i + Q^i = \tilde{P} y^i. \tag{4.3.6}$$

Observe that $\frac{\partial (P y^i)}{\partial y^i} = (n+1)P$ by the homogeneity of P and $\frac{\partial (\alpha s^i{}_0)}{\partial y^i} = 0$. Differentiating (4.3.6) with y^i yields

$$\tilde{P} - P = \frac{1}{n+1} \frac{\partial G_\alpha^k}{\partial y^k}.$$

Inserting this into (4.3.6) gives rise to

$$\alpha s^i{}_0 = Q^i = \frac{1}{n+1} \frac{\partial G_\alpha^k}{\partial y^k} y^i - G_\alpha^i.$$

Note that the right-hand side is quadratic in $y \in T_x M$ and the left-hand side is irrational since α is irrational. Consequently, we have $s^i{}_0 = 0$ and $G_\alpha^i = \frac{1}{n+1} \frac{\partial G_\alpha^k}{\partial y^k} y^i$, which mean that β is closed and α is locally projectively flat. The converse is obvious. This finishes the proof. $\qquad\square$

Remark 4.3.1. When $\dim M \geq 3$, Beltrami's Theorem tells us that α is a locally projectively flat if and only if α is of constant sectional curvature (see Theorem 4.3.3 below). Thus locally projectively flat Randers metrics are completely determined. Note that Beltrami's Theorem is not true in general in Finsler geometry. In [Yu], C. Yu classified locally projectively flat (α, β)-metrics via β-deformation based on Shen's result in [Sh6]. Further, we give an equivalent characterization for locally projectively flat general (α, β)-metrics ([FX]). The classification of locally projectively flat Finsler metrics is to be further studied.

Example 4.3.1. Let $\varphi = \varphi(y)$ be a Minkowski norm on $\mathbb{R}^n$ and

$$\mathcal{U} := \{y = (y^i) \in \mathbb{R}^n | \varphi(y) < 1\}$$

be a strongly convex domain in $\mathbb{R}^n$. Assume that $F(x, y)$ is the Funk metric on $\mathcal{U}$ defined by (2.1.12), i.e.,

$$F(x, y) = \varphi(F(x, y)x + y).$$

By Proposition 2.1.1, it is easy to check that (4.3.3) holds with the projective factor $P = \frac{1}{2}F$ (cf. Example 4.1.2). Thus, F is projectively flat on $\mathcal{U}$. Further, the Hilbert metric $\bar{F} = \frac{1}{2}(F(x, y) + F(x, -y))$ is also projectively flat on $\mathcal{U}$. In particular, the Finsler metric defined by (2.1.13) or (2.1.14) is projectively flat on $\mathcal{U}$.

Theorem 4.3.1 ([Ber3]). *If a Finsler metric F is locally projectively flat, then it is of scalar flag curvature with*

$$\mathbf{K}(x, y) = F^{-2}(y)(P^2 - P_{x^k} y^k),$$

where $P = P(x, y)$ is the projective factor of F.

Proof. Since F is locally projectively flat, we have $G^i = Py^i$. Plugging this into (3.2.17) gives rise to

$$R^i{}_k = \zeta\delta^i_k + \theta_k y^i,$$

where

$$\zeta := P^2 - P_{x^j}y^j, \quad \theta_k := 3(P_{x^k} - PP_{y^k}) + \zeta_{y^k}. \tag{4.3.7}$$

For any $y \in T_xM\backslash\{0\}$, let $\theta_y = \theta_k(x,y)dx^k$. Then

$$R_y(u) = R^i{}_k(y)u^k\frac{\partial}{\partial x^i} = \zeta(y)u + \theta_y(u)y \tag{4.3.8}$$

for any $u \in T_xM$. Note that $g_y(R_y(u),y) = 0$. Then

$$\theta_y(u) = -F^{-2}(y)\zeta(y)g_y(y,u).$$

Inserting this into (4.3.8) yields

$$R_y(u) = F^{-2}(y)\zeta(y)\left(F^2(y)u - g_y(y,u)y\right),$$

which means that F is of scalar flag curvature with $\mathbf{K}(x,y) = \frac{\zeta(y)}{F^2(y)}$. $\quad\square$

4.3.2 *Projectively Invariant Quantities*

In this section, we introduce the projectively geometric quantities, for example, the Weyl curvature and the Douglas curvature, etc. Let $\psi : (M,\bar{F}) \to (M,F)$ be a projective transformation of M, where $\bar{F} = \psi^*F$. Since $\bar{F}$ is projectively equivalent to F, we have

$$\bar{G}^i = G^i + Py^i,$$

where $P = P(x,y)$ is a positively homogeneous function of degree one in y. Plugging this into (3.2.17) and using the homogeneity of P yield

$$\bar{R}^i{}_k = R^i{}_k + \zeta\delta^i_k + \theta_k y^i, \tag{4.3.9}$$

where ζ and θ_k are defined in (4.3.7). Obviously, ζ is a positively homogeneous function of degree two in y and θ_k are positively homogeneous tensors of degree one in y with $\theta_k y^k = -\zeta$. Thus,

$$\overline{\mathrm{Ric}} = \mathrm{Ric} + (n-1)\zeta. \tag{4.3.10}$$

Let

$$W^i{}_k := A^i{}_k - \frac{1}{n+1}\frac{\partial A^t{}_k}{\partial y^t}y^i, \quad A^i{}_k := R^i{}_k - \frac{\mathrm{Ric}}{n-1}\delta^i_k. \tag{4.3.11}$$

By (4.3.9)–(4.3.11), we have

$$\bar{A}^i{}_k = A^i{}_k + \theta_k y^i.$$

Differentiating this with respect to y^i gives rise to

$$\frac{\partial \bar{A}^i{}_k}{\partial y^i} = \frac{\partial A^i{}_k}{\partial y^i} + (n+1)\theta_k.$$

Consequently, one obtains that $\bar{W}^i{}_k = W^i{}_k$. Define $W_y = W^i{}_k(x,y)\frac{\partial}{\partial x^i}\otimes dx^k$ for any $y \in T_xM\backslash\{0\}$. It is invariant under the projective transformation of M. We call the tensor W_y the *Weyl projective curvature tensor* and the family $W := \{W_y | y \in T_xM\backslash\{0\}, x \in M\}$ the *Weyl projective curvature* of F. Summing up, we have

Proposition 4.3.4. *The Weyl projective curvature is invariant under the projective transformation of M, that is to say, it is a projectively invariant quantity.*

For any $y \in T_xM\backslash\{0\}$, the *Berwald curvature tensor* of F is defined by

$$B_y = B^i{}_{jkl}(x,y)\frac{\partial}{\partial x^i}\otimes dx^j \otimes dx^k \otimes dx^l, \quad B^i{}_{jkl} := \frac{\partial^3 G^i}{\partial y^j \partial y^k \partial y^l}. \quad (4.3.12)$$

The family $B := \{B_y | y \in T_xM\backslash\{0\}, x \in M\}$ is called the *Berwald curvature* of F. Obviously $B^i{}_{jkl}$ are symmetric with respect to the lower indices j, k, l and the E-curvature tensor

$$E_{ij} = \frac{1}{2}[G^k]_{y^i y^j y^k} = \frac{1}{2}B^k{}_{ijk}. \quad (4.3.13)$$

This is the origin of the E-curvature $E = \{E_y | y \in T_xM, x \in M\}$ (see Definition 4.2.4). From this, we can define a new curvature tensor, called the *Douglas curvature tensor* of F, by $D_y = D^i{}_{jkl}(x,y)\frac{\partial}{\partial x^i}\otimes dx^j\otimes dx^k\otimes dx^l$, where

$$D^i{}_{jkl} := B^i{}_{jkl} - \frac{2}{n+1}\left(E_{jk}\delta^i_l + E_{kl}\delta^i_j + E_{lj}\delta^i_k + E_{jk\cdot l}y^i\right). \quad (4.3.14)$$

Plugging (4.3.12)–(4.3.13) into (4.3.14) gives

$$D^i{}_{jkl} = \frac{\partial^3}{\partial y^j \partial y^k \partial y^l}\left(G^i - \frac{1}{n+1}\frac{\partial G^t}{\partial y^t}y^i\right). \quad (4.3.15)$$

Obviously, $D^i{}_{jkl}$ are symmetric with respect to the lower indices j, k, l. Moreover, we have

$$D^i{}_{jkl}y^j = D^i{}_{jkl}y^k = D^i{}_{jkl}y^l = 0.$$

Similarly, we call the family $D := \{D_y | y \in T_x M \backslash \{0\}, x \in M\}$ the *Douglas curvature* of F. The following is a trivial fact.

Proposition 4.3.5. $D = B$ *if and only if* $E = 0$.

Similar to the proof of Proposition 4.3.4, we have the following.

Proposition 4.3.6. *The Douglas curvature is a projectively invariant quantity.*

A Finsler metric F on a manifold M is called a *Douglas metric* if the Douglas curvature $D = 0$. The following gives an equivalent characterization of locally projectively flat Finsler metrics.

Theorem 4.3.2 (Douglas). *Let (M, F) be an $n(\geq 3)$-dimensional Finsler manifold. Then F is locally projectively flat if and only if $W = D = 0$.*

Proof. The necessity follows from Propositions 4.3.4 and 4.3.6. The proof of the sufficiency is left to the readers. Also see §13.5 in [Sh2] for more details. □

In particular, when $F(x, y) = \sqrt{g_{ij}(x)y^i y^j}$ is a Riemannian metric, by (4.3.11), we have

$$A^i_{\ k} = \left(R_j^{\ i}_{\ kl} - \frac{1}{n-1} R_{jl} \delta^i_k \right) y^j y^l,$$

$$W^i_{\ k} = \left\{ R_j^{\ i}_{\ kl} - \frac{1}{n-1} (R_{jl} \delta^i_k - R_{jk} \delta^i_l) \right\} y^j y^l,$$

where R_{ij} are the Ricci tensor of F. Define a $(1,3)$-type tensor $W = W_j^{\ i}_{\ kl} \frac{\partial}{\partial x^i} \otimes dx^j \otimes dx^k \otimes dx^l$, where

$$W_j^{\ i}_{\ kl} = \frac{1}{3} \left\{ \frac{\partial^2 W^i_{\ k}}{\partial y^j \partial y^l} - \frac{\partial^2 W^i_{\ l}}{\partial y^j \partial y^k} \right\} = R_j^{\ i}_{\ kl} - \frac{1}{n-1} (R_{jl} \delta^i_k - R_{jk} \delta^i_l).$$

This is exactly the Weyl projective curvature tensor in Riemannian geometry. From this and Theorem 4.3.2, it is easy to check the following well known Beltrami's Theorem.

Theorem 4.3.3 (Beltrami). *An $n(\geq 3)$-dimensional Riemannian manifold (M, g) is locally projectively flat if and only if it is of constant sectional curvature.*

4.4 Relations Among Geometric Invariant Quantities

In previous sections, we introduce Riemannian geometric quantities, such as the Riemannian curvature and the Ricci curvature; non-Riemannian geometric quantities, for example, the Chern curvature, the Landsberg curvature and the S-curvature, etc.; and projectively invariant quantities, such as the Weyl curvature and the Douglas curvature. In fact, these geometric quantities are not independent each other. There are some relations among these quantities. In this section we give some identities involving these geometric invariant quantities, from which we obtain some characterizations of Finsler metrics with some curvature properties.

Lemma 4.4.1 ([MS]). *The following equations hold:*

$$C_{ijk|p|q}y^p y^q + C_{ijt}R^t{}_k = -\frac{1}{3}g_{it}R^t_{k\cdot j} - \frac{1}{3}g_{jt}R^t{}_{k\cdot i} - \frac{1}{6}g_{it}R^t{}_{j\cdot k} - \frac{1}{6}g_{jt}R^t{}_{i\cdot k},$$

$$\text{(4.4.1)}$$

$$I_{k|p|q}y^p y^q + I_t R^t{}_k = -\frac{1}{3}\left\{2R^i{}_{k\cdot i} + R^i{}_{i\cdot k}\right\}. \tag{4.4.2}$$

Proof. It follows from (3.3.14) that

$$R_t{}^i{}_{kl}y^l = R^i{}_{kl\cdot t}y^l + L^i{}_{kt|l}y^l = R^i{}_{k\cdot t} + R^i{}_{tk} + L^i{}_{tk|l}y^l.$$

Thus,

$$R_{jikl}y^l = g_{il}\left(R^l{}_{k\cdot j} + R^l{}_{jk}\right) + L_{ijk|l}y^l.$$

From this and the first equality in (3.3.7), one obtains

$$L_{ijk|l}y^l + C_{ijl}R^l{}_k = -\frac{1}{2}g_{il}\left(R^l{}_{k\cdot j} + R^l{}_{jk}\right) - \frac{1}{2}g_{jl}\left(R^l{}_{k\cdot i} + R^l{}_{ik}\right). \tag{4.4.3}$$

Contracting this with g^{ij} gives

$$J_{k|l}y^l + I_l R^l{}_k = -\left(R^l{}_{k\cdot l} + R^l{}_{lk}\right). \tag{4.4.4}$$

Note that $L_{ijk|l}y^l = C_{ijk|p|q}y^p y^q$ and $J_{k|l}y^l = I_{k|p|q}y^p y^q$. Plugging (3.2.18) into (4.4.3) and (4.4.4) respectively yields (4.4.1) and (4.4.2). □

Lemma 4.4.2 ([CMS]). *The following Ricci identities for S-curvature hold:*

$$S_{\cdot k|l}y^l - S_{|k} = I_{k|p|q}y^p y^q + I_l R^l{}_k, \tag{4.4.5}$$

$$S_{\cdot k||l}y^l - S_{|k} = -\frac{1}{3}\left\{2R^i{}_{k\cdot i} + R^i{}_{i\cdot k}\right\}. \tag{4.4.6}$$

Proof. (4.4.6) follows from (4.4.2) and (4.4.5) directly. Thus, it suffices to prove (4.4.5). Take a local frame $\{e_i\}_{i=1}^n$ for π^*TM and the corresponding local coframe $\{\omega^i, \omega^{n+i}\}$ for $T^*(TM_0)$. Write

$$d\tau = \tau_{|k}\omega^k + \tau_{\cdot k}\omega^{n+k}. \tag{4.4.7}$$

Note that $\tau_{\cdot k} = \tau_{y^k} = I_k$. From this, (4.4.7), (3.2.9) and (3.2.8), one obtains

$$0 = d^2\tau = \left(\tau_{|k|l}\omega^l + \tau_{|k\cdot l}\omega^{n+l}\right) \wedge \omega^k + \left(I_{k|l}\omega^l + I_{k\cdot l}\omega^{n+l}\right) \wedge \omega^{n+k} + I_k\Omega^k,$$

which implies that the following Ricci identities:

$$\tau_{|k|l} - \tau_{|l|k} = I_t R^t{}_{kl}, \tag{4.4.8}$$

$$\tau_{|k\cdot l} - I_{l|k} = -I_t L^t{}_{kl}. \tag{4.4.9}$$

Contracting (4.4.9) with y^k and using (4.2.8), (4.2.4), one obtains

$$S_{\cdot k} = (\tau_{|l}y^l)_{\cdot k} = I_{k|l}y^l + \tau_{|k} = J_k + \tau_{|k}.$$

Consequently,

$$S_{\cdot k|l} = \tau_{|k|l} + J_{k|l}. \tag{4.4.10}$$

From (4.4.10), (4.4.8) and $J_l y^l = 0$, we have

$$\begin{aligned}
S_{\cdot k|l}y^l - S_{|k} &= \left(S_{\cdot k|l} - S_{\cdot l|k}\right)y^l \\
&= \left(\tau_{|k|l} - \tau_{|l|k}\right)y^l + \left(J_{k|l} - J_{l|k}\right)y^l \\
&= I_{k|p|q}y^p y^q + I_t R^t{}_k.
\end{aligned}$$

This finishes the proof. $\qquad\square$

Lemma 4.4.3. *Let (M, F) be an n-dimensional Finsler manifold. If F is of scalar flag curvature $\mathbf{K} = \mathbf{K}(x, y)$, then*

$$I_{i|p|q}y^p y^q = -\frac{1}{3}F^2\left\{(n+1)\mathbf{K}_{\cdot i} + 3\mathbf{K}I_i\right\}; \tag{4.4.11}$$

$$M_{ijk|p|q}y^p y^q + \mathbf{K}F^2 M_{ijk} = 0. \tag{4.4.12}$$

Proof. Since F is of scalar flag curvature, we have by (4.1.3)

$$R^i{}_k = \mathbf{K}\left(F^2\delta^i_k - g_{kp}y^p y^i\right).$$

Differentiating this with respect to y^l yields

$$R^i{}_{k\cdot l} = \mathbf{K}_{\cdot l}F^2 h^i_k + \mathbf{K}\left(2g_{lp}y^p\delta^i_k - g_{kp}y^p\delta^i_l - g_{kl}y^i\right), \tag{4.4.13}$$

where $h^i_k = \delta^i_k - F^{-2}g_{kp}y^p y^i$. Then (4.4.11) follows from (4.4.2) and (4.4.13).

Further, let $h_{ij} = g_{ik}h_j^k$. By (4.4.1) and (4.4.13), we get

$$C_{ijk|p|q}y^p y^q = -\frac{1}{3}F^2 \left(\mathbf{K}_{.i}h_{jk} + \mathbf{K}_{.j}h_{ik} + \mathbf{K}_{.k}h_{ij} + 3\mathbf{K}C_{ijk}\right). \qquad (4.4.14)$$

Recall that the Matsumoto torsion is define by

$$M_{ijk} = C_{ijk} - \frac{1}{n+1}\left(I_i h_{jk} + I_j h_{ik} + I_k h_{ij}\right). \qquad (4.4.15)$$

Observe that $h_{ij|p} = 0$. Taking the covariant derivatives on (4.4.15) twice and using (4.4.11), (4.4.14) yield (4.4.12). $\qquad \square$

Theorem 4.4.1 ([CMS]). *Let (M, F) be an n-dimensional Finsler manifold of scalar flag curvature $\mathbf{K} = \mathbf{K}(x, y)$. If F is of almost isotropic S-curvature, then F is of almost isotropic flag curvature.*

Proof. Since the S-curvature is almost isotropic, there are a scalar function $c = c(x)$ and a closed 1-form $\eta = \eta_i(x)y^i$ on M such that

$$S = (n+1)(cF + \eta). \qquad (4.4.16)$$

Thus,

$$S_{.k|l}y^l - S_{|k} = (n+1)\left(c_{|l}y^l F_{.k} - c_{|k}F\right),$$

where we used the closeness of η, which is equivalent to $\eta_{i|j} = \eta_{j|i}$ for any $1 \le i, j \le n$.

On the other hand, plugging (4.4.13) into (4.4.6) yields

$$S_{.k|l}y^l - S_{|k} = -\frac{1}{3}(n+1)\mathbf{K}_{.k}F^2.$$

Consequently, one obtains

$$c_{|l}y^l F_{.k} - c_{|k}F = -\frac{1}{3}\mathbf{K}_{.k}F^2.$$

Note that $c_{|l} = c_{x^l}$. The above equation can be rewritten as

$$\left(\frac{1}{3}\mathbf{K} - \frac{1}{F}c_{x^l}y^l\right)_{y^k} = 0,$$

which means that there is a scalar function σ on M such that $\mathbf{K} = \frac{3c_{x^l}y^l}{F} + \sigma$. $\qquad \square$

Corollary 4.4.1. *Let (M, F) be an n-dimensional Finsler manifold of scalar flag curvature $\mathbf{K} = \mathbf{K}(x, y)$. If F is of constant S-curvature, then F is of isotropic flag curvature, in particular, F is of constant flag curvature when $n \ge 3$.*

Theorem 4.4.2. *Let (M, F) be an n-dimensional Finsler manifold of constant flag curvature $\mathbf{K} \neq 0$. If the mean Landsberg curvature $J = 0$, then F is Riemannian.*

Proof. It follows from (3.2.18) and (4.4.13) that

$$R^i{}_{kl} = \frac{1}{3}\mathbf{K}_{.l}F^2 h^i_k - \frac{1}{3}\mathbf{K}_{.k}F^2 h^i_l + \mathbf{K}\left(g_{lp}\delta^i_k - g_{kp}\delta^i_l\right)y^p. \quad (4.4.17)$$

Since $\mathbf{K}$ is a nonzero constant by the assumption, the above equality becomes

$$R^i{}_{kl} = \mathbf{K}\left(g_{lp}\delta^i_k - g_{kp}\delta^i_l\right)y^p. \quad (4.4.18)$$

Inserting this into (3.3.14) leads to

$$R_j{}^i{}_{kl} = \mathbf{K}\left(g_{jl}\delta^i_k - g_{jk}\delta^i_l\right) + L^i{}_{kj|l} - L^i{}_{lj|k} - L^i{}_{ks}L^s{}_{lj} + L^i{}_{ls}L^s{}_{kj}. \quad (4.4.19)$$

Plugging (4.4.19) into (3.3.12) yields

$$2\mathbf{K}\left(C_{jlt}\delta^i_k - C_{jkt}\delta^i_l\right) = P_j{}^i{}_{kt|l} - P_j{}^i{}_{lt|k} - P_j{}^i{}_{ks}L^s{}_{lt} + P_j{}^i{}_{ls}L^s{}_{kt}$$

$$-L^i{}_{kj|l\cdot t} + L^i{}_{lj|k\cdot t} + L^i{}_{ks}L^s{}_{lj\cdot t} + L^i{}_{ks\cdot t}L^s{}_{lj}$$

$$-L^i{}_{ls\cdot t}L^s{}_{kj} - L^i{}_{ls}L^s{}_{kj\cdot t}. \quad (4.4.20)$$

Contracting this with y_i and y^l respectively yields

$$L_{jkt|l}y^l + \mathbf{K}F^2 C_{jkt} = 0, \quad (4.4.21)$$

which means that

$$J_{k|l}y^l + \mathbf{K}F^2 I_k = 0. \quad (4.4.22)$$

Since $J_k = 0$, we have $I_k = 0$. By Deicke's Theorem, we conclude that F is Riemannian. $\qquad\square$

Note that the Berwald metrics are Landsberg metrics. By Theorem 4.4.2, Berwald metrics of nonzero constant flag curvature must be Riemannian.

Theorem 4.4.3 ([Sz], [Ma3]). *On an n-dimensional Finsler manifold (M, F), F is of scalar flag curvature if and only if the Weyl curvature $W = 0$.*

Proof. Assume that F is of scalar flag curvature $\mathbf{K} = \mathbf{K}(x, y)$. Then

$$R^i{}_k = \mathbf{K}(F^2 \delta^i_k - F F_{y^k} y^i), \quad R := \frac{1}{n-1} \text{Ric} = \mathbf{K} F^2. \quad (4.4.23)$$

Thus, $A^i{}_k = -\mathbf{K} F F_{y^k} y^i$ and

$$W^i{}_k = A^i{}_k - \frac{1}{n+1} \frac{\partial A^t{}_k}{\partial y^t} y^i = -\mathbf{K} F F_{y^k} y^i + \mathbf{K} F F_{y^k} y^i = 0.$$

Conversely, assume that $W^i{}_k = 0$ and let $\xi_k := \frac{1}{n+1} \frac{\partial A^t{}_k}{\partial y^t}$. Then $A^i{}_k = \xi_k y^i$ and $\xi_k y^k = -R$. Hence, by the definition of $A^i{}_k$ given in (4.3.11), we have

$$R^i{}_k = R \delta^i_k + \xi_k y^i.$$

Contracting this with g_{ij} yields

$$R_{jk} = R g_{jk} + \xi_k y_j,$$

where $y_j = g_{ij} y^i = F F_{y^j}$. Since $R_{jk} = R_{kj}$, we have $\xi_k y_j = \xi_j y_k$. Contracting this with y^j again gives $\xi_k = -F^{-2} R y_k$. Consequently,

$$R^i{}_k = R \delta^i_k - F^{-2} R y_k y^i = F^{-2} R (F^2 \delta^i_k - F F_{y^k} y^i).$$

Thus, F is of scalar flag curvature with $\mathbf{K} = \frac{R}{F^2}$. $\qquad\square$

4.5 Some Rigidity Theorems

We say that a Finsler manifold (M, F) is *forward complete* (resp., *backward complete*) if the geodesic $\gamma : [0, \ell) \to M$ can be extended to $[0, +\infty)$ (resp., $\gamma : (-\ell, 0] \to M$ can be extended to $(-\infty, 0]$). (M, F) is called *complete* if it is both forward complete and backward complete.

Let (M, F) be a complete Finsler manifold and d_F be the distance on M induced by F. Let $\mathbf{M}$ be the Matsumoto torsion. The *norm of* $\mathbf{M}$ at a point $x \in M$ is defined by

$$\|\mathbf{M}\|_x := \sup_{y,u,v,w \in T_x M_0} \frac{F(x, y) |\mathbf{M}_y(u, v, w)|}{\sqrt{g_y(u, u) g_y(v, v) g_y(w, w)}}.$$

We say that the Matsumoto torsion $\mathbf{M}$ *grows sub-exponentially* at a rate of $k > 0$ if for any point $x \in M$

$$M(x, r) := \sup_{\min(d_F(x,z), d_F(z,x)) \leq r} \|\mathbf{M}\|_x = o(e^{kr}), \quad (r \to +\infty).$$

We say that the Matsumoto torsion $\mathbf{M}$ *grows sub-linearly* if for any $x \in M$,

$$\lim_{r \to +\infty} r^{-1} M(x, r) = 0.$$

Similarly, we define the *norm* of the (mean) Cartan torsion I at a point $x \in M$ as follows.

$$\|I\|_x := \sup_{y,u \in T_x M_0} \frac{F(x,y)|I_y(u)|}{\sqrt{g_y(u,u)}},$$

$$\|C\|_x := \sup_{y,u,v,w \in T_x M_0} \frac{F(x,y)|C_y(u,v,w)|}{\sqrt{g_y(u,u)g_y(v,v)g_y(w,w)}}.$$

Let

$$I(x,r) := \sup_{\min(d_F(x,z),d_F(z,x)) \leq r} \|I\|_x, \quad C(x,r) := \sup_{\min(d_F(x,z),d_F(z,x)) \leq r} \|C\|_x.$$

We also can define the growth rates for I and C as above.

Theorem 4.5.1 ([MS]). *Let (M,F) be an $n(\geq 3)$-dimensional complete Finsler manifold of scalar flag curvature $\mathbf{K} = \mathbf{K}(x,y)$ with $\mathbf{K} \leq -1$. Suppose that the Matsumoto torsion grows sub-exponentially at rate of $k = 1$. Then F is a Randers metric.*

Proof. It suffices to prove that $\mathbf{M} = \mathbf{0}$ by Theorem 2.2.2. We argue this by contradiction. Assume that $\mathbf{M} \neq \mathbf{0}$. Then $\mathbf{M}_y(u,u,u) \neq 0$ for some $y, u \in T_x M_0$ with $F(x,y) = 1$. Let $\gamma(t)$ be the unit speed geodesic with $\gamma(0) = x$ and $\dot{\gamma}(0) = y$, and $U = U^i(t) \frac{\partial}{\partial x^i}$ be the parallel vector field along γ with $U(0) = u$. Let

$$\mathcal{M} := \mathcal{M}_{\dot{\gamma}}(U(t), U(t), U(t)) = \mathcal{M}_{ijk}(\gamma, \dot{\gamma})U^i(t)U^j(t)U^k(t).$$

By (4.4.12), we have

$$\mathcal{M}''(t) + \mathbf{K}(t)\mathcal{M}(t) = 0,$$

where $\mathbf{K}(t) := \mathbf{K}(\gamma(t), \dot{\gamma}(t)) \leq -1$. Let (a_1, b_1) be the maximal interval on which $\mathcal{M}(t) \neq 0$ and $f(t) := \frac{\mathcal{M}'(t)}{\mathcal{M}(t)}$ for $t \in (a_1, b_1)$. The above equality can be rewritten as

$$f' + f^2 = -\mathbf{K}(t) \geq 1. \tag{4.5.1}$$

Consider the ODE

$$\widetilde{\mathcal{M}}''(t) - \widetilde{\mathcal{M}}(t) = 0. \tag{4.5.2}$$

Then the solution of (4.5.2) with $\widetilde{\mathcal{M}}(0) = \mathcal{M}(0)$ and $\widetilde{\mathcal{M}}'(0) = \mathcal{M}'(0)$ is given by

$$\widetilde{\mathcal{M}}(t) = \mathcal{M}(0)\cosh(t) + \mathcal{M}'(0)\sinh(t).$$

Similarly, let (a_2, b_2) be the maximal interval on which $\widetilde{\mathcal{M}}(t) \neq 0$ and let $\tilde{f}(t) := \frac{\widetilde{\mathcal{M}}'(t)}{\widetilde{\mathcal{M}}(t)}$ for $t \in (a_2, b_2)$. Then (4.5.2) is rewritten as

$$\tilde{f}' + \tilde{f}^2 = 1. \tag{4.5.3}$$

We claim that $\varphi(t) := |\mathcal{M}(t)/\widetilde{\mathcal{M}}(t)|$ attains its minimum $\varphi(0) = 1$ at $t = 0$. To show this, let

$$h(t) := (f(t) - \tilde{f}(t)) \exp\left(\int (f(t) + \tilde{f}(t))dt\right). \tag{4.5.4}$$

By (4.5.1) and (4.5.3), we have

$$h'(t) = \left\{ f' + f^2 - \tilde{f}' - \tilde{f}^2 \right\} \exp\left(\int (f(t) + \tilde{f}(t))dt\right) \geq 0.$$

Note that $h(0) = 0$. Thus, $h(t) \leq 0$ for $t < 0$ and $h(t) \geq 0$ for $t > 0$. Since $h(t)$ has the same sign as $f(t) - \tilde{f}(t)$, we conclude that

$$\frac{\varphi'(t)}{\varphi(t)} = f(t) - \tilde{f}(t) = \begin{cases} \geq 0, & \text{if } t > 0, \\ \leq 0, & \text{if } t < 0. \end{cases}$$

This implies that $\varphi'(t) \geq 0$ for $t > 0$ and $\varphi'(t) \leq 0$ for $t < 0$. Therefore, $\varphi(t)$ attains its minimum at $t = 0$. Consequently, $\varphi(t) \geq \varphi(0) = 1$, namely, $|\mathcal{M}(t)| \geq |\widetilde{\mathcal{M}}(t)|$ for $\max(a_1, a_2) < t < \min(b_1, b_2)$. Clearly, $(a_2, b_2) \subset (a_1, b_1)$. Since $d_F(\gamma(-t), x) < t$ and $d_F(x, \gamma(t)) \leq t$ for any $t > 0$, one obtains

$$M(x, r) \geq \max\{\mathcal{M}(t) \mid |t| < r\}.$$

Suppose that $\mathcal{M}'(0) = 0$ or it has the same sign as $\mathcal{M}(0)$. Since $\widetilde{\mathcal{M}}(t) \neq 0$ for all $t > 0$, we have $b_2 = +\infty$ and

$$M(x, r) \geq |\mathcal{M}(r)| \geq |\mathcal{M}(0)| \cosh(r) + |\mathcal{M}'(0)| \sinh(r).$$

Suppose that $\mathcal{M}'(0)$ has the opposite sign as $\mathcal{M}(0)$. Since $\widetilde{\mathcal{M}}(t) \neq 0$ for all $t < 0$, we have $a_2 = -\infty$ and

$$M(x, r) \geq |\mathcal{M}(-r)| \geq |\mathcal{M}(0)| \cosh(r) + |\mathcal{M}'(0)| \sinh(r).$$

In either case, we have

$$\lim_{r \to \infty} \inf \frac{M(x, r)}{e^r} \geq \frac{1}{2}(|\mathcal{M}(0)| + |\mathcal{M}'(0)|) > 0.$$

However, $M(x, r)$ grows sub-exponentially at rate of $k = 1$. This is a contradiction. Thus $\mathbf{M} = 0$, which means that F is a Randers metric. $\quad\square$

Let (M, F) be a closed Finsler manifold. Then F is complete and the Matsumoto torsion is bounded. By Theorems 4.5.1 and 4.3.1, we have the following corollary.

Corollary 4.5.1. *Let (M, F) be an $n(\geq 3)$-dimensional closed Finsler manifold. Suppose that F is of scalar flag curvature with negative flag curvature, then it is a Randers metric. In particular if F is a locally projectively flat Finsler metric with negative flag curvature, then it is a locally projectively flat Randers metric.*

Theorem 4.5.2 ([ChS]). *Let (M, F) be a complete Finsler manifold with isotropic flag curvature $\mathbf{K} = \mathbf{K}(x)$.*

(1) *If $\mathbf{K} \leq -1$ and the mean Cartan torsion I grows sub-exponentially at rate of $k = 1$, then F is Riemannian;*

(2) *If $\mathbf{K} \leq 0$ and the Cartan torsion C (resp., I) grows sub-linearly, then F is Landsbergian (resp., weakly Landsbergian). Further, F is Riemannian on any open subset where $\mathbf{K} < 0$.*

Proof. Let $\gamma(t)$ be the unit speed geodesic on M with $\gamma(0) = x \in M$, $\dot{\gamma}(0) = y \in T_x M \backslash \{0\}$ and $U = U(t)$ the parallel vector field along γ with $U(0) = u \in T_x M \backslash \{0\}$. Set

$$I(t) := I_{\dot{\gamma}}(U(t)) = I_i(\gamma(t), \dot{\gamma}(t)) U^i(t).$$

By assumption, we have $\mathbf{K}_{.k} = 0$. It follows from (4.4.11) that

$$I''(t) + \mathbf{K}(t) I(t) = 0, \qquad (4.5.5)$$

where $\mathbf{K}(t) = \mathbf{K}(\gamma(t))$.

(1) Suppose that $\mathbf{K} \leq -1$ and $I(x, r) = o(e^r)$ for any $x \in M$. By the same arguments as in the proof of Theorem 4.5.1, we can show that $I(0) = I_i(x, y) u^i = 0$, which implies that $I_i = 0$ since y, u are arbitrary. Thus, $I = 0$. By Deicke's Theorem, F is Riemannian.

(2) Suppose that $\mathbf{K} \leq 0$ and I grows sub-linearly. We first show that $J = 0$ by contradiction. Thus F is weakly Landsbergian. Assume that $J_y(u) \neq 0$ for some $y, u \in T_x M \backslash \{0\}$. Let $I(t)$ be defined as above such that $I(0) = I_y(u)$ and $I'(0) = J_y(u) \neq 0$. Consequently, we have $I(\varepsilon) \neq 0$ and $I'(\varepsilon) \neq 0$ for sufficiently small $\varepsilon > 0$. Thus, we may assume that $I(0) \neq 0$ and $I'(0) \neq 0$.

Consider the function $\tilde{I}(t) := I(0) + I'(0)t$ for $t \in (a, b)$, where (a, b) is the maximal interval containing 0 on which $\tilde{I}(t)$ is not equal to zero. Let

$$f(t) := \frac{I'(t)}{I(t)}, \quad \tilde{f}(t) := \frac{\tilde{I}'(t)}{\tilde{I}(t)}.$$

We have

$$f'(t) + f(t)^2 = 0, \quad \tilde{f}'(t) + \tilde{f}(t)^2 = 0.$$

Let $h(t)$ be defined by (4.5.4). As in the proof of Theorem 4.5.1, we have $h'(t) \geq 0$. Note that $h(0) = 0$. Hence, $h(t) \geq 0$ for $t \geq 0$ and $h(t) \leq 0$ for $t \leq 0$, which imply that $\log \left| \frac{I(t)}{\tilde{I}(t)} \right|$ attains its minimum at $t = 0$, that is,

$$\log \left| \frac{I(t)}{\tilde{I}(t)} \right| \geq \log \left| \frac{I(0)}{\tilde{I}(0)} \right| = 0 \Leftrightarrow |I(t)| \geq |\tilde{I}(t)|.$$

Thus, we have

$$|I(t)| \geq |I(0) + I'(0)t|, \quad t \in (a, b).$$

If $I'(0)$ has the same sign as $I(0)$, then

$$I(x, t) \geq |I(t)| \geq |I(0)| + |I'(0)|t, \quad t > 0.$$

Hence, $b = +\infty$. Letting $t \to \infty$ yields that $I(t)$ grows at least linearly. This is impossible. If $I'(0)$ has the opposite sign as $I(0)$, then

$$I(x, t) \geq |I(t)| \geq |I(0)| - |I'(0)|t, \quad t < 0.$$

Thus, $a = -\infty$. Letting $t \to -\infty$ yields that $I(t)$ grows at least linearly. This is impossible. Therefore, $I'(0) = 0$, that is, $J_y(u) = 0$. This is a contradiction. In either case, we have $J = 0$. Thus, (4.4.11) is reduced to $\mathbf{K}I_k = 0$. By Deicke's Theorem, F is Riemannian on any subset where $\mathbf{K} < 0$.

Now, assume that $\mathbf{K} \leq 0$ and the Cartan torsion grows sub-linearly. By (4.4.14),

$$C_{ijk|p|q}y^p y^q + \mathbf{K}F^2 C_{ijk} = 0.$$

By a similar argument, one can show that $L_{ijk} = C_{ijk|l}y^l = 0$. Thus, F is Landsbergian. $\qquad\square$

The following Akbar-Zadeh's theorem can be regarded as a corollary of Theorem 4.5.2. Also see Corollary 7.2.5 in [ChS] or Theorem 9.1.2 in [Sh1] for more details.

Theorem 4.5.3 ([AZ]). *Let (M, F) be a complete Finsler space of constant flag curvature $\mathbf{K}$. Assume that the (mean) Cartan torsion is bounded.*

(1) *If $\mathbf{K} = 0$, then F is a (weak) Landsberg metric;*
(2) *If $\mathbf{K} < 0$, then F is a Riemannian metric.*

Proof. Let $\gamma = \gamma(t)$, the unit speed geodesic with $\gamma(0) = x \in M$ and $\dot{\gamma}(0) = y \in T_x M \setminus \{0\}$, and $U = U(t)$ the parallel vector field along γ with $U(0) = u \in T_x M \setminus \{0\}$. Set

$$I(t) := I_{\dot{\gamma}}(U(t)), \quad J(t) := J_{\dot{\gamma}}(U(t)),$$

$$C(t) := C_{\dot{\gamma}}(U(t), U(t), U(t)), \quad L(t) = L_{\dot{\gamma}}(U(t), U(t), U(t)).$$

It follows from (3.2.13) and (4.2.4) that

$$L(t) = C'(t), \quad J(t) = I'(t).$$

By (4.4.21)–(4.4.22), we have $L'(t) + \mathbf{K}C(t) = 0$ and $J'(t) + \mathbf{K}I(t) = 0$, equivalently,

$$C''(t) + \mathbf{K}C(t) = 0, \quad I''(t) + \mathbf{K}I(t) = 0.$$

Their general solutions are given by

$$C(t) = \mathfrak{s}_{\mathbf{K}}(t)L(0) + \mathfrak{s}'_{\mathbf{K}}(t)C(0), \quad I(t) = \mathfrak{s}_{\mathbf{K}}(t)J(0) + \mathfrak{s}'_{\mathbf{K}}(t)I(0),$$

where $\mathfrak{s}_{\mathbf{K}}(t)$ is the unique solution of the following equation:

$$y''(t) + \mathbf{K}y(t) = 0, \quad y(0) = 0, \quad y'(0) = 1.$$

Case 1. If $\mathbf{K} = 0$, then $C(t) = L(0)t + C(0)$. Note that C is bounded. We have $L(0) = 0$, i.e., $L_{ijk}(y)u^i u^j u^k = 0$. Thus, $L = 0$, that is, F is Landsbergian, since y, u are arbitrary. Similarly, if I is bounded, then F is weakly Landsbergian.

Case 2. If $\mathbf{K} < 0$, then

$$C(t) = \sinh(t)L(0) + \cosh(t)C(0).$$

Since C is bounded, we get $L(0) = C(0) = 0$, which implies that $C = 0$ as in case 1. Thus, F is Riemannian. $\square$

Every closed Finsler manifold (M, F) is complete and the Cartan torsion, the (mean) Cartan torsion on M are bounded. Thus we have the following.

Corollary 4.5.2 ([AZ]). *Let (M, F) be a closed Finsler space of constant flag curvature* **K**.

(1) *If* **K** $= 0$, *then F is a (weak) Landsberg metric;*
(2) *If* **K** < 0, *then F is a Riemannian metric.*

Similar to the arguments in the proof of Theorem 4.5.2 or Theorem 4.5.3, one obtains the following result.

Theorem 4.5.4 ([Wu1]). *Any closed weakly Landsberg manifold with the negative flag curvature must be Riemannian.*

Theorem 4.5.5 ([Sh5]). *Let (M, F) be a complete Finsler manifold with flag curvature* **K** ≤ 0. *Suppose that F has constant S-curvature and bounded mean Cartan torsion. Then F is weakly Landsbergian with $R_y(I_y) = 0$. Moreover, F is Riemannian on any open subset where* **K** < 0.

Proof. Let $\{e_i\}_{i=1}^n$ be a local frame for TM and $\{e_i = (x, y; e_i)\}_{i=1}^n$ be the corresponding local frame for $\pi^* TM$ with dual frame $\{\omega^i = (x, y; \omega^i)\}_{i=1}^n$, where $\{\omega^i\}_{i=1}^n$ is a local frame for T^*M dual to $\{e_i\}_{i=1}^{i=n}$. The dual of the mean Cartan tensor $I_y = I_i \omega^i$ is still denoted by I_y with $I_y = I^i(x, y)e_i$, where $I^i = g^{ij} I_j$. Similarly, the dual to the Landsberg tensor $J_y = J_i(x, y)\omega^i$ is denoted by $J_y = J^i(x, y)e_i$, where $J^i = g^{ij} J_j$. Thus,

$$R_y(I_y) = R^i{}_k I^k e_i.$$

By the assumption, $S = (n + 1)cF$ for some constant c. Thus,

$$S_{\cdot k|l} y^l - S_{|k} = (n + 1)c \left(F_{\cdot k|l} y^l - F_{|k} \right) = 0.$$

It follows from (4.4.5) that

$$J^i{}_{|l} y^l + R^i{}_l I^l = 0. \tag{4.5.6}$$

Let $\gamma : \mathbb{R} \to M$ be a geodesic and

$$I(t) := I^i(\gamma(t), \dot{\gamma}(t))e_i|_\gamma, \quad J(t) := J^i(\gamma(t), \dot{\gamma}(t))e_i|_\gamma.$$

Then $D_{\dot{\gamma}} I(t) = J(t)$ and $D_{\dot{\gamma}} J(t) = D_{\dot{\gamma}} D_{\dot{\gamma}} I(t)$. From these and (4.5.6), we have

$$D_{\dot{\gamma}} D_{\dot{\gamma}} I(t) + R_{\dot{\gamma}}(I(t)) = 0. \tag{4.5.7}$$

Consider the function $\varphi(t) := g_{\dot\gamma}(I(t), I(t))$. Then

$$\varphi''(t) = 2g_{\dot\gamma}(D_{\dot\gamma}D_{\dot\gamma}I(t), I(t)) + 2g_{\dot\gamma}(D_{\dot\gamma}I(t), D_{\dot\gamma}I(t))$$
$$= -2g_{\dot\gamma}(R_{\dot\gamma}(I(t)), I(t)) + 2g_{\dot\gamma}(J(t), J(t)) \geq 0, \qquad (4.5.8)$$

where we used $\mathbf{K} \leq 0$. Consequently, $\varphi(t)$ is convex and nonpositive. Suppose $\varphi'(t_0) \neq 0$ for some t_0. If $\varphi'(t_0) < 0$, then

$$\varphi(t) \geq \varphi(t_0) - \varphi'(t_0)(t_0 - t), \quad t < t_0.$$

If $\varphi'(t_0) > 0$, then

$$\varphi(t) \geq \varphi(t_0) + \varphi'(t_0)(t - t_0), \quad t > t_0.$$

One can see that $\lim_{t \to +\infty} \varphi(t) = \infty$ or $\lim_{t \to -\infty} \varphi(t) = \infty$. This implies that $I(t)$ is unbounded. This contradicts with the assumption. Therefore, $\varphi'(t) = 0$ and hence $\varphi''(t) = 0$, which implies that $R_{\dot\gamma}I(t) = 0$ and $J(t) = 0$ by (4.5.8). Since γ is arbitrary, we conclude that $R_y(I_y) = 0$ and $J_y = 0$. This prove the first claim.

Further, assume that F has negative flag curvature at a point $x \in M$. Since I_y is orthogonal to y with respect to g_y, and $R_y(I_y) = 0$, we obtain that $I_y = 0$ for all $y \in T_x M \backslash \{0\}$. By Deicke's Theorem, F is Riemannian.

$\square$

Chapter 5

Theory of Geodesics

Geodesics are one of the most important research objects on Finsler manifolds. In this chapter, we will introduce the basic theory of geodesics in Finsler geometry, including the exponential map, the first and second variation formulae of arc length, the Jacobi field, the conjugate point, the cut point and the minimality of geodesics, etc.

5.1 Geodesics and Exponential Map

In this section, we shall introduce the first variation formula of arc length for geodesics and the Gauss lemma. As applications, we will study the local minimizing property of arc length for geodesics.

5.1.1 *Geodesics*

Let (M, F) be a Finsler manifold and $\pi : TM_0 \to M$ be the natural projection from the tangent bundle TM to M. Assume that G is the spray induced by F and $\hat{\gamma}$ is the integral curve of G on TM_0, i.e., $\dot{\hat{\gamma}} = G|_{\hat{\gamma}}$. In local coordinate system (x^i, y^i), G is expressed by (3.1.15) and $\hat{\gamma}(t) = (x(t), y(t))$ satisfies

$$\dot{x}^i(t) = y^i(t), \quad \dot{y}^i(t) + 2G^i(x(t), y(t)) = 0, \tag{5.1.1}$$

which mean that $\gamma(t) = \pi \circ \hat{\gamma}(t) = (x^i(t))$ satisfies (3.1.21), that is, γ is a geodesic on (M, F). By the ODE theory, γ smoothly depends on $y \in TM_0$ and only C^1 at the zero section. Moreover, there is a unique geodesic $\gamma : (-\varepsilon, \varepsilon) \to M$ with $\gamma(0) = x$ and $\dot{\gamma}(0) = y$ for some sufficiently small $\varepsilon > 0$. Conversely, given a C^∞ geodesic $\gamma = \gamma(t)$ on (M, F), $\hat{\gamma}(t) = (\gamma(t), \dot{\gamma}(t))$ is then an integral of G, which is called the *canonical lift* of γ on TM_0.

Proposition 5.1.1. *If a C^∞ curve $\gamma : [a, b] \to M$ on a Finsler manifold (M, F) is a geodesic, then it is of constant speed. In particular, its arc length $L(\gamma) = (b - a)F(\dot{\gamma}(0))$.*

Proof. Since γ is a geodesic, it satisfies (3.1.21). From this and (3.1.3), we get

$$\frac{d}{dt}\left(F^2(\gamma(t), \dot{\gamma}(t))\right) = \frac{d}{dt}\left(g_{ij}(\gamma, \dot{\gamma})\dot{\gamma}^i(t)\dot{\gamma}^j(t)\right)$$

$$= \frac{\partial g_{ij}}{\partial x^k}\dot{\gamma}^k\dot{\gamma}^i\dot{\gamma}^j + 2C_{ijk}\ddot{\gamma}^k\dot{\gamma}^i\dot{\gamma}^j + 2g_{ij}\ddot{\gamma}^i\dot{\gamma}^j$$

$$= 2g_{il}N^l{}_j\dot{\gamma}^i\dot{\gamma}^j - 4g_{ij}G^i\dot{\gamma}^j = 0,$$

where we used (3.1.11) in the last equality. Consequently, $F(\gamma, \dot{\gamma})$ is a constant. $\square$

Proposition 5.1.1 implies that the parameter t is the arc length of γ if and only if $F(\dot{\gamma}) = 1$. A geodesic $\gamma(t)$ with $F(\dot{\gamma}) = 1$ is called a *normal geodesic*. A normal geodesic is a geodesic parameterized by arc length. To characterize the minimality of arc length for geodesics, we introduce the variation of a curve.

Definition 5.1.1. Let $\gamma : [a, b] \to M$ be a (piecewise) smooth curve on a Finsler manifold (M, F). If there is a sufficiently small positive number ε and a (piecewise) smooth map $H : [a, b] \times (-\varepsilon, \varepsilon) \to M$ with $H(t, 0) = \gamma(t)$ for all $a \le t \le b$, then H is called a *(piecewise) smooth variation* of γ. If H satisfies $H(a, s) = \gamma(a)$ and $H(b, s) = \gamma(b)$ for all $s \in (-\varepsilon, \varepsilon)$, then H is called a *variation preserving endpoints fixed*.

For any variation H of γ, we define two vector fields along H by

$$\widetilde{T}(t, s) := \frac{\partial H}{\partial t}, \quad \widetilde{V}(t, s) := \frac{\partial H}{\partial s},$$

which are respectively tangent vector fields of variation curves $\gamma_s :$ $[a, b] \to M, \gamma_s(t) = H(t, s)$ for all $s \in (-\varepsilon, \varepsilon)$ and the transversal curves $\gamma_t : (-\varepsilon, \varepsilon) \to M, \gamma_t(s) = H(t, s)$ for all $t \in [a, b]$. In particular, $T(t) := \widetilde{T}(t, 0) = \dot{\gamma}(t)$, which is just the tangent vector of $\gamma(t)$, and $V(t) := \widetilde{V}(t, 0)$, which is said to be a *variation vector field* along γ. Obviously, $V(a) = V(b) = 0$ if H is a variation preserving endpoints fixed. The following first variation formula for geodesics is well known ([BCS]).

Theorem 5.1.1 (First Variation Formula). *Let $\gamma : [a, b] \to M$ be a piecewise smooth curve on a Finsler manifold (M, F) with constant speed ℓ and $H : [a, b] \times (-\varepsilon, \varepsilon) \to M$ be a piecewise smooth variation of γ. Assume that $V(t)$ is the variation vector field along γ and $L(s) := L_F(\gamma_s)$ is the arc length of γ_s parametrized by $\gamma_s(t) = H(t, s)$. Then, for any partition $a = t_0 < t_1 \cdots < t_k = b$, we have the following first variation formula:*

$$L'(0) = \frac{1}{\ell} \left\{ g_T(T, V) \Big|_a^b + \sum_{i=1}^{k-1} g_T \left(T(t_i^-) - T(t_i^+), V(t_i) \right) \right.$$
$$\left. - \int_a^b g_T(V, D_T^T T) dt \right\}. \tag{5.1.2}$$

In particular, if H is a smooth variation, then

$$L'(0) = \frac{1}{\ell} \left\{ g_T(T, V) \Big|_a^b - \int_a^b g_T(V, D_T^T T) dt \right\}. \tag{5.1.3}$$

Proof. By definition,

$$L(s) = \int_a^b F(\gamma_s, \dot\gamma_s) dt = \sum_{i=1}^{k} \int_{t_{i-1}}^{t_i} F(\gamma_s, \dot\gamma_s) dt.$$

Then

$$L'(0) = \frac{dL}{ds} \Big|_{s=0} = \sum_{i=1}^{k} \int_{t_{i-1}}^{t_i} \left\{ \frac{1}{2F} \frac{d}{ds} [F^2(\gamma_s, \dot\gamma_s)] \right\} \Big|_{s=0} dt$$

$$= \frac{1}{2\ell} \sum_{i=1}^{k} \int_{t_{i-1}}^{t_i} \left\{ [F^2]_{x^j} V^j + [F^2]_{y^j} \frac{dV^j}{dt} \right\} dt$$

$$= \frac{1}{2\ell} \sum_{i=1}^{k} \int_{t_{i-1}}^{t_i} \left\{ [F^2]_{x^j} - [F^2]_{x^l y^j} \dot\gamma^l - [F^2]_{y^j y^l} \ddot\gamma^l \right\} V^j dt$$

$$+ \frac{1}{2\ell} \sum_{i=1}^{k} [F^2]_{y^j} V^j |_{t_{i-1}}^{t_i}$$

$$= -\frac{1}{\ell} \int_a^b g_{lj}(\dot\gamma) \left\{ \ddot\gamma^l + 2G^l(\gamma, \dot\gamma) \right\} V^j dt + \frac{1}{\ell} \sum_{i=1}^{k} g_{jl}(\dot\gamma) \dot\gamma^l V^j |_{t_{i-1}}^{t_i},$$

$$\tag{5.1.4}$$

where we used (3.1.14) in the last equality. Note that $D_{\dot\gamma}^{\dot\gamma} \dot\gamma = \left\{ \ddot\gamma^i(t) + 2G^i(\gamma, \dot\gamma) \right\} \frac{\partial}{\partial x^i}$. Thus (5.1.4) implies (5.1.2). $\qquad\square$

Definition 5.1.2. A piecewise smooth curve $\gamma : [a, b] \to M$ from a point $p = \gamma(a)$ to another point $q = \gamma(b)$ is called a *shortest path* if

$$d_F(p, q) = L_F(\gamma) = \int_a^b F(\gamma(t), \dot{\gamma}(t))dt.$$

Proposition 5.1.2. *Any shortest path on a Finsler manifold (M, F) of constant speed is a smooth geodesic.*

Proof. Let $\gamma : [a, b] \to M$ be a shortest path from $p = \gamma(a)$ to $q = \gamma(b)$ with constant speed $F(\gamma, \dot{\gamma}) = \ell > 0$. Consider a piecewise smooth variation $H : [a, b] \times (-\varepsilon, \varepsilon) \to M$ of γ preserving the endpoints fixed. The variation vector field V satisfies that $V(a) = V(b) = 0$. We may assume that H is C^∞ on each $[t_{i-1}, t_i] \times (-\varepsilon, \varepsilon)$ for some partition $a = t_0 < t_1 < \cdots < t_{k-1} < t_k = b$. Denote by $L(s) = L_F(\gamma_s)$ the length of γ_s parametrized by $\gamma_s(t) = H(t, s)$. By the assumption, we have $L(s) \geq L(0) = L_F(\gamma)$ for $|s| < \varepsilon$. Thus $L'(0) = 0$. By (5.1.2), we have

$$\sum_{i=1}^{k-1} g_T \left(T(t_i^-) - T(t_i^+), V(t_i) \right) - \int_a^b g_T(V, D_T^T T)dt = 0. \qquad (5.1.5)$$

Take $V(t) = f(t)D_T^T T$ in (5.1.5), where $f(t)$ is an arbitrary C^∞ function such that $f(t) > 0$ on $[t_{i-1}, t_i]$ and $f(t_i) = 0$ for $i = 1, \ldots, k$. Thus $D_T^T T = 0$ on each $[t_{i-1}, t_i]$, i.e., γ is a geodesic on each $[t_{i-1}, t_i]$. From this, (5.1.5) is reduced to

$$\sum_{i=1}^{k-1} g_T(T(t_i^-) - T(t_i^+), V(t_i)) = 0. \qquad (5.1.6)$$

To obtain the continuity of T, it suffices to construct a piecewise smooth vector field V along γ such that $V(a) = V(b) = 0$ and $V(t_i) = T(t_i^-) - T(t_i^+)$ for $1 \leq i \leq k - 1$. In fact, let f_i be C^∞ functions on $\mathbb{R}$ such that $f_i(a) = f_i(b) = 0$, $f_i > 0$ on $I_i := (t_i - \delta, t_i + \delta)$ for $1 \leq i \leq k-1$ and zero outside I_i, where δ is a positive number with $\delta < \frac{1}{2} \min_{1 \leq i \leq k}\{|t_i - t_{i-1}|\}$. Denoted by V_i the piecewise smooth parallel vector fields along γ with $V_i(t_i) = T(t_i^-) - T(t_i^+)$. Define $V(t) = f_i(t)V_i(t)$ on I_i for $1 \leq i \leq k-1$ and zero elsewhere in $[a, b]$. Such a vector field V satisfies the above requirement. From (5.1.6), we get $T(t_i^+) = T(t_i^-)$. By the uniqueness of geodesics, γ is a smooth geodesic. $\square$

Proposition 5.1.2 shows that the shortest path on a Finsler manifold (M, F) from the point p to another point q must be a geodesic. The converse is not true in general, even in Riemannian geometry. However, it is true locally. This will be discussed in Section 5.4 below.

5.1.2 *Exponential Map*

Let $\gamma_y : [0, \ell] \to M$ be a geodesic on M with $\gamma(0) = x$ and $\dot{\gamma}(0) = y \in T_x M \backslash \{0\}$. By the homogeneity of G, we have for any $\delta > 0$ and $y \in T_x M \backslash \{0\}$,

$$\gamma_{\delta y}(t) = \gamma_y(\delta t),$$

which means that the curves $\gamma_y(\delta t)(0 \leq t \leq \ell/\delta)$ and $\gamma_{\delta y}(t)(0 \leq t \leq \ell)$ are the same geodesics with $\gamma_y(0) = x$ and $\dot{\gamma}_y(0) = \delta y$. Thus, we may assume that γ_y is defined on $[0, 1]$. Denote by $\mathcal{U}_x(M)$ the set of all tangent vectors $y \in T_x M$ for any $x \in M$ such that γ_y is defined on $[0, 1]$. Define a map $\exp_x : \mathcal{U}_x M \to M$ by

$$\exp_x(y) := \begin{cases} \gamma_y(1), & \text{if } y \neq 0, \\ x, & \text{if } y = 0, \end{cases}$$

which is called the *exponential map* at x. By definition, for each $y \in T_x M \backslash \{0\}$, the curve $t \to \exp_x(ty)(t \in [0, 1])$ is a geodesic emanating from x. Moreover, $\exp_x$ is C^∞ on $\mathcal{U}_x M \backslash \{0\}$.

For any $y \in \mathcal{U}_x M$, we identity $T_y(\mathcal{U}_x M)$ with $T_x M$ in a natural way. The differential map of $\exp_x$ at $y \in \mathcal{U}_x M$ is a linear map

$$d(\exp_x)|_y : T_x M \to T_z M, \quad z = \exp_x(y)$$

and

$$d(\exp_x)|_0 : T_x M \to T_x M$$

is an endomorphism provided that $\exp_x$ is differentiable at the origin of $T_x M$.

Theorem 5.1.2 ([Wh]). *Let (M, F) be a Finsler manifold. Then the exponential map $\exp$ is C^1 at the zero section of TM and for any $x \in M$, $d(exp_x)|_0 : T_x M \to T_x M$ is the identity map at the origin of $T_x M$.*

The proof is omitted. See [BCS] for more details. Now, we discuss the regularity of the exponential map at the zero section of TM.

Assume that $\exp_x$ is C^2 at $y = 0$ for all $x \in M$. For any $y \in T_x M$ with $F(y) < r$ (sufficiently small), let $f(y) := \exp_x(y)$ and $\gamma(t) = \exp_x(ty)$ be a geodesic with $\dot{\gamma}(0) = y$ for $0 \le t \le 1 + \varepsilon (\varepsilon > 0)$. Then $\gamma(t) = f(ty)$. By the geodesic equation (3.1.21) and letting $t = 0$, we have

$$y^j y^k \frac{\partial^2 f^i}{\partial y^j \partial y^k}(0) + 2G^i(y) = 0,$$

which imply that $G^i(y)$ are quadratic in y. By definition, such a Finsler metric is a Berwald metric. Conversely, If F is a Berwald metric, then $G^i(y) = \frac{1}{2}\Gamma^i_{jk}(x)y^j y^k$ are quadratic in $y \in T_x M$ for all $x \in M$. Thus, the geodesic $\gamma = \gamma(t)$ can be characterized by

$$\ddot{\gamma} + \Gamma^i_{jk}(\gamma)\dot{\gamma}^j \dot{\gamma}^k = 0.$$

By the ODE theory, $\gamma(t)$ smoothly depends on the initial conditions $\gamma(0) = x$ and $\dot{\gamma}(0) = y$ for sufficiently small t. Hence, $\exp_x$ is C^∞ at $y = 0$ for all $x \in M$. We have proved the following.

Theorem 5.1.3 ([AZ]). *Let (M, F) be a Finsler manifold. Then the exponential map* $\exp$ *is C^2 at the zero section if and only if F is a Berwald metric.*

The following lemma shows that the exponential map preserves the length of the tangent vector with respect to the Finsler metric F.

Lemma 5.1.1 (Gauss Lemma). *Let (M, F) be a Finsler manifold. Assume that $\exp_x y$ is defined for a fixed $y \in T_x M \backslash \{0\}$. Then, for any $v \in T_y(T_x M) \cong T_x M$, we have*

$$g_{d(\exp_x)_y(y)}\big(d(\exp_x)_y(y), d(\exp_x)_y(v)\big) = g_y(y, v). \qquad (5.1.7)$$

Proof. Note that $\exp_x y$ is defined. There is a small $\varepsilon > 0$ such that $\exp_x t(y + sv)$ is well defined for any $(t, s) \in [0, 1] \times (-\varepsilon, \varepsilon)$. Let $H : [0, 1] \times (-\varepsilon, \varepsilon) \to M$ defined by

$$H(t, s) := \exp_x t(y + sv).$$

Obviously, $\gamma_s(t) = H(t, s)$ is a geodesic for each s. Such a variation is said to be a *geodesic variation* of the geodesic $\gamma(t) = H(t, 0) = \exp_x(ty)$. The variation vector field

$$V(t) = \frac{\partial H}{\partial s}\bigg|_{s=0} = d(\exp_x)_{ty}(tv) = td(\exp_x)_{ty}(v).$$

It is easy to see that $T(t) = \dot{\gamma}(t) = d(\exp_x)_{ty}(y)$. We may assume that $F(\dot{\gamma}) = \ell(\text{constant})$ by Proposition 5.1.1. From (5.1.3), we obtain

$$L'(0) = \frac{1}{\ell}g_T(T,V)\Big|_0^1 = \frac{1}{\ell}\left\{g_{d(\exp_x)_y(y)}\big(d(\exp_x)_y(y), d(\exp_x)_y(v)\big)\right\}.$$

$$(5.1.8)$$

On the other hand, since $\gamma_s(t)$ is a geodesic with constant speed for each s, we have $F(\dot{\gamma}_s(t)) = F(\dot{\gamma}_s(0)) = F(y + sv)$. Thus, $L(s) = F(y + sv)$. Consequently,

$$L'(0) = \frac{d}{ds}F(y + sv)\Big|_{s=0} = F_{y^i}(y)v^i = \frac{1}{F(y)}g_y(y,v). \qquad (5.1.9)$$

Since $F(y) = F(\dot{\gamma}(0)) = \ell$, by (5.1.8) and (5.1.9), one obtains (5.1.7). $\qquad\square$

Fix a point $x \in M$, define the *tangent ball* and the *tangent sphere* of radius r by

$$\mathbf{B}_r(x) = \{y \in T_xM | F(x,y) < r\}, \quad \mathbf{S}_r(x) = \{y \in T_xM | F(x,y) = r\}.$$

If r is sufficiently small, then $\exp_x$ is well defined on $\mathbf{B}_r(x)$ and $\mathbf{S}_r(x)$, respectively. We call $\exp_x(\mathbf{B}_r(x))$ a *geodesic ball* and $\exp_x(\mathbf{S}_r(x))$ a *geodesic sphere* centered at x with a radius r.

For any $y \in \mathbf{S}_r(x)$, we have $F_{y^i}dy^i = 0$. Since $F_{y^i} = \frac{1}{F(y)}g_{ij}y^j$, we have $g_y(y,v) = 0$ for any vector v tangent to $\mathbf{S}_r(x)$. With respect to the Riemannian metric $g_y = g_{ij}(x,y)dy^i \otimes dy^j$ induced by F on $T_xM\backslash\{0\}$, the radial vector ty is orthogonal to $\mathbf{S}_{tr}(x)$ for any $t > 0$. By the Gauss lemma, one obtains the following special form of the Gauss lemma. This means that the radial geodesics γ are orthogonal to trajectories of the geodesic spheres with respect to $g_{\dot{\gamma}}$.

Corollary 5.1.1. *Let $T(t)$ be a tangent vector field of a radial geodesic $\exp_x(ty)(0 \le t \le 1)$ and $v(t)$ be tangent to $\mathbf{S}_{tr}(x)$. Then*

$$g_T\left(T, d(\exp_x)_y(v)\right) = 0.$$

Now, we begin to prove the locally minimizing property of arc length for a geodesic.

Theorem 5.1.4. *Let (M,F) be an n-dimensional Finsler manifold. For any $x \in M$ and $y \in \mathbf{B}_\delta(x)$, where $\delta > 0$ sufficiently small, let $\gamma(t) = \exp_x(ty) : [0,1] \to \exp_x(\mathbf{B}_\delta(x))$ be a radial geodesic starting from x and*

$\widetilde{\gamma} : [a, b] \to M$ *be a piecewise smooth curve on* M *from* $\widetilde{\gamma}(a) = \gamma(0) = x$ *to* $\widetilde{\gamma}(b) = \gamma(1)$. *Then*

$$L_F(\widetilde{\gamma}) \geq L_F(\gamma),$$

in which the equality holds if and only if $\widetilde{\gamma}([a, b]) = \gamma([0, 1])$.

Proof. We prove this according to the following two cases.

Case 1. $\widetilde{\gamma} \subseteq \exp_x(\mathbf{B}_\delta(x))$. In this case, $\sigma := \exp_x^{-1} \circ \widetilde{\gamma} : [a, b] \to \mathbf{B}_\delta(x)$ is a piecewise smooth curve from the origin o to $y \in \mathbf{B}_\delta(x)$. Take a piecewise smooth curve $v : [a, b] \to \mathbf{S}_1(x)$ and a nonnegative piecewise smooth function $r : [a, b] \to \mathbb{R}$ such that $\sigma(t) = r(t)v(t)$. Thus, $\widetilde{\gamma}(t) = \exp_x(r(t)v(t))$. Consequently, $r(a) = 0$, $r(b) = F(y)$ and $y = F(y)v(b)$. Moreover,

$$\dot{\widetilde{\gamma}}(t) = d(\exp_x)_{r(t)v(t)} (\dot{r}(t)v(t) + r(t)\dot{v}(t)).$$

Since $F(v(t)) = 1$, we have $g_{v(t)}(v(t), \dot{v}(t)) = 0$, which implies that

$$g_{r(t)v(t)}(v(t), \dot{v}(t)) = 0 \quad \text{for} \quad r(t) > 0.$$

When $r(t) > 0$, by (1.1.5) and the Gauss lemma, one obtains

$$F(\dot{\widetilde{\gamma}}(t))F\Big(d(\exp_x)_{r(t)v(t)}(r(t)v(t))\Big)$$

$$\geq g_{d(\exp_x)_{r(t)v(t)}}\Big(d(\exp_x)_{r(t)v(t)}(r(t)v(t)),$$

$$d(\exp_x)_{r(t)v(t)}(\dot{r}(t)v(t) + r(t)\dot{v}(t))\Big)$$

$$= g_{r(t)v(t)}\Big(r(t)v(t), \dot{r}(t)v(t) + r(t)\dot{v}(t)\Big)$$

$$= g_{v(t)}\Big(r(t)v(t), \dot{r}(t)v(t)\Big) = r(t)\dot{r}(t). \qquad (5.1.10)$$

On the other hand, by the Gauss lemma again, we have

$$F\big(d(\exp_x)_{r(t)v(t)}(r(t)v(t))\big) = F\left(r(t)v(t)\right) = r(t). \qquad (5.1.11)$$

Combining (5.1.10) with (5.1.11) yields $F(\dot{\widetilde{\gamma}}(t)) \geq \dot{r}(t)$. Obviously, this is true for $r(t) = 0$. Consequently,

$$L_F(\widetilde{\gamma}) = \int_a^b F(\dot{\widetilde{\gamma}}(t))dt \geq \int_a^b \dot{r}(t)dt = r(b) - r(a) = F(y) = L_F(\gamma)$$

for $r(t) \geq 0$.

Case 2. $\widetilde{\gamma}$ is not completely included in $\exp_x(\mathbf{B}_\delta(x))$. In this case, there is a $c \in [a,b]$ such that $\widetilde{\gamma}(c) \in \partial(\exp_x(\mathbf{B}_\delta(x)))$ and $\widetilde{\gamma}|_{[a,c]} \subseteq \exp_x(\mathbf{B}_\delta(x))$. Thus,

$$L_F(\widetilde{\gamma}) \geq L_F(\widetilde{\gamma}|_{[a,c]}) \geq \delta > L_F(\gamma).$$

Together Case 1 with Case 2, we always have $L_F(\widetilde{\gamma}) \geq L_F(\gamma)$.

From the above arguments, $L_F(\widetilde{\gamma}) = L_F(\gamma)$ if and only if only Case 1 occurs and the inequality in (5.1.10) becomes an equality. Thus, $d(\exp_x)_{r(t)v(t)}(\dot{r}(t)v(t) + r(t)\dot{v}(t))$ and $d(\exp_x)_{r(t)v(t)}(r(t)v(t))$ are linearly dependent. That is to say, $d(\exp_x)_{r(t)v(t)}(\dot{v}(t)) = 0$ for $r(t) > 0$, which implies that $\dot{v}(t) = 0$, i.e., $v(t)$ is a constant vector, when $r(t)$ is sufficiently small. Hence, $v(t) = v(b) = \frac{y}{F(y)}$. Thus $\widetilde{\gamma}([a,b]) = \gamma([0,1])$ by a reparameterization of $\gamma(t)$. This finishes the proof. $\qquad\square$

A geodesic $\gamma : [0,1] \to M$ is said to be *minimizing* if its length furnishes the absolute minimum among all piecewise smooth curves from $\gamma(0)$ to $\gamma(1)$.

5.1.3 *Geodesic Completeness*

Recall that a Finsler metric F on a smooth manifold M is forward (resp., backward) complete if every geodesic $\gamma : [0,\ell) \to M$ (resp. $\gamma : (-\ell,0] \to M$) can be extended to a geodesic on $[0,\infty)$ (resp., $(-\infty,0]$) (see Section 4.5). Then F is forward complete if and only if $\overleftarrow{F}(y) := F(-y)$ is backward complete.

There are nonreversible Finsler metrics which are only forward complete. For example, the Funk metric on a strongly convex domain in $\mathbb{R}^n$ is forward complete, but not complete. While the Klein metric is complete. An important fact is that every closed Finsler manifold is complete. The proofs of these facts are left to the readers (cf. [Sh1]). Further, if (M,F) is connected and the exponential map $\exp_x$ is defined on all of T_xM for any $x \in M$, then we say that (M,F) is *geodesic complete*.

Let d_F be the distance induced by F and $\{x_i\}$ be a sequence on a metric space (M, d_F). For any $\varepsilon > 0$, if there exists a positive number N such that $d_F(x_i, x_j) < \varepsilon$ (resp., $d_F(x_j, x_i) < \varepsilon$) when $j > i \geq N$, then we say that $\{x_i\}$ is a *forward (resp., backward) Cauchy sequence*. Similarly, we say that the sequence $\{x_i\}$ is *forward (resp., backward) convergent* to x if

$$\lim_{i \to \infty} d_F(x_i, x) = 0 \quad (\text{resp., } \lim_{i \to \infty} d_F(x, x_i) = 0)$$

for any $x \in M$. The metric space (M, d_F) is called *forward* (resp., *backward*) complete if every forward (resp., backward) Cauchy sequence is forward (resp., backward) convergent. It is easy to see that the forward completeness is equivalent to the backward completeness if the reversibility Λ of F is finite. In this case, (M, F) is complete.

Theorem 5.1.5 (Hopf-Rinow). *Let (M, F) be a connected Finsler manifold. Then the following statements are equivalent.*

   (i) *The Finsler manifold (M, F) is forward complete.*
  (ii) *The metric space (M, d_F) is forward complete.*
 (iii) *The Finsler manifold (M, F) is geodesic complete.*
  (iv) *There is a point $x \in M$ such that $\exp_x$ is defined on all of $T_x M$.*
   (v) *Every closed and forward bounded subset on (M, d_F) is compact.*

   Furthermore, if any of the above holds, then for any pair of points $x, z \in M$, there exists a globally minimizing geodesic from x to z.

The following result is well known. It is an important application of the covering space theory.

Theorem 5.1.6 (Cartan–Hadamard). *Let (M, F) be a simply connected and forward complete Finsler manifold with nonpositive flag curvature. Then the exponential map $\exp_x : T_x M \to M$ is a C^1 diffeomorphism for every $x \in M$.*

For the proofs of Theorems 5.1.5–5.1.6, the readers refer to [BCS] or [Sh1].

5.2   Second Variation of Length

Let (M, F) be a Finsler manifold and $\gamma : [a, b] \to M$ be a smooth geodesic on M with $F(\dot{\gamma}) = \ell$. Consider a smooth variation $H : [a, b] \times (-\varepsilon, \varepsilon) \to M$ of γ and $\widetilde{T}(t, s) = \frac{\partial H}{\partial t}$, $\widetilde{V}(t, s) = \frac{\partial H}{\partial s}$ (cf. Section 5.1.1). With respect to the Chern connection of F, we have

$$D_{\widetilde{T}}^{\widetilde{T}} \widetilde{V} - D_{\widetilde{V}}^{\widetilde{T}} \widetilde{T} = [\widetilde{T}, \widetilde{V}] = 0, \tag{5.2.1}$$

$$D_{\widetilde{V}}^{\widetilde{T}} D_{\widetilde{T}}^{\widetilde{T}} \widetilde{V} - D_{\widetilde{T}}^{\widetilde{T}} D_{\widetilde{V}}^{\widetilde{T}} \widetilde{V} = \Omega(\widetilde{V}, \widetilde{T}) \widetilde{V} \tag{5.2.2}$$

from (3.1.18) and (3.2.2). Assume that $V(t) = \widetilde{V}(t, 0)$ is the variation vector field along γ and $L(s) := L_F(\gamma_s)$ is the arc length of γ_s parametrized by

$\gamma_s(t) = H(t, s)$. Then

$$L'(s) = \int_a^b \frac{1}{2F} \frac{d}{ds}(F^2(\gamma_s, \dot{\gamma}_s))dt = \int_a^b \frac{1}{F(\widetilde{T})} g_{\widetilde{T}}\left(D_{\widetilde{V}}^{\widetilde{T}}\widetilde{T}, \widetilde{T}\right) dt$$

$$= \int_a^b \frac{1}{F(\widetilde{T})} g_{\widetilde{T}}\left(D_{\widetilde{T}}^{\widetilde{T}}\widetilde{V}, \widetilde{T}\right) dt$$

by (5.2.1). Further, by (5.2.2),

$$L''(s) = \int_a^b \frac{1}{F(\widetilde{T})} \left\{ g_{\widetilde{T}}\left(D_{\widetilde{V}}^{\widetilde{T}}D_{\widetilde{T}}^{\widetilde{T}}\widetilde{V}, \widetilde{T}\right) + \|D_{\widetilde{T}}^{\widetilde{T}}\widetilde{V}\|_{g_{\widetilde{T}}}^2 \right\} dt$$

$$- \int_a^b \frac{1}{F^3(\widetilde{T})} \left(g_{\widetilde{T}}\left(D_{\widetilde{T}}^{\widetilde{T}}\widetilde{V}, \widetilde{T}\right)\right)^2 dt$$

$$= \int_a^b \frac{1}{F(\widetilde{T})} \left\{ g_{\widetilde{T}}\left(D_{\widetilde{T}}^{\widetilde{T}}D_{\widetilde{V}}^{\widetilde{T}}\widetilde{V}, \widetilde{T}\right) + g_{\widetilde{T}}\left(\Omega(\widetilde{V}, \widetilde{T})\widetilde{V}, \widetilde{T}\right) + \|D_{\widetilde{T}}^{\widetilde{T}}\widetilde{V}\|_{g_{\widetilde{T}}}^2 \right\} dt$$

$$- \int_a^b \frac{1}{F^3(\widetilde{T})} \left(g_{\widetilde{T}}\left(D_{\widetilde{T}}^{\widetilde{T}}\widetilde{V}, \widetilde{T}\right)\right)^2 dt. \tag{5.2.3}$$

Now, we calculate each term of the RHS in the above equality. First of all, observe that

$$g_{\widetilde{T}}\left(D_{\widetilde{T}}^{\widetilde{T}}D_{\widetilde{V}}^{\widetilde{T}}\widetilde{V}, \widetilde{T}\right) = \widetilde{T}g_{\widetilde{T}}\left(D_{\widetilde{V}}^{\widetilde{T}}\widetilde{V}, \widetilde{T}\right) - g_{\widetilde{T}}\left(D_{\widetilde{V}}^{\widetilde{T}}\widetilde{V}, D_{\widetilde{T}}^{\widetilde{T}}\widetilde{T}\right)$$

$$- 2C_{\widetilde{T}}\left(D_{\widetilde{T}}^{\widetilde{T}}\widetilde{T}, D_{\widetilde{V}}^{\widetilde{T}}\widetilde{V}, \widetilde{T}\right)$$

and $\widetilde{T}(t, 0) = T(t) = \dot{\gamma}$ satisfying $\nabla_T^T T = 0$. Thus,

$$\left.\int_a^b \frac{1}{F(\widetilde{T})} g_{\widetilde{T}}\left(D_{\widetilde{T}}^{\widetilde{T}}D_{\widetilde{V}}^{\widetilde{T}}\widetilde{V}, \widetilde{T}\right) dt\right|_{s=0} = \frac{1}{\ell} g_T(D_V^T V, T)|_a^b. \tag{5.2.4}$$

Similarly, since

$$g_{\widetilde{T}}\left(D_{\widetilde{T}}^{\widetilde{T}}\widetilde{V}, \widetilde{T}\right) = \widetilde{T}g_{g_{\widetilde{T}}}(\widetilde{V}, \widetilde{T}) - g_{\widetilde{T}}\left(\widetilde{V}, D_{\widetilde{T}}^{\widetilde{T}}\widetilde{T}\right) - 2C_{\widetilde{T}}\left(D_{\widetilde{T}}^{\widetilde{T}}\widetilde{T}, \widetilde{V}, \widetilde{T}\right),$$

we have

$$\left.\int_a^b \frac{1}{F^3(\widetilde{T})} \left(g_{\widetilde{T}}\left(D_{\widetilde{T}}^{\widetilde{T}}\widetilde{V}, \widetilde{T}\right)\right)^2 dt\right|_{s=0} = \frac{1}{\ell^3} \int_a^b \left(T(g_T(V, T))\right)^2 dt. \tag{5.2.5}$$

Also, $\Omega(\tilde{V}, \tilde{T})\tilde{V} = R_{\tilde{T}}(\tilde{V}, \tilde{T})\tilde{V}$ from (3.2.5). Thus,

$$\int_a^b \frac{1}{F(\tilde{T})} g_{\tilde{T}}\left(\Omega(\tilde{V}, \tilde{T})\tilde{V}, \tilde{T}\right) dt\bigg|_{s=0} = \frac{1}{\ell} \int_a^b g_T\left(R_T(V,T)V, T\right)$$

$$= -\frac{1}{\ell} \int_a^b g_T\left(R_T(V), V\right). \quad (5.2.6)$$

Plugging (5.2.4)–(5.2.6) into (5.2.3) yields

$$L''(0) = \frac{1}{\ell} g_T(D_V^T V, T)\big|_a^b + \frac{1}{\ell} \int_a^b \left\{ g_T\left(D_T^T V, D_T^T V\right) - g_T\left(R_T(V), V\right) \right\} dt$$

$$-\frac{1}{\ell^3} \int_a^b \left(T(g_T(V,T))\right)^2 dt. \quad (5.2.7)$$

Note that

$$\int_a^b g_T\left(D_T^T V, D_T^T V\right) dt = g_T\left(D_T^T V, V\right)\big|_a^b - \int_a^b g_T\left(D_T^T D_T^T V, V\right) dt.$$

$$(5.2.8)$$

(5.2.7) may be reexpressed by

$$L''(0) = \frac{1}{\ell}\left\{ g_T(D_V^T V, T)\big|_a^b + g_T\left(D_T^T V, V\right)\big|_a^b \right\}$$

$$-\frac{1}{\ell} \int_a^b g_T\left(D_T^T D_T^T V + R_T(V), V\right) dt$$

$$-\frac{1}{\ell^3} \int_a^b \left(T(g_T(V,T))\right)^2 dt. \quad (5.2.9)$$

If we write $V = f(t)T + V^\perp$ by an orthogonal decomposition of V along γ with respect to g_T, then $D_T^T V = f'(t)T + D_T^T V^\perp$, which implies that $T(g_T(V,T)) = \ell^2 f'(t)$ and

$$g_T(D_T^T V, D_T^T V) = \ell^2 (f'(t))^2 + g_T(D_T^T V^\perp, D_T^T V^\perp).$$

Thus, (5.2.7) can be rewritten as

$$L''(0) = \frac{1}{\ell} g_T(D_V^T V, T)\big|_a^b$$

$$+ \frac{1}{\ell} \int_a^b \left\{ g_T\left(D_T^T V^\perp, D_T^T V^\perp\right) - g_T\left(R_T(V^\perp), V^\perp\right) \right\} dt, \quad (5.2.10)$$

where we used that $R_T(T) = 0$. Summing up, one obtains the second variation formula as follows.

Theorem 5.2.1 (Second Variation Formula). *Let $\gamma : [a, b] \to M$ be a smooth geodesic on a Finsler manifold (M, F). Then, for any variation $H : [a, b] \times (-\varepsilon, \varepsilon) \to M$ of γ, the second variation formula is given by one of (5.2.7) and (5.2.9)–(5.2.10).*

In particular, if H keeps the two endpoints fixed and is a normal variation with respect to g_T, i.e., $g_T(V, T) = 0$, then

$$L''(0) = \frac{1}{\ell} \int_a^b \left\{ g_T \left(D_T^T V, D_T^T V \right) - g_T \left(R_T(V), V \right) \right\} dt \quad (5.2.11)$$

or

$$L''(0) = -\frac{1}{\ell} \int_a^b g_T \left(D_T^T D_T^T V + R_T(V), V \right) dt. \quad (5.2.12)$$

The above variation formula can be extended to the case of a piecewise smooth curve as in Theorem 5.1.1. With the same notations as in Theorem 5.1.1, we have the corresponding second variation formula. In particular, if γ is a piecewise smooth curve and H is a piecewise smooth variation of γ which keeps the two endpoints fixed and is normal with respect to $g_T = g_{\dot\gamma}$, then

$$L''(0) = \frac{1}{\ell} \sum_i g_T \left(\Delta_{t_i}(D_T^T V), V_{t_i} \right) - \frac{1}{\ell} \int_a^b g_T \left(D_T^T D_T^T V + R_T(V), V \right) dt, \quad (5.2.13)$$

where

$$\Delta_{t_i}(D_T^T V) = \lim_{t \to t_i + 0} D_T^T V - \lim_{t \to t_i - 0} D_T^T V.$$

As an application of Theorem 5.2.1, we obtain the well known Bonnet–Myers theorem as follows ([BCS]).

Theorem 5.2.2 (Bonnet-Myers). *Let (M, F) be an n-dimensional forward complete and connected Finsler manifold with Ric$\geq K > 0$. Then M is compact and its diameter d is at most $\pi \sqrt{(n-1)/K}$. Moreover, the fundamental group $\pi(M, x)$ is finite.*

Proof. Since M is forward complete, for any two points $p, q \in M$, there is a minimal normal geodesic $\gamma : [0, \ell] \to M$ from $p = \gamma(0)$ to $q = \gamma(\ell)$ by Hopf–Rinow Theorem. Obviously, $\ell = d_F(p, q)$. Take parallel $g_{\dot\gamma}$-orthonormal frame fields $e_i(t)$ $(1 \leq i \leq n)$ along γ with $e_n(t) = \dot{\gamma}$. Let

$$V_i = \left(\sin \frac{\pi t}{\ell} \right) e_i(t), \quad 1 \leq i \leq n-1.$$

Then $V_i(0) = V_i(\ell) = 0$. For each $1 \leq i \leq n-1$, there exists a normal variation $H_i : [0, \ell] \times (-\varepsilon, \epsilon) \to M$ of γ preserving the end points p, q fixed such that its variation field is given by V_i. By the second variation formula (5.2.12), one obtains

$$L_i''(0) = \int_0^\ell \left(\sin \frac{\pi t}{\ell} \right)^2 \left(\frac{\pi^2}{\ell^2} - g_{\dot{\gamma}}(R_{\dot{\gamma}}(e_i), e_i) \right) dt,$$

Consequently,

$$\sum_{i=1}^{n-1} L_i''(0) = \int_0^\ell \left(\sin \frac{\pi t}{\ell} \right)^2 \left((n-1)\frac{\pi^2}{\ell^2} - \mathrm{Ric}(\dot{\gamma}) \right) dt$$

$$\leq \left((n-1)\frac{\pi^2}{\ell^2} - K \right) \int_0^\ell \left(\sin \frac{\pi t}{\ell} \right)^2 dt$$

$$= \frac{\ell}{2} \left((n-1)\frac{\pi^2}{\ell^2} - K \right).$$

If $\ell > \pi\sqrt{(n-1)/K}$, then $\sum_{i=1}^{n-1} L_i''(0) < 0$, which implies that there is an index i such that $L_i''(0) < 0$. Thus the length of γ attains the maximum among the family of curves $\{H_i^{(s)} : [0, \ell] \to M | H_i^{(s)}(0) = p, H_i^{(s)}(\ell) = q\}$, which is impossible because of the minimality of γ. Hence, $d_F(p, q) = \ell \leq \pi\sqrt{(n-1)/K}$ and the diameter $d \leq \pi\sqrt{(n-1)/K}$ by the arbitrariness of p and q. By the Hopf–Rinow theorem, M is compact.

Let $\tilde{M}$ be a simply connected covering space (i.e., a universal cover) of M with a smooth projection $\rho : \tilde{M} \to M$. Let $\tilde{F} = \rho^* F$. Then $(\tilde{M}, \tilde{F})$ is a smooth forward complete Finsler manifold by a standard argument. Since $(\tilde{M}, \tilde{F})$ is locally isometric to (M, F), the Ricci curvature of $(\tilde{M}, \tilde{F})$ satisfies $\tilde{\mathrm{Ric}} \geq K$. By previous arguments, $(\tilde{M}, \tilde{F})$ is compact, which means that every closed subset $\rho^{-1}(x) \subset \tilde{M}$ is compact for any $x \in M$, hence finite. On the other hand, by hypothesis, M is connected, so all its fundamental groups $\pi(M, x)$, where the x denotes the base point, are isomorphic. Since $\tilde{M}$ is a universal cover of M, any specific $\pi(M, x)$ is bijective with the collection of isolated points $\rho^{-1}(x)$. Thus, $\pi(M, x)$ is finite. $\square$

Remark 5.2.1. For the first claim in Theorem 5.2.2, we have a simpler proof. In fact, fix a point $x \in M$ and consider a normal geodesic $\gamma : [0, \ell_\gamma) \to$

M emanating from x, here $\ell_\gamma \in (0, \infty)$ is taken as the supremum of $t > 0$ such that $d_F(x, \gamma(t)) = t$. Define a C^∞ vector field V on an open set $\mathcal{U} \subset M \backslash \{x\}$ such that $V(\gamma(t)) = \dot{\gamma}(t)$ for each γ and $t \in (0, \ell_\gamma)$, and all integral curves of V are geodesics, i.e., V is a geodesic field on $\mathcal{U}$. By Proposition 4.1.2, the Riemannian metric g_V has the Ricci curvature $\widehat{\mathrm{Ric}} \geq K$. Note that $F(V) = \sqrt{g_V(V, V)}$. Therefore, the Riemannian Bonnet–Myers theorem on (M, g_V) implies that $\ell_\gamma \leq \pi\sqrt{(n-1)/K}$.

5.3 Jacobi Fields and Conjugate Points

To study the singularity of the exponential map, we need introduce the notion of conjugate point. We first define the Jacobi field along a geodesic and then give its geometric explanation.

Let (M, F) be an n-dimensional Finsler manifold and $\gamma : [a, b] \to M$ be a smooth geodesic on M. Assume that $H : [a, b] \times (-\varepsilon, \varepsilon) \to M$ is a geodesic variation of γ such that the curve $\gamma_s = H(\cdot, s)$ is a geodesic for each $s \in (-\varepsilon, \varepsilon)$. We denote by $\widetilde{T}$, $\widetilde{V}$ as in Section 5.2 and by $V(t) = \widetilde{V}(t, 0)$ the variation vector field along γ. Then $D_{\widetilde{T}}^{\widetilde{T}}\widetilde{T} = 0$ and (5.2.1) holds. By (5.2.1)-(5.2.2), we have

$$D_{\widetilde{T}}^{\widetilde{T}} D_{\widetilde{T}}^{\widetilde{T}} \widetilde{V} = D_{\widetilde{T}}^{\widetilde{T}} D_{\widetilde{V}}^{\widetilde{T}} \widetilde{T} = R_{\widetilde{T}}(\widetilde{T}, \widetilde{V})\widetilde{T}.$$

Restricting this to γ yields

$$D_{\dot{\gamma}}^{\dot{\gamma}} D_{\dot{\gamma}}^{\dot{\gamma}} V = R_{\dot{\gamma}}(\dot{\gamma}, V)\dot{\gamma} = -R_{\dot{\gamma}}(V). \tag{5.3.1}$$

Choose local $g_{\dot{\gamma}}$-orthonormal frame fields $e_1(t), \ldots, e_n(t)$ such that $e_n(t) = \dot{\gamma}$ and $e_i(t)(1 \leq i \leq n-1)$ are parallel vector fields along γ with respect to the Chern connection. Let $V(t) = V^i(t)e_i(t)$. Then (5.3.1) is rewritten as $V''(t) + R_{\dot{\gamma}}(V) = 0$, i.e.,

$$(V^i)'' + R^i{}_k(\dot{\gamma})V^k = 0. \tag{5.3.2}$$

By the ODE theory, there is a unique smooth solution $V = V(t)$ for (5.3.1) or (5.3.2) with $V_0 = V(t_0)$ and $V'(t_0) = (D_{\dot{\gamma}}^{\dot{\gamma}}V)(t_0)$.

A vector field V is called a *Jacobi field* along γ if V satisfies the Jacobi field equation (5.3.1) or (5.3.2). Thus, the set of all Jacobi fields along γ constitutes an $2n$-dimensional linear space. The above arguments show that the variational vector field V of a geodesic variation is a Jacobi field. Conversely, we have the following.

Proposition 5.3.1. *Let $\gamma : [0, l] \to M$ be a geodesic on an n-dimensional Finsler manifold (M, F). Assume that $J(t)$ is a Jacobi field along γ. Then $J(t)$ is the variational vector field for some geodesic variation of γ.*

Proof. Let $c(s)$ be a smooth curve with $\dot{c}(0) = J(0)$ for $s \in (-\varepsilon, \varepsilon)$. We take parallel vector fields $T(s)$ and $W(s)$ along $c(s)$ with $T(0) = \dot{\gamma}(0)$ and $W(0) = D_T^T J(0)$, where $T(t) = \dot{\gamma}(t)$. Then $H : [0, l] \times (-\varepsilon, \varepsilon) \to M$ defined by

$$H(t, s) = \exp_{c(s)} t\left(T(s) + sW(s)\right) \qquad (5.3.3)$$

gives a smooth geodesic variation of γ. Let $\widetilde{T}(t, s) = \frac{\partial H}{\partial t}$ and $\widetilde{V}(t, s) = \frac{\partial H}{\partial s}$. Obviously, $V(t) = \widetilde{V}(t, 0)$, which is a Jacobi field along γ. Next, we prove that $J(t) = V(t)$. It suffices to prove that they satisfy the same initial data. In fact,

$$V(0) = \left.\frac{\partial H(0, s)}{\partial s}\right|_{s=0} = \dot{c}(0) = J(0).$$

Moreover, since

$$V(t) = \left.\frac{\partial H(t, s)}{\partial s}\right|_{s=0} = \dot{c}(0) + \left[\left(d\exp_{c(s)}\right)_{tT(s)} tW(s)\right]_{s=0}$$

$$= J(0) + t\left(d\exp_{c(0)}\right)_{tT(0)} W(0),$$

we have

$$V'(0) = \left.D_T V(t)\right|_{t=0} = \left[\left(d\exp_{c(0)}\right)_{tT(0)} W(0)\right]_{t=0} = W(0) = D_T^T J(0).$$

This finishes the proof. $\qquad\square$

From the proof of Proposition 5.3.1, we have the following corollary.

Corollary 5.3.1. *Let $\gamma : [0, l] \to M$ be a geodesic on a Finsler manifold (M, F). Then, for any $v, w \in T_{\gamma(0)}M$, there exists a unique Jacobi field $J = J(t)$ along γ with $J(0) = v$ and $D_{\dot{\gamma}}^{\dot{\gamma}} J(0) = w$. Further,*

$$J(t) = v + t\left(d\exp_{\gamma(0)}\right)_{t\dot{\gamma}(0)}(w) \qquad (5.3.4)$$

is the variational field of the geodesic variation defined by (5.3.3).

Let $J(t)$ be a Jacobi field along a geodesic γ. Then

$$\frac{d^2}{dt^2} g_{\dot{\gamma}}(J, \dot{\gamma}) = g_{\dot{\gamma}}(R_{\dot{\gamma}}(\dot{\gamma}, J)\dot{\gamma}, \dot{\gamma}) = -g_{\dot{\gamma}}(R_{\dot{\gamma}}(J), \dot{\gamma}) = -g_{\dot{\gamma}}(J, R_{\dot{\gamma}}(\dot{\gamma})) = 0,$$

which means that $g_{\dot{\gamma}}(J, \dot{\gamma})$ is a linear function in t. If

$$g_{\dot{\gamma}}(J(t_1), \dot{\gamma}(t_1)) = 0, \quad g_{\dot{\gamma}}(J(t_2), \dot{\gamma}(t_2)) = 0, \quad t_1 \neq t_2,$$

then $g_{\dot{\gamma}}(J, \dot{\gamma}) = 0$. In this case, $J(t)$ is said to be *normal* along γ. For any Jacobi field J, it is easy to check that the normal vector field $J^{\perp} := J - g_{\dot{\gamma}}(J, \dot{\gamma})\dot{\gamma}$ is also a Jacobi field and vice versa. Thus, it suffices to consider the normal Jacobi fields when we study the Jacobi fields.

Definition 5.3.1. Let $\gamma : [a, b] \to M$ be a geodesic on a Finsler manifold (M, F) with $p = \gamma(a)$ and $q = \gamma(b)$. The point q is called a *conjugate point* of p along γ if there is a nonzero Jacobi field $J = J(t)$ along γ which vanishes at p and q.

Obviously, q is a conjugate point of p along γ if and only if p is a conjugate point of q along γ. We also say that the points p and q are *conjugate* along γ. Let $\varphi : M_1 \to M_2$ be a differentiable map between two differential manifolds M_1 and M_2. Recall that p is a *critical point* of φ if the linear map $d\varphi : T_p M_1 \to T_{\varphi(p)} M_2$ is singular, namely, there is a nonzero $v \in T_p M_1$ such that $d\varphi_p(v) = 0$. In this case, $\varphi(p)$ is said to be a *critical value* of φ. The following gives a geometric explanation of the conjugate point.

Proposition 5.3.2. $q = \gamma(l)$ *is a conjugate point of* $p = \gamma(0)$ *along the geodesic* $\gamma : [0, l] \to (M, F)$ *if and only if* q *is a critical value of the exponential map* $\exp_p : T_p M \to M$.

Proof. Assume that $q = \gamma(l)$ is a conjugate point of $p = \gamma(0)$ along $\gamma(t) = \exp_x(tv)$ for any $v \in T_p M$ and $t \in [0, l]$. Then there is a nonzero Jacobi field J along γ with $J(0) = J(l) = 0$. By (5.3.4), there is a nonzero vector $w = D_{\dot{\gamma}}^{\dot{\gamma}} J(0) \in T_{lv}(T_p M)$ such that $(d\exp_p)_{lv}(w) = 0$. Thus, $q = \exp_x(lv)$ is a critical value of $\exp_p$. Conversely, we may assume that $q = \exp_p(lv)$ for some $v \in T_p M$ is a critical value of $\exp_x$. Then there is a nonzero vector $w \in T_{lv}(T_p M)$ such that $(d\exp_p)_{lv}(w) = 0$. We identify the tangent space $T_{lv}(T_p M)$ with the tangent space $T_p M$ in a standard way and define a geodesic variation $H(t, s) = \exp_p t(v + sw)$ of $\gamma(t) = \exp_p(tv)$ with $H(0, s) = p$ and

$H(t,0) = \gamma(t)$. Its variation field $J(t)$ is a Jacobi field with $J(0) = 0$. On the other hand,

$$J(l) = \left.\frac{\partial H(t,s)}{\partial s}\right|_{(l,0)} = l(d\exp_p)_{lv}(w) = 0.$$

Consequently, q is a conjugate point of p.   □

The following result shows that the Jacobi field J along a geodesic $\gamma : [a,b] \to (M,F)$ is uniquely determined by its values at a and b if there is no conjugate point pair on γ.

Proposition 5.3.3. *Let $\gamma : [a,b] \to M$ be a geodesic on a Finsler manifold (M,F) with $p = \gamma(a)$ and $q = \gamma(b)$. If q is not a conjugate point of p along γ, then, for any $v \in T_pM$ and $w \in T_qM$, there is a unique Jacobi field J along γ such that $J(a) = v$ and $J(b) = w$.*

Proof. Since q is not a conjugate point of p, $J(t) \equiv 0$ when $v = w = 0$ by definition. Now we consider the case when $v \neq 0$ or $w \neq 0$. By Corollary 5.3.1, there is a unique Jacobi field V along γ with $V(a) = 0$ and $D_{\dot\gamma}^{\dot\gamma} V(a) = u$ for any nonzero $u \in T_pM$. From this, we can choose Jacobi fields $V_1,\ldots,V_n$ along γ such that $V_i(a) = 0 (1 \leq i \leq n)$ and $D_{\dot\gamma}^{\dot\gamma}V_1(a),\ldots,D_{\dot\gamma}^{\dot\gamma}V_n(a)$ are linearly independent. Then $V_1(b),\ldots,V_n(b)$ are linearly independent. Otherwise, there are constants $k_1,\ldots,k_n$, which are not all zero, with $\sum_i k_i V_i(b) = 0$. Hence $V := \sum_i k_i V_i$ is a Jacobi field with $V(a) = V(b) = 0$. Note that V is a nonzero vector field since $D_{\dot\gamma}^{\dot\gamma}V(a) = \sum_i k_i D_{\dot\gamma}^{\dot\gamma}V_i(a) \neq 0$. Consequently, q is a conjugate point of p, which is impossible by the assumption. Thus, for any $w \in T_qM$, there are constants $k_1,\ldots,k_n$ such that $w = \sum_i k_i V_i(b)$. Let $J_1 = \sum_i k_i V_i$. Then J_1 is a Jacobi field along γ with $J_1(a) = 0$ and $J_1(b) = w$.

On the other hand, for any nonzero vector $\tilde{u} \in T_qM$, there is a unique Jacobi field $\tilde{V}$ with $\tilde{V}(b) = 0$ and $D_{\dot\gamma}^{\dot\gamma}\tilde{V}(b) = \tilde{u}$. In the same way as above, there exist Jacobi fields $\tilde{V}_1,\ldots,\tilde{V}_n$ such that $\tilde{V}_i(b) = 0 (1 \leq i \leq n)$ and $D_{\dot\gamma}^{\dot\gamma}\tilde{V}_1(b),\ldots,D_{\dot\gamma}^{\dot\gamma}\tilde{V}_n(b)$ are linearly independent. Then $\tilde{V}_1(a),\ldots,\tilde{V}_n(a)$ are linearly independent. For any $v \in T_pM$, there are constants $\tilde{k}_1,\ldots,\tilde{k}_n$ such that $v = \sum_i \tilde{k}_i \tilde{V}_i(a)$. Let $J_2 = \sum_i \tilde{k}_i \tilde{V}_i$. Then J_2 is a Jacobi field along γ with $J_2(a) = v$ and $J_2(b) = 0$. Consequently, $J = J_1 + J_2$ is a Jacobi field with $J(a) = v$ and $J(b) = w$.

Assume that $\bar{J}$ and $\tilde{J}$ are two different Jacobi fields with $\bar{J}(a) = \tilde{J}(a) = v$ and $\bar{J}(b) = \tilde{J}(b) = w$. Then $J := \bar{J} - \tilde{J}$ is a nonzero Jacobi field with $J(a) = J(b) = 0$. Thus, $J \equiv 0$ and hence $\bar{J} = \tilde{J}$.   □

Finally, we discuss the Jacobi fields and conjugate points on a Finsler manifold (M, F) of nonpositive or nonnegative flag curvature. First of all, we discuss Jacobi fields and conjugate points on a Finsler manifold (M, F) of constant flag curvature.

Example 5.3.1. Let (M, F) be a Finsler manifold of constant flag curvature $\mathbf{K} = c$. Then the Riemann curvature R satisfies that

$$R_y(u) = c\Big(g_y(y, y)u - g_y(y, u)y\Big) \tag{5.3.5}$$

by (4.1.1). Let $\gamma : [0, l] \to M$ be a geodesic with $F(\dot\gamma) = 1$. As mentioned earlier, it suffices to consider the normal Jacobi fields. Assume that $J = J(t)$ is a normal Jacobi field along γ. It follows from the Jacobi field equation and (5.3.5) that

$$D_{\dot\gamma}^{\dot\gamma} D_{\dot\gamma}^{\dot\gamma} J + cJ = 0. \tag{5.3.6}$$

Choose $g_{\dot\gamma}$-orthonormal frame fields $e_1(t), \ldots, e_n(t)$ along γ such that $e_n(t) = \dot\gamma(t)$ and $e_1(t), \ldots, e_n(t)$ are parallel along γ. Let $J(t) = \sum_{i=1}^n J^i(t)e_i(t)$. Then $J^n = 0$ and (5.3.6) is reduced to

$$(J^i(t))'' + cJ^i(t) = 0, \quad 1 \leq i \leq n - 1, \tag{5.3.7}$$

whose solutions are given by

$$J^i(t) = \begin{cases} \lambda^i \frac{\sin(\sqrt{c}t)}{\sqrt{c}} + \mu^i \cos(\sqrt{c}t), & \text{if } c > 0, \\ \lambda^i t + \mu^i, & \text{if } c = 0, \\ \lambda^i \frac{\sinh(\sqrt{-c}t)}{\sqrt{-c}} + \mu^i \cosh(\sqrt{-c}t), & \text{if } c < 0, \end{cases} \tag{5.3.8}$$

where $\lambda^i, \mu^i (1 \leq i \leq n - 1)$ are constants. Let

$$\mathfrak{s}_c(t) = \begin{cases} \frac{\sin(\sqrt{c}t)}{\sqrt{c}}, & \text{if } c > 0, \\ t, & \text{if } c = 0, \\ \frac{\sinh(\sqrt{-c}t)}{\sqrt{-c}}, & \text{if } c < 0. \end{cases} \tag{5.3.9}$$

From (5.3.8), all normal Jocobi fields J along γ can be expressed by

$$J(t) = A(t)\mathfrak{s}_c(t) + B(t)\mathfrak{s}_c'(t), \tag{5.3.10}$$

where $A(t) = \sum_{i=1}^{n-1} \lambda^i e_i(t)$ and $B(t) = \sum_{i=1}^{n-1} \mu^i e_i(t)$ are arbitrary two parallel vector fields along γ and orthogonal to $\dot\gamma$ with respect to $g_{\dot\gamma}$. In particular, all normal Jacobi fields J along γ with $J(0) = 0$ are given by

$$J(t) = A(t)\mathfrak{s}_c(t). \tag{5.3.11}$$

Obviously, J is a nonzero Jacobi field if and only if all λ^i are nonzero for $1 \leq i \leq n-1$. The roots of $J(t) = 0$ are $t = \frac{k\pi}{\sqrt{c}}(k = 0, 1, 2, \ldots)$ for $c > 0$ and $t = 0$ for $c \leq 0$. Therefore $q_k = \gamma(k\pi/\sqrt{c})(k = 0, 1, 2, \ldots)$ are conjugate points of $p = \gamma(0)$ along γ when $c > 0$ and there is no conjugate point pair on (M, F) when $c \leq 0$.

Example 5.3.1 shows that (5.3.11) gives all normal Jacobi fields J with $J(0) = 0$ along a geodesic γ on a Finsler manifold (M, F) of constant flag curvature $\mathbf{K} = c$. In the case when $c > 0$, there are infinitely many conjugate point pairs along γ. While there is no conjugate point pairs along γ when $c \leq 0$. In fact, this is also true for any Finsler manifold (M, F) of nonpositive flag curvature.

Theorem 5.3.1. *There is no conjugate point pair on a Finsler manifold (M, F) of nonpositive flag curvature.*

Proof. We prove this by contradiction. Let $\gamma : [0, l] \to M$ be a geodesic. Assume that there is a nonzero Jacobi field $J = J(t)$ along γ on (M, F) such that $J(0) = J(l) = 0$. Since J satisfies (5.3.1), equivalently, $D_{\dot{\gamma}}^{\dot{\gamma}} D_{\dot{\gamma}}^{\dot{\gamma}} J = -R_{\dot{\gamma}}(J)$, we have

$$\frac{d}{dt} g_{\dot{\gamma}}(D_{\dot{\gamma}}^{\dot{\gamma}} J, J) = -g_{\dot{\gamma}}(R_{\dot{\gamma}}(J), J) + g_{\dot{\gamma}}\left(D_{\dot{\gamma}}^{\dot{\gamma}} J, D_{\dot{\gamma}}^{\dot{\gamma}} J\right)$$

$$= -\mathbf{K}(\Pi, \dot{\gamma})\left(g_{\dot{\gamma}}(\dot{\gamma}, \dot{\gamma}) g_{\dot{\gamma}}(J, J) - g_{\dot{\gamma}}(\dot{\gamma}, J)^2\right) + g_{\dot{\gamma}}\left(D_{\dot{\gamma}}^{\dot{\gamma}} J, D_{\dot{\gamma}}^{\dot{\gamma}} J\right),$$

where we used (3.1.19) and (4.1.1). By the Chauchy–Schwarz inequality and the nonpositivity of $\mathbf{K}$, the right-hand side of the above equality is nonnegative. This implies that $g_{\dot{\gamma}}(D_{\dot{\gamma}}^{\dot{\gamma}} J, J)$ is nondecreasing on $[0, l]$. Since $J(0) = J(l) = 0$, we have $g_{\dot{\gamma}}(D_{\dot{\gamma}}^{\dot{\gamma}} J, J) = 0$. Consequently,

$$\frac{d}{dt} g_{\dot{\gamma}}(J, J) = 2g_{\dot{\gamma}}(D_{\dot{\gamma}}^{\dot{\gamma}} J, J) + 2C_{\dot{\gamma}}(D_{\dot{\gamma}}^{\dot{\gamma}} \dot{\gamma}, J, J) = 0, \tag{5.3.12}$$

that is, $g_{\dot{\gamma}}(J, J)$ is a linear function in t. Note that $J(0) = J(l) = 0$. Thus, $J \equiv 0$, which is impossible. $\qquad\square$

5.4 First Conjugate Point and Cut Point

In this section, we shall study the minimizing property of arc length for a geodesic among "nearby" curves that share its endpoints. Roughly speaking, the arc length of a geodesic γ emanating from p attains its minimum among all "nearby" piecewise smooth curves that share its endpoints when $\gamma(t)$ meets the first conjugate point of p along γ. While this minimizing property fails when it gets pass the first conjugate point. So, the cut point of p along a geodesic must occur either before, or exactly at the first conjugate point. For the sake of simplicity, we always assume that the geodesic is smooth throughout this section.

Let $\mathfrak{V}_\gamma$ be a set of piecewise smooth vector fields V along a geodesic $\gamma : [a, b] \to M$. Define $I_\gamma : \mathfrak{V}_\gamma \times \mathfrak{V}_\gamma \to \mathbb{R}$ by

$$I_\gamma(V, W) := \int_a^b \left\{ g_{\dot\gamma}\left(D_{\dot\gamma}^{\dot\gamma} V, D_{\dot\gamma}^{\dot\gamma} W\right) - g_{\dot\gamma}\left(R_{\dot\gamma}(V), W\right) \right\} dt. \quad (5.4.1)$$

I_γ is said to be the *index form* along γ. Since $R_{\dot\gamma}$ is self-joint with respect to $g_{\dot\gamma}$, I_γ is symmetric, i.e., $I_\gamma(V, W) = I_\gamma(W, V)$. Note that the index form may be defined for a piecewise smooth geodesic γ. See §5.4 in [BCS] for more details.

Lemma 5.4.1. *If V is a Jacobi field along γ, then, for any $W \in \mathfrak{V}_\gamma$,*

$$I_\gamma(V, W) = g_{\dot\gamma}\left(D_{\dot\gamma}^{\dot\gamma} V, W\right) \Big|_a^b.$$

Proof. Note that

$$\frac{d}{dt} g_{\dot\gamma}\left(D_{\dot\gamma}^{\dot\gamma} V, W\right) = g_{\dot\gamma}\left(D_{\dot\gamma}^{\dot\gamma} D_{\dot\gamma}^{\dot\gamma} V, W\right) + g_{\dot\gamma}\left(D_{\dot\gamma}^{\dot\gamma} V, D_{\dot\gamma}^{\dot\gamma} W\right). \quad (5.4.2)$$

The proof follows from the integration by parts and the Jacobi field equation. □

Lemma 5.4.2 (Index Lemma). *Let $\gamma : [0, l] \to M$ be a geodesic on an n-dimensional Finsler manifold (M, F) and $J = J(t)$ be a Jacobi field along γ. Suppose that there is no conjugate point of $\gamma(0)$ along γ. Then, for any piecewise smooth vector field V along γ with $V(0) = J(0)$ and $V(l) = J(l)$, we have $I_\gamma(V, V) \geq I_\gamma(J, J)$. The equality holds if and only if $V = J$.*

Proof. We prove this lemma according to two cases.

Case 1. $V(0) = J(0) = 0$. Choose a $g_{\dot\gamma(l)}$-orthonormal basis $\{e_i\}_{i=1}^n$ in $T_{\gamma(l)}M$. For each i, there is a unique Jacobi field J_i along γ with $J_i(0) = 0$ and $J_i(l) = e_i$ by Proposition 5.3.3. Then $J_1, \ldots, J_n$ are linearly independent on $M \backslash \{\gamma(0)\}$. Since $J_i(0) = 0$, by Corollary 5.3.1, there are smooth vector fields $A_i(t) = (d\exp_{\gamma(0)})_{t\dot\gamma(0)}(D_{\dot\gamma}^{\dot\gamma} J_i(0))$ along γ with $A_i(0) = D_{\dot\gamma}^{\dot\gamma} J_i(0)$ such that $J_i = t A_i$. We claim that $\{A_i(t)\}_{i=1}^n$ are linearly independent on $[0, l]$. In fact, it is easy to see that $A_1(t), \ldots, A_n(t)$ are linearly independent on $(0, l]$. For the case when $t = 0$, we assume that there are some constants λ^i $(1 \leq i \leq n)$, such that $\sum_i \lambda_i A_i(0) = 0$, equivalently, $\sum_i \lambda_i D_{\dot\gamma}^{\dot\gamma} J_i(0) = 0$. Then $W(t) := \sum_i \lambda_i J_i(t)$ is a Jacobi field with $W(0) = D_{\dot\gamma}^{\dot\gamma} W(0) = 0$, which implies that $W(t) = 0$ by the uniqueness of the Jacobi field. Since $J_1(t), \ldots, J_n(t)$ are linearly independent on $(0, l]$, we get $\lambda_i = 0$ for all $1 \leq i \leq n$. Thus, $\{A_i(0)\}_{i=1}^n$ are linearly independent. The claim follows.

Now, we assume that $V(t) = \sum_i \tilde{f}_i(t) A_i(t)$ with $\tilde{f}_i(0) = 0$, where $\tilde{f}_i(t)$ are piecewise smooth functions in t. Observe that $\tilde{f}_i(t) = t f_i(t)$ for some piecewise function $f_i(t)$ $(1 \leq i \leq n)$. In fact,

$$\tilde{f}_i(t) = \tilde{f}_i(t) - \tilde{f}_i(0) = \int_0^1 \frac{d}{ds} \tilde{f}_i(ts) ds = t \int_0^1 \tilde{f}_i'(ts) ds.$$

Thus, $V(t) = \sum_i f_i(t) J_i(t)$. Put $\tilde{J}(t) := \sum_i f_i(l) J_i(t)$. Then $\tilde{J}$ is a Jacobi field along γ with $\tilde{J}(0) = J(0) = 0$ and $\tilde{J}(l) = V(l) = J(l)$. Therefore, $J(t) = \tilde{J}(t) = \sum_i f_i(l) J_i(t)$ by Proposition 5.3.3. Note that $R_{\dot\gamma}$ is self-adjoint with respect to $g_{\dot\gamma}$. We have

$$\frac{d}{dt}\left(g_{\dot\gamma}(D_{\dot\gamma}^{\dot\gamma} J_i, J_j) - g_{\dot\gamma}(J_i, D_{\dot\gamma}^{\dot\gamma} J_j)\right) = g_{\dot\gamma}\left(D_{\dot\gamma}^{\dot\gamma} D_{\dot\gamma}^{\dot\gamma} J_i, J_j\right) - g_{\dot\gamma}\left(J_i, D_{\dot\gamma}^{\dot\gamma} D_{\dot\gamma}^{\dot\gamma} J_j\right)$$
$$= -g_{\dot\gamma}\left(R_{\dot\gamma}(J_i), J_j\right) + g_{\dot\gamma}\left(J_i, R_{\dot\gamma}(J_j)\right) = 0,$$

which means that

$$g_{\dot\gamma}(D_{\dot\gamma}^{\dot\gamma} J_i, J_j) = g_{\dot\gamma}(J_i, D_{\dot\gamma}^{\dot\gamma} J_j) \tag{5.4.3}$$

since $J_i(0) = 0$. By Lemma 5.4.1 and $J(0) = 0$, one obtains

$$I_\gamma(J, J) = g_{\dot\gamma}\left(D_{\dot\gamma}^{\dot\gamma} J(l), J(l)\right) = \sum_{i,j} f_i(l) f_j(l) g_{\dot\gamma}\left(D_{\dot\gamma}^{\dot\gamma} J_i(l), J_j(l)\right). \tag{5.4.4}$$

Next, we calculate $I(V, V)$. Since

$$D_{\dot\gamma}^{\dot\gamma} V = \sum_i f_i' J_i + \sum_i f_i D_{\dot\gamma}^{\dot\gamma} J_i := A + B,$$

we have

$$I_\gamma(V, V) = \int_0^l \left\{ g_{\dot\gamma}(A, A) + 2 g_{\dot\gamma}(A, B) + g_{\dot\gamma}(B, B) - g_{\dot\gamma}\left(R_{\dot\gamma}(V), V\right) \right\} dt.$$

$$\tag{5.4.5}$$

By integration by parts and the Jacobi field equation,

$$\int_0^l g_{\dot\gamma}(B,B)dt = \sum_{i,j} \int_0^l f_i f_j g_{\dot\gamma}\left(D_{\dot\gamma}^{\dot\gamma} J_i, D_{\dot\gamma}^{\dot\gamma} J_j\right) dt$$

$$= \sum_{i,j} \int_0^l f_i f_j \left(\frac{d}{dt}g_{\dot\gamma}(D_{\dot\gamma}^{\dot\gamma} J_i, J_j) - g_{\dot\gamma}(D_{\dot\gamma}^{\dot\gamma} D_{\dot\gamma}^{\dot\gamma} J_i, J_j)\right) dt$$

$$= \sum_{i,j} f_i(l) f_j(l) g_{\dot\gamma}\left(D_{\dot\gamma}^{\dot\gamma} J_i(l), J_j(l)\right)$$

$$+ \sum_{i,j} \int_0^l f_i f_j g_{\dot\gamma}\left(R_{\dot\gamma}(J_i), J_j\right) dt$$

$$- \sum_{i,j} \int_0^l \left\{f_i' f_j g_{\dot\gamma}(D_{\dot\gamma}^{\dot\gamma} J_i, J_j) + f_i f_j' g_{\dot\gamma}(D_{\dot\gamma}^{\dot\gamma} J_i, J_j)\right\} dt$$

$$= I_\gamma(J,J) + \int_0^l g_{\dot\gamma}(R_{\dot\gamma}(V),V)dt - 2\int_0^l g_{\dot\gamma}(A,B)dt, \quad (5.4.6)$$

where we used (5.4.3)-(5.4.4) in the last equality. Plugging (5.4.6) in (5.4.5) yields

$$I_\gamma(V,V) = I_\gamma(J,J) + \int_0^l g_{\dot\gamma}(A,A)dt \geq I_\gamma(J,J).$$

Obviously, $I_\gamma(V,V) = I_\gamma(J,J)$ if and only if $A = \sum_i f_i' J_i = 0$ for $t \in [0,l]$, which means that $f_i(t) = f_i(l)$ since $J_1,\ldots,J_n$ are linearly independent on $(0,l]$. Thus, $A = 0$ if and only if $V = J$.

Case 2. $V(0) = J(0) \neq 0$. In this case, let $\tilde J(t) = 0$ and $W(t) = V(t) - J(t)$. Then $W(0) = \tilde J(0) = 0$ and $W(l) = \tilde J(l) = 0$. By Case 1, we have $I_\gamma(W,W) \geq I_\gamma(\tilde J, \tilde J) = 0$ and the equality holds if and only if $W = \tilde J = 0$. Note that

$$I_\gamma(V,J) = I_\gamma(J,V) = g_{\dot\gamma}\left(D_{\dot\gamma}^{\dot\gamma} J, V\right)\Big|_0^l = g_{\dot\gamma}\left(D_{\dot\gamma}^{\dot\gamma} J, J\right)\Big|_0^l = I_\gamma(J,J)$$

by Lemma 5.4.1. Thus,

$$0 \leq I_\gamma(W,W) = I_\gamma(V,V) - 2I_\gamma(V,J) + I_\gamma(J,J) = I_\gamma(V,V) - I_\gamma(J,J),$$

which means that $I_\gamma(V,V) \geq I_\gamma(J,J)$, and the equality holds if and only if $V = J$. $\qquad\square$

Definition 5.4.1. Let (M, F) be a forward complete Finsler manifold. For any $p \in M$ and a unit vector $y \in S_pM$ (indicatrix of F at p), let $\gamma_y(t) = \exp_p(ty)$ be a geodesic that passes through p with initial velocity y. We define a positive number c_y by

$$c_y := \sup\left\{r \in \mathbb{R}^+ \mid \text{no point } \gamma_y(t)(0 < t \leq r) \text{ is conjugate to } p\right\}.$$

c_y is called the *conjugate value* of y and put

$$c_p := \inf_{y \in S_pM} c_y, \quad c_M := \inf_{p \in M} c_p.$$

c_p and c_M are respectively called the *conjugate radius* of p and the *conjugate radius* of M. While the *conjugate locus* of p is defined as

$$\mathbf{C}_p := \{\gamma_y(c_y) \mid y \in S_p(M) \text{ with } c_y < \infty\}.$$

Obviously, $c_y \in (0, \infty]$. If $c_y < \infty$, the point $\gamma_y(c_y)$ is known as the *first conjugate point* of p along $\gamma_y(t)$. In other words, c_y is the first positive number such that there is a nonzero Jacobi field J along $\gamma(t) = \exp_p(ty)(0 \leq t \leq r)$ satisfying $J(0) = J(r) = 0$. Otherwise, we say that p has no conjugate point along γ_y.

Proposition 5.4.1. *Let* $\gamma : [0, l] \to M$ *be a geodesic on an n-dimensional Finsler manifold* (M, F).

(1) *If* $\gamma(0)$ *has no conjugate point along* γ, *then* $I_\gamma(V, V) > 0$ *for any nonzero piecewise smooth vector field* V *along* γ *with* $V(0) = V(l) = 0$.
(2) *If there is a conjugate point of* $\gamma(0)$ *along* γ *and* $\gamma(l)$ *is the first conjugate point of* $\gamma(0)$ *along* γ, *then, for any nonzero piecewise smooth vector field* V *along* γ *with* $V(0) = V(l) = 0$, $I_\gamma(V, V) \geq 0$ *and the equality holds if and only if* V *is a Jacobi field.*

Proof. (1) If $\gamma(0)$ has no conjugate point along γ, then the Jacobi field $J(t)$ along γ with $J(0) = J(l) = 0$ must vanish by Proposition 5.3.3. It follows from Lemma 5.4.2 that $I_\gamma(V, V) > 0$ for any nonzero vector field V along γ with $V(0) = V(l) = 0$.

(2) First, we assume that $0 < r < l$. Then $I_\gamma(V, V) > 0$ on $[0, r]$. Now, we assume that $r = l$ and $\gamma(t) = \exp_{\gamma(0)}(ty)$ for $t \in [0, r]$, where $y = \frac{1}{F(\dot{\gamma}(0))}\dot{\gamma}(0)$. Let $r_i = l\left(1 - (i + 1)^{-1}\right)$ and $\{V_i(t) \mid 0 < t < r_i\}$ be a sequence of piecewise smooth vector fields along $\gamma_i = \exp_{\gamma(0)}(ty)$ for $t \in [0, r_i]$ such that $V_i(0) = V_i(r_i) = 0$ and $\lim_{i \to \infty} V_i = V$. Then $I_{\gamma_i}(V_i, V_i) > 0$ from (1). Letting $i \to \infty$ yields $I_\gamma(V, V) \geq 0$.

Let $W(t)$ be a piecewise smooth vector field along γ with $W(0) = W(l) = 0$. Then, for any $\lambda \in \mathbb{R}$,

$$I_\gamma(V + \lambda W, V + \lambda W) = I_\gamma(V, V) + 2\lambda I_\gamma(V, W) + \lambda^2 I_\gamma(W, W).$$

Thus, for any nonzero vector field V along γ with $V(0) = V(l) = 0$, $I_\gamma(V, V) = 0$ if and only if $I_\gamma(V, W) = 0$ for any nonzero vector field W along γ with $W(0) = W(l) = 0$. On the other hand, by (5.4.1)–(5.4.2), one obtains

$$I_\gamma(V, W) = -\int_0^l g_{\dot\gamma}\left(D_{\dot\gamma}^{\dot\gamma} D_{\dot\gamma}^{\dot\gamma} V + R_{\dot\gamma}(V), W\right) dt.$$

In particular, taking

$$W(t) = f(t)\left(D_{\dot\gamma}^{\dot\gamma} D_{\dot\gamma}^{\dot\gamma} V + R_{\dot\gamma}(V)\right),$$

where $f : [0, l] \to M$ is a smooth positive function with $f(0) = f(l) = 0$, one obtains

$$0 = I_\gamma(V, W) = \int_0^l f(t)\|D_{\dot\gamma}^{\dot\gamma} D_{\dot\gamma}^{\dot\gamma} V + R_{\dot\gamma}(V)\|_{g_{\dot\gamma}}^2 \, dt,$$

which means that V is a Jacobi field. This ends the proof. $\square$

Let $\overline{\mathfrak{V}}_\gamma := \{V \in \mathfrak{V}_\gamma | V(0) = V(l) = 0\}$. It follows from the proof of Proposition 5.4.1 that $I_\gamma(V, V) = 0$ if and only if V is in the null space of I_γ on $\overline{\mathfrak{V}}_\gamma$, i.e., $I_\gamma(V, W) = 0$ for any $W \in \overline{\mathfrak{V}}_\gamma$. Thus, by Proposition 5.4.1, we get the following result.

Corollary 5.4.1. *I_γ has nontrivial null space on the set $\overline{\mathfrak{V}}_\gamma$ if and only if $\gamma(l)$ is the first conjugate point of $\gamma(0)$ along γ.*

Now, we give a geometric characterization of the first conjugate point as follows.

Theorem 5.4.1. *Let $\gamma : [0, l] \to M$ be a geodesic on a Finsler manifold (M, F) and $H : [0, l] \times (-\varepsilon, \varepsilon) \to M$ be a smooth variation of γ keeping the two endpoints fixed.*

(i) *If there is no conjugate point of $\gamma(0)$ along γ, then there is a positive $\delta(< \varepsilon)$ such that for any nonzero $s \in (-\delta, \delta)$ we have $L(s) > L(0)$.*

(ii) *If there is a conjugate point of $\gamma(0)$ along γ, then there is a positive number $\delta(< \varepsilon)$ such that for any nonzero $s \in (-\delta, \delta)$ we have $L(s) < L(0)$.*

Proof. (i) Let V be the variational field of H. By the assumption, V is a non-vanishing vector field with $V(0) = V(l) = 0$. Since $\gamma(l)$ is not a conjugate point of $\gamma(0)$, we have $I_\gamma(V, V) > 0$ by Proposition 5.4.1(1). Note that the variation H is arbitrary. In particular, for any normal variation H whose variation field is determined by V with $V(0) = V(l) = 0$, we have $I_\gamma(V, V) > 0$. From this and (5.2.11), one obtains that $L''(0) > 0$, namely, the functional $L(s)$ attains the minimum at $s = 0$. Consequently, there is a positive $\delta(< \varepsilon)$ such that $L(s) > L(0)$ for any nonzero $s \in (-\delta, \delta)$.

(ii) It suffices to construct a piecewise smooth vector field V along γ which is orthogonal to $\dot\gamma$ with respect to $g_{\dot\gamma}$ such that $I_\gamma(V, V) < 0$. Assume that $t_0 \in (0, l)$ such that $\gamma(t_0)$ is the first conjugate point of $\gamma(0)$. Then there is a nonzero Jacobi field $J_1(t)$ along $\gamma|_{[0,t_0]}$ with $J_1(0) = J_1(t_0) = 0$ and hence $g_{\dot\gamma}(J_1, \dot\gamma) = 0$. Since $J_1(t_0) = 0$, we have $D^{\dot\gamma}_{\dot\gamma} J_1(t_0) \neq 0$. Otherwise, $J_1(t) = 0$ by Corollary 5.3.1. This is a contradiction.

Choose a sufficiently small $\varepsilon > 0$ such that $[t_0 - \varepsilon, t_0 + \varepsilon] \subset [0, l]$ and there is no conjugate point of $\gamma(t_0 - \varepsilon)$ along $\gamma|_{[t_0-\varepsilon,t_0+\varepsilon]}$. Thus $J_1(t_0 - \varepsilon) \neq 0$ since $\gamma(t_0 - \varepsilon)$ is not a conjugate point of $\gamma(0)$. There is a unique Jacobi field J_2 along $\gamma|_{[t_0-\varepsilon,t_0+\varepsilon]}$ such that

$$J_2(t_0 - \varepsilon) = J_1(t_0 - \varepsilon), \quad J_2(t_0 + \varepsilon) = 0$$

by Proposition 5.3.3. Moreover, $g_{\dot\gamma}(J_2, \dot\gamma) = 0$. Let

$$\bar{J}_1(t) = \begin{cases} J_1(t), & t \in [t_0 - \varepsilon,\, t_0], \\ 0, & t \in [t_0,\, t_0 + \varepsilon] \end{cases}$$

Since $D^{\dot\gamma}_{\dot\gamma} J_1(t_0) \neq 0$, $\bar{J}_1(t)$ is not differentiable at $t = t_0$. Consequently, $\bar{J}_1(t)$ is not a Jacobi field along $\gamma|_{[t_0-\varepsilon,\, t_0+\varepsilon]}$, which implies that $\bar{J}_1 \neq J_2$. Define a piecewise smooth vector field V along γ by

$$V(t) = \begin{cases} J_1(t), & t \in [0,\, t_0 - \varepsilon], \\ J_2(t), & t \in [t_0 - \varepsilon, t_0 + \varepsilon], \\ 0, & t \in [t_0 + \varepsilon,\, l]. \end{cases}$$

Then, by Lemmas 5.4.1–5.4.2,

$$I_\gamma(V, V) = I_{\gamma|_{[0,t_0+\varepsilon]}}(V, V) = I_{\gamma|_{[0,t_0-\varepsilon]}}(J_1, J_1) + I_{\gamma|_{[t_0-\varepsilon,t_0+\varepsilon]}}(J_2, J_2)$$

$$< I_{\gamma|_{[0,t_0-\varepsilon]}}(J_1, J_1) + I_{\gamma|_{[t_0-\varepsilon,t_0+\varepsilon]}}(\bar{J}_1, \bar{J}_1)$$

$$= I_{\gamma|_{[0,t_0-\varepsilon]}}(J_1, J_1) + I_{\gamma|_{[t_0-\varepsilon,t_0]}}(J_1, J_1)$$

$$= I_{\gamma|_{[0,t_0]}}(J_1, J_1) = g_{\dot\gamma}(D^{\dot\gamma}_{\dot\gamma} J_1, J_1)|_0^{t_0} = 0.$$

This finishes the proof. $\qquad\square$

Likewise, we also can define the *cut value* i_y of $y \in S_p M$ for any $p \in M$ by

$$i_y := \sup \{r \in \mathbb{R}^+ | \text{the segment } \gamma_y|_{[0,r]} \text{ is global minimizing}\},$$

where $\gamma_y(t)$ is a unit speed geodesic. Note that $i_y \in (0, \infty]$. If $i_y < \infty$, the point $\gamma_y(i_y)$ is called the *cut point* of p along γ_y. If $i_y = \infty$, we say that the cut point does not exist in the direction y. Put

$$i_p := \inf_{y \in S_p M} i_y, \qquad i_M := \inf_{p \in M} i_p.$$

We call i_p and i_M the *injectivity radius* at p and the *injectivity radius* of M respectively. Whereas

$$\mathcal{C}_p := \{\gamma_y(i_y) | y \in S_p M \text{ with } i_y < \infty\}$$

is called the *cut locus* of p, which is a closed subset of M with zero Hausdorff measure ([BCS]).

Proposition 5.4.2. *Let (M, F) be a forward complete Finsler manifold. For any $p \in M$ and $y \in S_p M$, let $\gamma : [0, \infty) \to M$ be a unit speed geodesic with $\gamma(0) = p$ and $\dot{\gamma}(0) = y$. Then*

(1) *$i_y \leq c_y$ and hence $i_p \leq c_p$ at each $p \in M$;*
(2) *for any $r < i_y$, the geodesic $\gamma|_{[0,r]}$ is the unique minimizer of arc length among all piecewise smooth curves that share its endpoints;*
(3) *there is at least another minimizing geodesic σ issuing from p to $\gamma(i_y)$ when $i_y < c_y$.*

For its proof, we refer to §8.2 in [BCS] and §12.2 in [Sh1]. We omit it here.

Chapter 6

Comparison Theorems

Comparison theorems are one of the basic tools to study the global geometric analysis and topology on manifolds. Essentially, they describe more general properties on manifolds via the relationships between the Jacobi fields and the curvature of manifolds. Under some circumstances, we study complete Riemannian manifolds by comparing the geometry of a general manifolds M with that of a simply connected model space of constant sectional curvature. However, things are little different in Finsler geometry. On one hand, the geometry and topology of Finsler manifolds are restricted by the non-Riemannian quantities as well as the Riemannian quantities. On the other hand, the model spaces become more complicated and even have not been completely described clearly. Fortunately, Finsler geometers overcome these obstructions and establish the theory of comparison geometry. Let us begin with some basic concepts, such as the gradient and the Laplacian etc., and then introduce some comparison theorems and their applications.

6.1 Gradient, Hessian and Finsler Laplacian

Let (M, F, m) be an n-dimensional Finsler measure space and $\mathcal{L} : TM \to T^*M$ be the *Legendre transformation* associated with F and its dual norm F^*. That is to say, for each $x \in M$, $\mathcal{L}_x$ sends $y \in T_xM$ to a unique element $\xi = \mathcal{L}_x(y) \in T_x^*M$ given by

$$\mathcal{L}_x(y) := \begin{cases} g_y(y, \cdot), & y \in T_xM \backslash \{0\}, \\ 0, & y = 0. \end{cases}$$

such that $F(x, y) = F^*(x, \xi)$ and $\xi(y) = F^2(y)$, where g_y is the fundamental tensor of F. Then $\mathcal{L} := \{\mathcal{L}_x : T_xM \to T_x^*M | x \in M\}$. In general, $\mathcal{L}_x$ is a nonlinear map from T_xM to T_x^*M. It is linear only when F comes from a

Riemannian metric. In local coordinates, $\mathcal{L}_x(y) = g_{ij}(y)y^j dx^i \in T_x^*M\backslash\{0\}$ if $y \neq 0$ and $\mathcal{L}_x(y) = 0$ if $y = 0$. Write $\xi = \mathcal{L}_x(y) = \xi_i dx^i$ for any $y \in T_xM\backslash\{0\}$, where $\xi_i = g_{ij}(y)y^j$ (the latter expression makes sense at 0 as $\mathcal{L}_x(0) = 0$). Denote by $g^{*ij}(\xi) = \frac{1}{2}[F^{*2}]_{\xi^i\xi^j}(\xi)$. Then

$$g^{*ij}(x,\xi) = g^{ij}(x,y), \qquad (6.1.1)$$

where $(g^{ij}(x,y))$ is the inverse matrix of $(g_{ij}(x,y))$.

For a C^2 function $u : M \to R$, the *gradient* ∇u of u is defined by $\nabla u := \mathcal{L}^{-1}(du)$. Obviously, $\nabla u = 0$ if $du = 0$. In a local coordinate system, we can reexpress ∇u as

$$\nabla u := \begin{cases} g^{ij}(\nabla u)\frac{\partial u}{\partial x^i}\frac{\partial}{\partial x^j}, & x \in M_u, \\ 0, & x \in M\backslash M_u, \end{cases} \qquad (6.1.2)$$

where $M_u = \{x \in M | du(x) \neq 0\}$. In general, ∇u is only continuous on M, but smooth on M_u.

Given a weakly differentiable vector field V on a Finsler measure space (M, F, m), the *divergence* of V with respect to m, denoted by $\text{div}_m(V)$, is defined by

$$\int_M \varphi(\text{div}_m V)dm = -\int_M d\varphi(V)dm$$

for any $\varphi \in C_0^\infty(M)$. If V is differentiable, then $\text{div}_m(V)$ is given by (2.4.1). The *Finsler Laplacian* $\boldsymbol{\Delta}_m$ of u on (M, F, m) is formally defined by $\boldsymbol{\Delta}_m u = \text{div}_m(\nabla u)$. Note that ∇u is weakly differentiable. The Finsler Laplacian should be understood in the weak sense by

$$\int_M \varphi\boldsymbol{\Delta}_m u\,dm = -\int_M d\varphi(\nabla u)dm \qquad (6.1.3)$$

for any $\phi \in C_0^\infty(M)$ ([Sh1]). In local coordinates $(x^i)_{i=1}^n$, we write $V = V^i\frac{\partial}{\partial x^i}$ and $dm = \sigma(x)dx$, where $dx = dx^1 \wedge \cdots \wedge dx^n$. Then $\text{div}_m(V)$ (simply, $\text{div}(V)$) is expressed by

$$\text{div}(V) = \frac{1}{\sigma}\frac{\partial(\sigma V^i)}{\partial x^i} \qquad (6.1.4)$$

and $\boldsymbol{\Delta}_m u$ (simply, $\boldsymbol{\Delta}u$) is expressed by

$$\boldsymbol{\Delta}u = \frac{1}{\sigma}\frac{\partial}{\partial x^i}\left(\sigma g^{ij}(\nabla u)\frac{\partial u}{\partial x^j}\right) \qquad (6.1.5)$$

on M_u. Obviously, the Finsler Laplacian is a nonlinear weakly differential operator on M. The *Hessian* of u on M_u is defined by

$$\nabla^2 u(X,Y) = g_{\nabla u}(D_X^{\nabla u} \nabla u, Y) \tag{6.1.6}$$

for any $X, Y \in \Gamma(TM)$, where D is the covariant differentiation with respect to the Chern connection ([WuX]). Equivalently,

$$\nabla^2 u(X,Y) = X g_{\nabla u}(\nabla u, Y) - g_{\nabla u}(\nabla u, D_X^{\nabla u} Y)$$
$$= XY(u) - D_X^{\nabla u} Y(u), \tag{6.1.7}$$

where we used (3.1.19). In particular, if F is Riemannian, then it is reduced to the one in Riemannian case. It is easy to check that $\nabla^2 u(X,Y)$ is symmetric with respect to X and Y. In fact, by (3.1.18)–(3.1.19) and $C_{\nabla u}(\nabla u, X, Y) = 0$, we have

$$g_{\nabla u}\left(D_X^{\nabla u} \nabla u, Y\right) = X\left(g_{\nabla u}(\nabla u, Y)\right) - g_{\nabla u}\left(\nabla u, D_X^{\nabla u} Y\right)$$
$$= XY(u) - g_{\nabla u}\left(\nabla u, D_Y^{\nabla u} X + [X,Y]\right)$$
$$= XY(u) - g_{\nabla u}\left(\nabla u, D_Y^{\nabla u} X\right) - [X,Y](u)$$
$$= Y\left(g_{\nabla u}(\nabla u, X)\right) - g_{\nabla u}\left(\nabla u, D_Y^{\nabla u} X\right)$$
$$= g_{\nabla u}\left(D_Y^{\nabla u} \nabla u, X\right). \tag{6.1.8}$$

For any $x \in M_u$, we choose a local frame $\{\frac{\partial}{\partial x^i}\}_{i=1}^n$ around x such that $g_{\nabla u}(\partial/\partial x^i, \partial/\partial x^j) = g_{ij}(\nabla u)$, whose inverse matrix is $(g^{ij}(\nabla u))$. Note that $\nabla^i u = g^{ij}(\nabla u)\frac{\partial u}{\partial x^j}$ and

$$\frac{\partial}{\partial x^i}\left(g^{kl}(\nabla u)\right)\frac{\partial u}{\partial x^l} = -g^{kp}(\nabla u)g^{ql}(\nabla u)\frac{\partial g_{pq}(\nabla u)}{\partial x^i}\frac{\partial u}{\partial x^l}$$
$$= -g^{kp}(\nabla u)g^{ql}(\nabla u)\left\{g_{pt}\Gamma_{iq}^t + g_{tq}\Gamma_{ip}^t + 2C_{pqt}N_i^t \right.$$
$$\left. + 2C_{pqt}\frac{\partial(\nabla^t u)}{\partial x^i}\right\}(\nabla u)\frac{\partial u}{\partial x^l}$$
$$= -(\nabla^q u)\Gamma_{iq}^k(\nabla u) - g^{kp}(\nabla u)\Gamma_{ip}^l(\nabla u)\frac{\partial u}{\partial x^l}$$
$$= -N_i^k(\nabla u) - g^{kp}(\nabla u)\Gamma_{ip}^l(\nabla u)\frac{\partial u}{\partial x^l}, \tag{6.1.9}$$

where we used (3.1.2) and $C_{ijk}(y)y^i = 0$. It follows from (6.1.5), (6.1.9) and (4.2.10) that

$$\mathbf{\Delta} u = (\nabla^i u)\frac{\partial}{\partial x^i}(\log \sigma) - N^i_{\ i}(\nabla u) + g^{ij}(\nabla u)\left(\frac{\partial^2 u}{\partial x^i \partial x^j} - \Gamma^k_{ij}(\nabla u)\frac{\partial u}{\partial x^k}\right)$$

$$= g^{ij}(\nabla u)\left(\frac{\partial^2 u}{\partial x^i \partial x^j} - \Gamma^k_{ij}(\nabla u)\frac{\partial u}{\partial x^k}\right) - S(\nabla u) \qquad (6.1.10)$$

on M_u. Observe that

$$D^{\nabla u}_{\partial/\partial x^i} \nabla u = D^{\nabla u}_{\partial/\partial x^i}\left(\nabla^k u \frac{\partial}{\partial x^k}\right)$$

$$= \frac{\partial}{\partial x^i}\left(g^{kl}(\nabla u)\frac{\partial u}{\partial x^l}\right)\frac{\partial}{\partial x^k} + (\nabla^j u)\Gamma^k_{ij}(\nabla u)\frac{\partial}{\partial x^k}$$

$$= g^{kl}(\nabla u)\left\{\frac{\partial^2 u}{\partial x^l \partial x^i} - \Gamma^j_{il}(\nabla u)\frac{\partial u}{\partial x^j}\right\}\frac{\partial}{\partial x^k}. \qquad (6.1.11)$$

Consequently, from (6.1.6), one obtains

$$u_{;ij} := \nabla^2 u\left(\partial/\partial x^i, \partial/\partial x^j\right) = g_{\nabla u}(D^{\nabla u}_{\partial/\partial x^i}\nabla u, \partial/\partial x^j)$$

$$= \frac{\partial^2 u}{\partial x^i \partial x^j} - \Gamma^k_{ij}(\nabla u)\frac{\partial u}{\partial x^k}, \qquad (6.1.12)$$

which implies that $\nabla^2 u = u_{;ij}dx^i \otimes dx^j$ is a symmetric covariant tensor field defined on M_u. Obviously, $u_{;ij}$ are the covariant derivatives of order two with respect to the Levi–Civita connection of $g_{\nabla u}$. Combining (6.1.10) with (6.1.12) yields the following result.

Lemma 6.1.1. $\mathbf{\Delta} u = \hat{\Delta} u - S(\nabla u)$ on M_u, where $\hat{\Delta} u = \mathrm{tr}_{\nabla u} \nabla^2 u$ is the Laplacian on the Riemannian manifold $(M, g_{\nabla u}, dV_{g_{\nabla u}})$, here $\mathrm{tr}_{\nabla u}$ means taking the trace with respect to the weighted Riemannian metric $g_{\nabla u}$.

Given a non-vanishing smooth vector field V, one can define the *weighted gradient* and the *weighted Laplacian* on a weighted Riemannian manifold (M, g_V), respectively, given by

$$\nabla^V u := \begin{cases} g^{ij}(x, V)\frac{\partial u}{\partial x^j}\frac{\partial}{\partial x^i} & \text{for } V \in T_x M \backslash \{0\}, \\ 0, & \text{for } V = 0, \end{cases} \qquad (6.1.13)$$

and $\mathbf{\Delta}^V u := \mathrm{div}_m(\nabla^V u)$ in the weak sense. It should be pointing out that we used the divergence with respect to the measure m rather than the volume measure of g_V. It is easy to see that $\nabla^{\nabla u} u = \nabla u$ and $\boldsymbol{\Delta}^{\nabla u} u = \boldsymbol{\Delta} u$. The latter follows from

$$\int_M \phi \boldsymbol{\Delta}^{\nabla u} u \, dm = -\int_M d\phi(\nabla^{\nabla u} u) dm = -\int_M d\phi(\nabla u) dm = \int_M \phi \boldsymbol{\Delta} u \, dm$$

for any $\phi \in C_0^\infty(M)$.

For the reverse structure $\overleftarrow{F}$ of F, we may define the gradient, divergence and Laplacian etc. with respect to $\overleftarrow{F}$. Since $\overleftarrow{g}(x,y) = g(x,-y)$, we have $\overleftarrow{\nabla} u = -\nabla(-u)$ and $\overleftarrow{\boldsymbol{\Delta}} u = -\boldsymbol{\Delta}(-u)$.

Likewise, we can define the Finsler $p(> 1)$-Laplacian. The *Finsler p-Laplacian* $\boldsymbol{\Delta}_{p,m}$ (simply, $\boldsymbol{\Delta}_p$) is formally defined by $\boldsymbol{\Delta}_{p,m} u := \operatorname{div}_m \left(F^{p-2}(x, \nabla u)\nabla u \right)$. To be more precisely, $\boldsymbol{\Delta}_p u$ is defined in the weak sense through the identity

$$\int_M \varphi \boldsymbol{\Delta}_p u \, dm = -\int_M F^{p-2}(x, \nabla u) d\varphi(\nabla u) dm, \qquad (6.1.14)$$

for all $\varphi \in C_0^\infty(M)$. In particular, when $p = 2$, $\boldsymbol{\Delta}_p$ is exactly the usual Finsler Laplacian $\boldsymbol{\Delta}$ (cf. [Xia5]–[Xia7], [YH1]–[YH2]).

6.2 Rauch Comparison Theorem

Rauch comparison theorem allows us to draw geometric conclusions from comparing length of the Jacobi field on a Finsler manifold (M, F) with the corresponding length of that on another Finsler manifold $(\tilde{M}, \tilde{F})$ whose flag curvature is suitably related to that of M. In particular, it tells us that Jacobi fields grow fastest on negatively curved spaces, less so on flat spaces, and considerably slower on positively curved spaces. We follow the approaches in [CE] and [BCS] to argue this.

For the sake of convenience, we use $(\ )'$ to abbreviate D_T^T or $\tilde{D}_{\tilde{T}}^{\tilde{T}}$ and $(\ )''$ to abbreviate $D_T^T D_T^T$ or $\tilde{D}_{\tilde{T}}^{\tilde{T}} \tilde{D}_{\tilde{T}}^{\tilde{T}}$. Also, we abbreviate

$$\|W\|_T := \sqrt{g_T(W, W)}, \quad \|\tilde{W}\|_{\tilde{T}} := \sqrt{\tilde{g}_{\tilde{T}}(\tilde{W}, \tilde{W})}$$

for any $W \in \Gamma(TM)$ and $\tilde{W} \in \Gamma(T\tilde{M})$ in the following.

Theorem 6.2.1 (Rauch). *Let (M, F) and $(\tilde{M}, \tilde{F})$ be n-dimensional and $\tilde{n}(\tilde{n} \geq n)$-dimensional Finsler manifolds respectively, and let $\gamma : [0, \ell] \to M$ and $\tilde{\gamma} : [0, \ell] \to \tilde{M}$ be normal geodesics with $\dot{\gamma}(t) = T(t)$ and $\dot{\tilde{\gamma}}(t) = \tilde{T}(t)$. Assume that, for each $t \in [0, \ell]$ and any $X \in T_{\gamma(t)}M$, $\tilde{X} \in T_{\tilde{\gamma}(t)}\tilde{M}$, the flag curvatures of flags Π and $\tilde{\Pi}$, respectively, spanned by T, X and $\tilde{T}, \tilde{X}$,* satisfy $\mathbf{K}(\Pi, T) \leq \tilde{\mathbf{K}}(\tilde{\Pi}, \tilde{T})$. *Assume further that* $\tilde{\gamma}(t)$ *does not have conjugate point pairs on* $[0, \ell]$. *Let* $J, \tilde{J}$ *be the Jacobi fields along* $\gamma, \tilde{\gamma}$ *such that* $J(0), \tilde{J}(0)$ *are respectively tangent to* $\gamma, \tilde{\gamma}$ *with* $g_T(J(0), J(0)) = \tilde{g}_{\tilde{T}}(\tilde{J}(0), \tilde{J}(0))$ *and*

$$g_T(T(0), J'(0)) = \tilde{g}_{\tilde{T}}(\tilde{T}(0), \tilde{J}'(0)), \quad g_T(J'(0), J'(0)) = \tilde{g}_{\tilde{T}}(\tilde{J}'(0), \tilde{J}'(0)).$$
$$\tag{6.2.1}$$

Then for any $t \in [0, \ell]$,

$$g_T(J, J) \geq \tilde{g}_{\tilde{T}}(\tilde{J}, \tilde{J}). \tag{6.2.2}$$

Proof. We first treat with the case when $J, \tilde{J}$ are perpendicular to $T, \tilde{T}$, respectively and $J(0) = \tilde{J}(0) = 0$. It is obvious that (6.2.2) holds if $\tilde{J} = 0$. Assume that $\tilde{J}$ is a non-vanishing Jacobi field. Consider the ratio $\|J\|_T^2 / \|\tilde{J}\|_{\tilde{T}}^2$, as a function in the parameter t of geodesics. This is well defined except at $t = 0$ since $\tilde{\gamma}$ has no conjugate points of $\tilde{\gamma}(0)$. Applying L'Hôspital's rule twice and the assumptions to deduce that

$$\lim_{t \to 0} \frac{\|J\|_T^2}{\|\tilde{J}\|_{\tilde{T}}^2} = \lim_{t \to 0} \frac{g_T(J', J)}{\tilde{g}_{\tilde{T}}(\tilde{J}', \tilde{J})} = \lim_{t \to 0} \frac{g_T(J', J')}{\tilde{g}_{\tilde{T}}(\tilde{J}', \tilde{J}')} = 1.$$

To prove that $\|J\|_T^2 \geq \|\tilde{J}\|_{\tilde{T}}^2$, it suffices to prove that J is nowhere zero on $(0, \ell]$ and

$$\frac{d}{dt}\left(\frac{\|J\|_T^2}{\|\tilde{J}\|_{\tilde{T}}^2}\right) \geq 0, \quad \text{equivalently}, \quad \frac{g_T(J', J)}{\|J\|_T^2} \geq \frac{\tilde{g}_{\tilde{T}}(\tilde{J}', \tilde{J})}{\|\tilde{J}\|_{\tilde{T}}^2} \tag{6.2.3}$$

for $t \in (0, \ell]$.

Fix $t_0 \in (0, \ell]$, such that J has no zero points on $(0, t_0]$. For any $t_1 \in (0, t_0]$, define vector fields on $(0, t_1]$

$$W(t) = \frac{J(t)}{\sqrt{g_T(J(t_1), J(t_1))}}, \quad \tilde{W}(t) = \frac{\tilde{J}(t)}{\sqrt{\tilde{g}_{\tilde{T}}(\tilde{J}(t_1), \tilde{J}(t_1))}}.$$

Then W and $\tilde{W}$ are Jacobi fields along $\gamma|_{(0,t_1]}$ and $\tilde{\gamma}|_{(0,t_1]}$ with $W(0) = \tilde{W}(0) = 0$ and $\|W\|_T(t_1) = \|\tilde{W}\|_{\tilde{T}}(t_1) = 1$. Moreover, by Lemma 5.4.1,

$$\left.\frac{g_T(J', J)}{\|J\|_T^2}\right|_{t_1} = g_T(W', W)\big|_{t_1} = I_{\gamma|_{[0,t_1]}}(W, W), \tag{6.2.4}$$

$$\left.\frac{\tilde{g}_{\tilde{T}}(\tilde{J}', \tilde{J})}{\|\tilde{J}\|_{\tilde{T}}^2}\right|_{t_1} = \tilde{g}_{\tilde{T}}(\tilde{W}', \tilde{W})\big|_{t_1} = \tilde{I}_{\tilde{\gamma}|_{[0,t_1]}}(\tilde{W}, \tilde{W}). \tag{6.2.5}$$

Now, we choose g_T-orthonormal frame fields $e_i(t)(1 \leq i \leq 1)$ along γ and $\tilde{g}_{\tilde{T}}$-orthonormal frame fields $\tilde{e}_i(t)(1 \leq i \leq \tilde{n})$ along $\tilde{\gamma}$ respectively by parallel translations such that

$$e_1(t_1) = T(t_1), \quad e_2(t_1) = W(t_1), \quad \tilde{e}_1(t_1) = \tilde{T}(t_1), \quad \tilde{e}_2(t_1) = \tilde{W}(t_1).$$

Obviously, $e_1(t) = T(t)$ and $\tilde{e}_1(t) = \tilde{T}(t)$. Since W and $\tilde{W}$ are normal Jacobi fields, we may set

$$W(t) = \sum_{i=2}^{n} W^i(t)e_i(t), \quad \tilde{W}(t) = \sum_{i=2}^{\tilde{n}} \tilde{W}^i(t)\tilde{e}_i(t).$$

By the assumptions, we have

$$W^2(0) = \cdots = W^n(0) = 0, \quad W^2(t_1) = 1, \quad W^3(t_1) = \cdots = W^n(t_1) = 0;$$
$$\tilde{W}^2(0) = \cdots = \tilde{W}^n(0) = 0, \quad \tilde{W}^2(t_1) = 1, \quad \tilde{W}^3(t_1) = \cdots = \tilde{W}^n(t_1) = 0.$$

Define a normal vector field along $\tilde{\gamma}|_{[0,t_1]}$ with respect to $\tilde{g}_{\tilde{T}}$ by

$$V(t) = \sum_{i=2}^{\tilde{n}} W^i(t)\tilde{e}_i(t).$$

Then $V(0) = \tilde{W}(0) = 0$ and $V(t_1) = \tilde{W}(t_1) = \tilde{e}_2(t_1)$. From the index lemma, one obtains that $\tilde{I}_{\tilde{\gamma}|_{[0,t_1]}}(V, V) \geq \tilde{I}_{\tilde{\gamma}|_{[0,t_1]}}(\tilde{W}, \tilde{W})$. On the other hand, by the construction of $V(t)$, we have $\|V\|_{\tilde{T}} = \|W\|_T$. Since $\mathbf{K}(\Pi, T) \leq \tilde{\mathbf{K}}(\tilde{\Pi}, \tilde{T})$ for flags $\Pi = \mathrm{span}\{W, T\}$ and $\tilde{\Pi} = \mathrm{span}\{V, \tilde{T}\}$, one obtains

$$\tilde{I}_{\tilde{\gamma}|_{[0,t_1]}}(\tilde{W}, \tilde{W}) \leq \tilde{I}_{\tilde{\gamma}|_{[0,t_1]}}(V, V) = \int_0^{t_1} \left\{ \tilde{g}_{\tilde{T}}(V', V') - \tilde{g}_{\tilde{T}}(\tilde{R}_{\tilde{T}}(V), V) \right\} dt$$

$$\leq \int_0^{t_1} \left\{ g_T(W', W') - g_T(R_T(W), W) \right\} dt = I_{\gamma|_{[0,t_1]}}(W, W).$$

From this, $J(0) = \tilde{J}(0) = 0$, and (6.2.4)–(6.2.5), (6.2.3) holds for any $t_1 \in (0, t_0]$. Next we prove that $t_0 = \ell$, i.e., J has no zero points on $(0, \ell]$. In fact, if there is the first point $t_0 \in (0, \ell]$ such that $J(t_0) = 0$, then J has no zero points on $(0, t_0)$. Following previous arguments, we get

$$g_T(J(t), J(t)) \geq \tilde{g}_{\tilde{T}}(\tilde{J}(t), \tilde{J}(t)), \quad t \in (0, t_0).$$

Taking the limit $t \to t_0$ yields

$$0 = g_T(J(t_0), J(t_0)) \geq \tilde{g}_{\tilde{T}}(\tilde{J}(t_0), \tilde{J}(t_0)) > 0,$$

which is impossible. Thus, $t_0 = \ell$ and hence (6.2.3) holds on $(0, \ell]$.

In general case, let

$$J = J^\perp + g_T(J,T)T, \quad \tilde{J} = \tilde{J}^\perp + \tilde{g}_{\tilde{T}}(\tilde{J},\tilde{T})\tilde{T},$$

where $J^\perp$ and $\tilde{J}^\perp$ are respectively the normal Jacobi fields along γ and $\tilde{\gamma}$ with respect to g_T and $\tilde{g}_{\tilde{T}}$. Since $J(0), \tilde{J}(0)$ are respectively tangent to $\gamma, \tilde{\gamma}$, we have $J^\perp(0) = \tilde{J}^\perp(0) = 0$. Consequently, we have $\|J^\perp\|_T \geq \|\tilde{J}^\perp\|_{\tilde{T}}$ from the previous case. On the other hand, since J and $\tilde{J}$ are Jacobi fields, $g_T(J,T) = at + b$ and $\tilde{g}_{\tilde{T}}(\tilde{J},\tilde{T}) = \tilde{a}t + \tilde{b}$. It follows from the first equation of (6.2.1) that $a = \tilde{a}$. Note that

$$g_T(J,J) = g_T(J^\perp,J^\perp) + g_T(J,T)^2, \quad \tilde{g}_{\tilde{T}}(\tilde{J},\tilde{J}) = \tilde{g}_{\tilde{T}}(\tilde{J}^\perp,\tilde{J}^\perp) + \tilde{g}_{\tilde{T}}(\tilde{J},\tilde{T})^2.$$

Since $\|J(0)\|_T = \|\tilde{J}(0)\|_{\tilde{T}}$ and $J^\perp(0) = \tilde{J}^\perp(0) = 0$, we have $g_T(J,T)(0) = \tilde{g}_{\tilde{T}}(\tilde{J},\tilde{T})(0)$, which implies that $b = \tilde{b}$. Thus, $g_T(J,T) = \tilde{g}_{\tilde{T}}(\tilde{J},\tilde{T})$ and hence (6.2.2) follows. □

In the case when $(\tilde{M}, \tilde{F})$ has constant flag curvature c, by Rauch Theorem and Example 5.3.1, one obtains the following results.

Corollary 6.2.1. *Let (M,F) be an n-dimensional Finsler manifold and $\gamma : [0,\ell] \to M$ be a normal geodesic. Suppose that*

(i) *the flag curvature $\mathbf{K}(\Pi,T) \leq c$ for any flag Π spanned by $T = \dot{\gamma}$ and $W \in T_{\gamma(t)}M$;*

(ii) *J is a normal Jacobi field along γ;*

(iii) *$J(0) = 0$.*

Then, for any $0 < t \leq \ell$ if $c \leq 0$ and any $0 < t < \frac{\pi}{\sqrt{c}}$ if $c > 0$, we have $\|J\|_T \geq \mathfrak{s}_c(t)\|J'(0)\|_T$ and

$$\frac{g_T(J',J)}{g_T(J,J)} \geq \frac{\mathfrak{s}_c'(t)}{\mathfrak{s}_c(t)}, \tag{6.2.6}$$

where $\mathfrak{s}_c(t)$ is defined as in Example 5.3.1.

Corollary 6.2.2. *Let (M,F) be an n-dimensional Finsler manifold and $\gamma : [0,\ell] \to M$ be a normal geodesic. Suppose that*

(i) *the flag curvature $\mathbf{K}(\Pi,T) \geq c$ for any flag Π spanned by $T = \dot{\gamma}$ and $W \in T_{\gamma(t)}M$;*

(ii) *J is a normal Jacobi field along γ;*

(iii) *$J(0) = 0$.*

Then, for any $0 < t \leq \ell$ if $c \leq 0$ and any $0 < t < \frac{\pi}{\sqrt{c}}$ if $c > 0$, we have $\|J\|_T \leq \mathfrak{s}_c(t)\|J'(0)\|_T$ *and*

$$\frac{g_T(J', J)}{g_T(J, J)} \leq \frac{\mathfrak{s}_c'(t)}{\mathfrak{s}_c(t)}, \qquad (6.2.7)$$

where $\mathfrak{s}_c(t)$ is defined as in Example 5.3.1.

6.3 Hessian and Laplacian Comparison Theorems

In this section, we shall discuss the Hessian and Laplacian comparison theorems for the distance function. These comparison theorems were first studied by Wu–Xin in [WuX] and later developed by Ohta–Sturm in [OS1]. For this purpose, let us first compute the Hessian of distance function.

Let (M, F, m) be an n-dimensional Finsler measure space. Consider the *distance function* from a fixed point $p \in M$,

$$r(x) := d_F(p, x), \quad x \in M.$$

According to Theorem 5.1.3, $\exp_p$ is C^∞ on $T_pM\backslash\{p\}$ and C^1 at the origin in T_pM. Then $r(x) = F_x(\exp_p^{-1}(x))$ is smooth on $M\backslash(\mathcal{C}_p \cup \{p\})$, where $\mathcal{C}_p$ is the cut locus of p.

For any $x \in M\backslash(\mathcal{C}_p \cup \{p\})$, the distance function r is smooth at x. Let $\gamma : [0, r(x)] \to M$ be a minimal normal geodesic from p to x and $\sigma : [0, \varepsilon) \to M$ be a geodesic with $\sigma(0) = x$ and $\dot\sigma(0) = X$ for any $X \in T_xM$, where $\varepsilon > 0$ sufficiently small. Note that r is smooth at $\sigma(s)$ for any sufficiently small s. Thus, $r(\sigma(s)) = d_F(p, \sigma(s))$. Let $\gamma_s : [0, r(\sigma(s))] \to M$ be a minimal geodesic from p to $\sigma(s)$. Then $H : [0, r] \times (-\varepsilon, \varepsilon) \to M, H(t, s) := \gamma_s(t)$ defines a geodesic variation of γ emanating from p with $H(t, 0) = \gamma(t)$ and $H(r, s) = \sigma(s)$, where $r = r(\sigma(s))$. Its variational field V satisfies $V(0) = 0$ and $V(r) = X$. By the first variation formula (see Theorem 5.1.1), we have

$$X(r) = L'(0) = g_T(T, V)\big|_0^r = g_T(T, X),$$

where $T = \dot\gamma$. On the other hand,

$$X(r) = dr(X) = g_{\nabla r}(\nabla r, X).$$

Thus, $T = \nabla r$ by Corollary 1.1.3 and hence $\dot\gamma = \nabla r$ with $F(\nabla r) = 1$. Consequently, ∇r is a unit speed geodesic field on $M\backslash(\mathcal{C}_p \cup \{p\})$. From this and (6.1.6) or (6.1.7), it is easy to check the following result.

Proposition 6.3.1. *Assume that $x \in M$ is not the cut point of p. Then*

$$\nabla^2 r(\nabla r, \nabla r) = \nabla^2 r(\nabla r, X) = 0,$$

for any $X \in T_x M$.

Proposition 6.3.1 tells us that we need not consider the radial components of X when we calculate $\nabla^2 r(X, X)$. With the above notations, we assume that $X \in T_x M$ with $g_{\nabla r}(\nabla r, X) = 0$. Since x is not a conjugate point of p, there is a unique Jacobi field J along γ such that

$$J(0) = 0, \quad J(r) = X \tag{6.3.1}$$

by Proposition 5.3.3. Note that J can be regarded as a variational field for some geodesic variation of γ from Proposition 5.3.1. Hence, $[J, \nabla r] = 0$, that is,

$$D_{\nabla r}^{\nabla r} J = D_J^{\nabla r} \nabla r. \tag{6.3.2}$$

From the Jacobi field equation (5.3.1) and $\dot{\gamma} = T = \nabla r$, we can see that $g_T(J, T)$ is a linear function in t. This means that $g_T(J, T) = 0$ along γ since $J(0) = 0$ and $g_T(T(r), J(r)) = g_T(T(r), X) = 0$. Thus, by (6.1.6), (6.3.1)–(6.3.2) and the Jacobi field equation,

$$\nabla^2 r(X, X) = \nabla^2 r(J, J)\big|_x = g_T(D_J^T T, J)\big|_x = g_T(D_T^T J, J)\big|_x$$

$$= \int_0^r \frac{d}{dt}\left(g_T(D_T^T J, J)\right) dt$$

$$= \int_0^r \left\{ g_T(D_T^T J, D_T^T J) + g_T(R_T(J), J) \right\} dt = I_\gamma(J, J). \tag{6.3.3}$$

Thus, one obtains the following.

Proposition 6.3.2. *Assume that $x \in M$ is not the cut point of p. Then, for any $X \in T_x M$ with $g_{\nabla r}(\nabla r, X) = 0$, there is a Jacobi field J along γ satisfying (6.3.1)–(6.3.2) such that $\nabla^2 r(X, X) = g_{\nabla r}(D_{\nabla r}^{\nabla r} J, J)\big|_x = I_\gamma(J, J)$.*

Now, we are ready to prove the following Hessian comparison theorem.

Theorem 6.3.1 (Hessian). *Let (M, F), $(\tilde{M}, \tilde{F})$ be two n-dimensional Finsler manifolds, and let $\gamma : [0, \ell] \to M$ and $\tilde{\gamma} : [0, \ell] \to \tilde{M}$ be normal geodesics such that $\gamma(\ell), \tilde{\gamma}(\ell)$ are not cut points of $\gamma(0), \tilde{\gamma}(0)$, respectively.*

Set $r = d_F(\gamma(0), \cdot)$ and $\tilde{r} = \tilde{d}_{\tilde{F}}(\tilde{\gamma}(0), \cdot)$. For any $X \in T_{\gamma(\ell)}M$ and $\tilde{X} \in T_{\tilde{\gamma}(\ell)}\tilde{M}$ satisfying

$$g_{\nabla r}(\nabla r, X) = \tilde{g}_{\tilde{\nabla}\tilde{r}}(\tilde{\nabla}\tilde{r}, \tilde{X}), \quad g_T(X, X) = \tilde{g}_{\tilde{T}}(\tilde{X}, \tilde{X}),$$

if the radial flag curvatures satisfy $\mathbf{K}(\Pi, \nabla r) \geq (\leq)\tilde{\mathbf{K}}(\tilde{\Pi}, \tilde{\nabla}\tilde{r})$, where $\Pi = \mathrm{span}\{\nabla r, X\}$ and $\tilde{\Pi} = \mathrm{span}\{\tilde{\nabla}\tilde{r}, \tilde{X}\}$, then

$$\nabla^2 r(X, X) \leq (\geq)\tilde{\nabla}^2\tilde{r}(\tilde{X}, \tilde{X}).$$

Proof. Let $T = \dot{\gamma} = \nabla r$ and $\tilde{T} = \dot{\tilde{\gamma}} = \tilde{\nabla}\tilde{r}$ as before. By the assumptions and Proposition 6.3.1, we assume that $g_T(T, X) = \tilde{g}_{\tilde{T}}(\tilde{T}, \tilde{X}) = 0$ and $g_T(X, X) = \tilde{g}_{\tilde{T}}(\tilde{X}, \tilde{X}) = 1$ without loss of generality. It follows from Proposition 6.3.2 that

$$\nabla^2 r(X, X) = I_\gamma(J, J), \quad \tilde{\nabla}^2\tilde{r}(\tilde{X}, \tilde{X}) = I_{\tilde{\gamma}}(\tilde{J}, \tilde{J}), \qquad (6.3.4)$$

where J and $\tilde{J}$ are respectively Jacobi fields along γ and $\tilde{\gamma}$ satisfying

$$J(0) = \tilde{J}(0) = 0, \quad J(\ell) = X, \quad \tilde{J}(\ell) = \tilde{X}.$$

Choose parallel $\tilde{g}_{\tilde{T}}$–orthonormal frame fields $\tilde{e}_1(t), \dots, \tilde{e}_n(t) = \tilde{T}$ along $\tilde{\gamma}$. Note that $\tilde{g}_{\tilde{T}}(\tilde{T}, \tilde{J}) = 0$. We may write

$$\tilde{J}(t) = \sum_{i=1}^{n-1} \lambda^i(t)\tilde{e}_i(t), \quad \lambda^i(0) = 0 (1 \leq i \leq n-1).$$

Similarly, we obtain parallel g_T–orthonormal frame fields $e_1(t), \dots, e_n(t) = T$ along γ via taking the parallel translation of g_T–orthonormal basis $\{e_1(\ell), \dots, e_n(\ell)\}$ in $T_{\gamma(\ell)}M$ such that

$$X = J(\ell) = \sum_{i=1}^{n-1} \lambda^i(\ell)e_i(\ell).$$

Define a vector field Z along γ by

$$Z(t) = \sum_{i=1}^{n-1} \lambda^i(t)e_i(t).$$

Then $Z(0) = J(0) = \tilde{J}(0) = 0$ and $Z(\ell) = X = J(\ell)$. By the index lemma (Lemma 5.4.2), we have

$$I_\gamma(J, J) \leq I_\gamma(Z, Z). \qquad (6.3.5)$$

Moreover,

$$g_T(Z,Z) = \tilde{g}_{\tilde{T}}(\tilde{J},\tilde{J}), \quad g_T(D_T^T Z, D_T^T Z) = \tilde{g}_{\tilde{T}}(D_{\tilde{T}}^{\tilde{T}}\tilde{J}, D_{\tilde{T}}^{\tilde{T}}\tilde{J}). \quad (6.3.6)$$

Therefore, by (6.3.4)–(6.3.6),

$$\nabla^2 r\,(X,X) = I_\gamma\,(J,J) \le I_\gamma\,(Z,Z)$$

$$= \int_0^\ell \left\{ g_T\left(D_T^T Z, D_T^T Z\right) - g_T\left(R_T(Z),Z\right)\right\} dt$$

$$\le \int_0^\ell \left\{ g_{\tilde{T}}\left(\tilde{D}_{\tilde{T}}^{\tilde{T}}\tilde{J}, \tilde{D}_{\tilde{T}}^{\tilde{T}}\tilde{J}\right) - \tilde{g}_{\tilde{T}}\left(\tilde{R}_{\tilde{T}}(\tilde{J}),\tilde{J}\right)\right\} dt$$

$$= I_{\tilde{\gamma}}\left(\tilde{J},\tilde{J}\right) = \tilde{\nabla}^2 \tilde{r}\left(\tilde{X},\tilde{X}\right).$$

This finishes the proof. □

In particular, when (M,F) is a Finsler manifold of constant flag curvature c, the normal Jacobi field J along the geodesic $\gamma : [0,r] \to M$ with $J(0) = 0$ is given by (5.3.11), i.e., $J(t) = A(t)\mathfrak{s}_c(t)$, where $A(t)$ is a parallel normal vector field along γ. Let

$$\mathfrak{ct}_c(t) = \frac{\mathfrak{s}_c'(t)}{\mathfrak{s}_c(t)} = \begin{cases} \sqrt{c}\cot(\sqrt{c}t), & \text{if } c > 0, \\ \frac{1}{t}, & \text{if } c = 0, \\ \sqrt{-c}\coth(\sqrt{-c}t), & \text{if } c < 0, \end{cases} \quad (6.3.7)$$

where $t \ge 0$ if $c \le 0$ and $t \in [0, \pi/\sqrt{c}]$ if $c > 0$. Note that $J(r) = X \in T_{\gamma(r)}M$ and $g_T(T, A(t)) = 0$. By Proposition 6.3.2, we have

$$\nabla^2 r(X,X) = g_T(J, D_T^T J)\big|_{t=r} = g_T\big(\mathfrak{s}_c(t)A(t), \mathfrak{s}_c'(t)A(t)\big)\big|_{t=r}$$

$$= \mathfrak{ct}_c(r)g_T(X,X).$$

Back to the general case and using Propositions 6.3.1–6.3.2, we get

$$\nabla^2 r = \mathfrak{ct}_c(r)(g_{\nabla r} - dr \otimes dr). \quad (6.3.8)$$

By Theorem 6.3.1, one obtains the following result, which improves Theorem 4.1 in [WuX].

Corollary 6.3.1. *Let (M,F) be an n-dimensional Finsler manifold and $r = d_F(p, \cdot)$, a distance function from a fixed point $p \in M$. Assume that,* *for any* $X \in T_{\gamma(\ell)}M$, *the radial flag curvature* $\mathbf{K}(\Pi, \nabla r) \geq (\leq)c$, *where* $\Pi = span\{\nabla r, X\}$. *Then*

$$\nabla^2 r(X, X) \leq (\geq)\mathfrak{ct}_c(r)\left(g_{\nabla r}(X, X) - g_{\nabla r}(\nabla r, X)^2\right) \qquad (6.3.9)$$

pointwise on $M\backslash(\mathcal{C}_p \cup \{p\})$.

Taking the trace on both sides of (6.3.9) with respect to $g_{\nabla r}$ and using Lemma 6.1.1, one obtains the following Laplacian comparison theorem for the distance function r.

Theorem 6.3.2. *Under the same assumptions as in Corollary 6.3.1, we have*

$$\mathbf{\Delta} r \leq (n-1)\mathfrak{ct}_c(r) + \|S\| \ (\mathbf{\Delta} r \geq (n-1)\mathfrak{ct}_c(r) - \|S\|) \qquad (6.3.10)$$

point-wise on $M\backslash(\mathcal{C}_p \cup \{p\})$, *where* $\|S\|$ *is the point-wise norm of S-curvature which is defined by*

$$\|S\|_x = \sup_{X \in T_x M\backslash\{0\}} \frac{|S(X)|}{F(X)}.$$

Under the condition for radial Ricci curvature, we have the following result.

Theorem 6.3.3. *Let* (M, F, m) *be an n-dimensional Finsler manifold and* $r(\cdot) = d_F(p, \cdot)$ *the distance function from a fixed point* $p \in M$. *Assume that the radial Ricci curvature* $\mathrm{Ric}(\nabla r) \geq (n-1)c$. *Then*

$$\mathbf{\Delta} r \leq (n-1)\mathfrak{ct}_c(r) + \|S\| \qquad (6.3.11)$$

point-wise on $M\backslash(\mathcal{C}_p \cup \{p\})$.

Proof. For any $q \in M\backslash(\mathcal{C}_p \cup \{p\})$, let $\gamma : [0, \ell] \to M$ be a normal minimal geodesic on (M, F) from $p = \gamma(0)$ to $q = \gamma(\ell)$. Then $t = d_F(p, \gamma(t)) = r(\gamma(t))$ and $\dot{\gamma} = \nabla r(\gamma(t))$. We choose local $g_{\nabla r}$-orthonormal parallel frame fields $e_1(t), \ldots, e_n(t) = \nabla r$ along γ. For $1 \leq i \leq n-1$, let J_i be the unique Jacobi field along γ such that $J_i(0) = 0, J_i(\ell) = e_i(\ell)$, and $W_i(t) = \frac{\mathfrak{s}_c(t)}{\mathfrak{s}_c(\ell)}e_i(t)$, where $\mathfrak{s}_c(t)$ is defined by (5.3.9). Clearly, we have

$W_i(0) = J_i(0) = 0$ and $W_i(\ell) = J_i(\ell)$. From Proposition 6.3.2 and the index lemma, we get

$$
\operatorname{tr}_{\nabla r} \nabla^2 r|_q = \sum_{i=1}^{n} \nabla^2 r(e_i(\ell), e_i(\ell)) = \sum_{i=1}^{n} I_\gamma(J_i, J_i) \leq \sum_{i=1}^{n} I_\gamma(W_i, W_i)
$$

$$
= \frac{1}{\mathfrak{s}_c(\ell)^2} \int_0^{\ell} \left\{ (n-1)\mathfrak{s}_c'(t)^2 - \operatorname{Ric}(\nabla r)\mathfrak{s}_c(t)^2 \right\} dt
$$

$$
\leq \frac{1}{\mathfrak{s}_c(\ell)^2} \int_0^{\ell} \left\{ (n-1)\mathfrak{s}_c'(t)^2 - (n-1)c\mathfrak{s}_c(t)^2 \right\} dt
$$

$$
= (n-1)\mathfrak{ct}_c(r(q)),
$$

which together with Lemma 6.1.1 yields (6.3.11). $\square$

Remark 6.3.1. The Hessian $\nabla^2 r$ with respect to the Finsler metric F is different from the Hessian $\operatorname{Hess}(r)$ with respect to the Riemannian metric $g_{\nabla r}$ even if ∇r is a geodesic field. Similarly, $\mathbf{\Delta} r \neq \hat{\Delta} r$, where $\hat{\Delta} r = \operatorname{tr}_{\nabla r} \nabla^2 r$, which is the Laplacian of r with respect to $g_{\nabla r}$.

Theorems 6.3.2–6.3.3 imply that the estimates of $\mathbf{\Delta} r$ are affected by not only Riemannian quantities, for example, flag curvature and Ricci curvature, but also non-Riemnanian quantity, such as, S-curvature. In [Oh1], S. Ohta introduced the weighted Ricci curvature in terms of the Ricci curvature and the S-curvature. Studies show that the weighted Ricci curvature might be more suitable to study the global analysis and topology of Finsler manifolds ([Oh3]–[Oh4], [OS1]–[OS2], [WX], [Xc], [Xia4]–[Xia10]).

Definition 6.3.1 ([Oh1]). Given a vector $y \in T_x M$, let $\gamma : (-\varepsilon, \varepsilon) \longrightarrow M$ be the geodesic such that $\gamma(0) = x$ and $\dot{\gamma}(0) = y$. We set $dm = e^{-\Psi(\gamma)} \operatorname{Vol}_{\dot{\gamma}}$ along γ, where $\operatorname{Vol}_{\dot{\gamma}}$ is the volume form of $g_{\dot{\gamma}}$. Define the *weighted Ricci curvature* involving a parameter $N \in [n, \infty]$ by

(1) $\operatorname{Ric}_n(y) := \begin{cases} \operatorname{Ric}(y) + (\Psi \circ \gamma)''(0), & \text{if} \quad (\Psi \circ \gamma)'(0) = 0, \\ -\infty, & \text{if} \quad (\Psi \circ \gamma)'(0) \neq 0, \end{cases}$

(2) $\operatorname{Ric}_N(y) := \operatorname{Ric}(y) + (\Psi \circ \gamma)''(0) - \frac{(\Psi \circ \gamma)'(0)^2}{N-n}$ for $N \in (n, \infty)$,

(3) $\operatorname{Ric}_\infty(y) := \operatorname{Ric}(y) + (\Psi \circ \gamma)''(0)$.

For $\lambda \geq 0$ and $N \in [n, \infty]$, define $\operatorname{Ric}_N(\lambda y) := \lambda^2 \operatorname{Ric}(y)$.

We say that $\operatorname{Ric}_N \geq K$ for some $K \in R$ if $\operatorname{Ric}_N(y) \geq KF^2(y)$ for all $(x, y) \in TM$. We remark that $(\Psi \circ \gamma)'(0) = S(x, y)$ (S-curvature), and

$$
(\Psi \circ \gamma)''(0) = \frac{d}{dt}[S(\gamma(t), \dot{\gamma}(t))]_{t=0} = S_{x^i} y^i - 2S_{y^i} G^i = S_{|i} y^i, \qquad (6.3.12)
$$

which is exactly the change rate of S-curvature along γ, denoted by $\dot{S}(y)$. Then $\mathrm{Ric}_N(y)$ and $\mathrm{Ric}_\infty(y)$ can be reexpressed by

$$\mathrm{Ric}_N(y) = \mathrm{Ric}(y) + \dot{S}(y) - \frac{S(y)^2}{N-n}, \quad \mathrm{Ric}_\infty(y) = \mathrm{Ric}(y) + \dot{S}(y). \quad (6.3.13)$$

For the reverse metric $\overleftarrow{F}$, the weighted Ricci curvature $\overleftarrow{\mathrm{Ric}}_N(y) = \mathrm{Ric}_N(-y)$. Thus, $\overleftarrow{\mathrm{Ric}}_N \geq K$ is equivalent to $\mathrm{Ric}_N \geq K$ for some $K \in \mathbb{R}$. In particular, if $(M, g, e^{-f}dV_g)$ is a weighted Riemannian manifold, where $f \in C^\infty(M)$ and dV_g is the Riemannian measure with respect to g, then $\tau(x, y) = f(x), S(x, y) = df_x(y)$ and $\dot{S}(x, y) = \mathrm{Hess}(f)_x(y)$. Thus,

$$\mathrm{Ric}_N = \mathrm{Ric} + \mathrm{Hess}(f) - \frac{df \otimes df}{N-n}, \quad \mathrm{Ric}_\infty = \mathrm{Ric} + \mathrm{Hess}(f), \quad (6.3.14)$$

where Ric and Hess mean the Ricci curvature and the Hessian of g. Ric_N is exactly the weighted Ricci curvature Ric_N^f for $N \in [n, \infty]$ on $(M, g, e^{-f}dV_g)$ ([Li]).

Lemma 6.3.1. *If V is a non-vanishing geodesic vector field on an open set $\mathcal{U}$ in an n-dimensional Finsler space (M, F), then Ric_N is just the weighted Ricci curvature $\hat{\mathrm{Ric}}_N$ of the weighted Riemannian metric $\hat{g} = g_V$ on $\mathcal{U}$ for any $N \in [n, \infty]$.*

Proof. Since V is a geodesic field, we have $\hat{R}^i{}_k(v) = R^i{}_k(v)$ and hence $\hat{\mathrm{Ric}}(v) = \mathrm{Ric}(v)$ for any $x \in \mathcal{U}$ and $v = V_x \in T_x\mathcal{U}$ by Lemma 4.1.2 and Proposition 4.1.2. Let $\gamma : (-\varepsilon, \varepsilon) \to \mathcal{U}$ be an integral curve of V, which is a geodesic such that $\gamma(0) = x, \dot{\gamma}(0) = v = V_x$ and $\dot{\gamma}(t) = V(\gamma(t))$. Along γ, we decompose the measure m as $dm = \sigma(t)dx$, where

$$\sigma(t) = e^{-\Psi(\gamma(t))}\sqrt{\det(g_{ij}(\gamma, \dot{\gamma}))},$$

equivalently,

$$\Psi(\gamma(t)) = \log\frac{\sqrt{\det(g_{ij}(\gamma, \dot{\gamma}))}}{\sigma(t)}.$$

On the other hand, $dm = \sigma(t)dx$ can be regarded as the weighted Riemannian measure on the weighted Riemannian manifold $(\mathcal{U}, g_V, e^{-\Psi}dV_{g_V})$ restricted to γ. On $(\mathcal{U}, g_V, e^{-\Psi}dV_{g_V})$, the weighted Ricci curvature $\hat{\mathrm{Ric}}_N$ is given by

$$\hat{\mathrm{Ric}}_N = \hat{\mathrm{Ric}} + \mathrm{Hess}(\Psi) - \frac{d\Psi \otimes d\Psi}{N-n}, \quad N \in (n, \infty)$$

and $\hat{\mathrm{Ric}}_n = \lim_{N \to n} \hat{\mathrm{Ric}}_N$ and $\hat{\mathrm{Ric}}_\infty = \lim_{N \to \infty} \hat{\mathrm{Ric}}_N$. Note that $\hat{\mathrm{Ric}}(v) = \mathrm{Ric}(v)$, $d\Psi(v) = \frac{d}{dt}|_{t=0}(\Psi \circ \gamma(t)) = S(v)$ and $\mathrm{Hess}(\Psi)(v, v) = \frac{d^2}{dt^2}(\Psi \circ \gamma(t))|_{t=0} = \dot{S}(v)$. Thus $\mathrm{Ric}_N(v) = \hat{\mathrm{Ric}}_N(v, v) = \hat{\mathrm{Ric}}_N(v)$. Since v is arbitrary, we have $\mathrm{Ric}_N = \hat{\mathrm{Ric}}_N$. $\qquad\square$

From Lemma 6.3.1 and the Bonnet–Myers theorem for weighted Riemannian manifolds (see Theorem 5 in [Qz] or Theorem 3 in [BL]), we have the following weighted version of the Bonnet–Myers theorem in the same way as in Remark 5.2.1.

Theorem 6.3.4 (Bonnet-Myers). *If (M, F, m) is forward complete and satisfies $\mathrm{Ric}_N \geq K > 0$ for $N \in [n, \infty)$, then we have*

$$\mathrm{diam}(M) \leq \pi \sqrt{\frac{N-1}{K}}.$$

In particular, M is compact.

To give a weighted version of the Laplacian comparison theorem, we need the following lemma.

Lemma 6.3.2. *Let (M, F, m) be an n-dimensional Finsler measure space. Assume that V is a nowhere vanishing vector field on M such that $D^V V$ is symmetric in the sense that $g_V(X, D_Y^V V) = g_V(Y, D_X^V V)$ for any vector fields X, Y on M. Set $f := \frac{1}{2} F^2(V)$. Then*

(i) $\nabla^V f = D_V^V V.$

(ii) *It holds*

$$\|\hat{D}V\|_{\hat{g}}^2 \geq \frac{(\mathrm{div}_m V)^2}{N} + \frac{N}{N-1} \left(\frac{\mathrm{div}_m V}{N} - \frac{V(f)}{F^2(V)} \right)^2 - \frac{1}{N-n} S^2(V)$$

$$(6.3.15)$$

for any $N \in (n, \infty)$, where $\hat{D}$ is the Levi-Civita connection of the weighted Riemannian metric $\hat{g} := g_V$.

Proof. (i) In the local coordinates (x^i), since

$$\frac{\partial f}{\partial x^i} = \frac{1}{2} \frac{\partial}{\partial x^i} (g_V(V, V)) = g_V \left(V, D_{\frac{\partial}{\partial x^i}}^V V \right) = g_V \left(D_V^V V, \frac{\partial}{\partial x^i} \right),$$

we have

$$\nabla^V f = g^{ij}(V) g_V \left(D_V^V V, \frac{\partial}{\partial x^i} \right) \frac{\partial}{\partial x^j} = D_V^V V.$$

(ii) Choose a local $\hat{g}$-orthonormal frame $\{e_i\}_{i=1}^n$ such that $V = \sum_i V_i e_i = V_1 e_1$, where $V_1 = F(V) > 0$ and $V_i = 0 (i \geq 2)$. Then

$$V(f) = g_V(V, \nabla^V f) = g_V(V, D_V^V V) = F^2(V) V_{1;1},$$

where ";" means the covariant derivative with respect to the Levi-Civita connection of $\hat{g}$. Thus

$$\mathrm{div}_m V = \sum_{i=1}^{n} V_{i;i} - S(V) = V_{1;1} + \sum_{j=2}^{n} V_{j;j} - S(V).$$

From this, one obtains

$$\frac{(\mathrm{div}_m V)^2}{N} + \frac{N}{N-1}\left(\frac{\mathrm{div}_m V}{N} - \frac{V(f)}{F^2(V)}\right)^2$$

$$= V_{1;1}^2 + \frac{1}{N-1}\left(\sum_{j=2}^{n} V_{j;j}\right)^2 - \frac{2S(V)}{N-1}\sum_{j=2}^{n} V_{j;j} + \frac{S(V)^2}{N-1}.$$

Since

$$-\frac{2}{N-1}S(V)\sum_{j=2}^{n} V_{j;j} \le \frac{1}{N-1}\left(\frac{n-1}{N-n}S(V)^2 + \frac{N-n}{n-1}\left(\sum_{j=2}^{n} V_{j;j}\right)^2\right),$$

we have

$$\frac{(\mathrm{div}_m V)^2}{N} + \frac{N}{N-1}\left(\frac{\mathrm{div}_m V}{N} - \frac{V(f)}{F^2(V)}\right)^2$$

$$\le V_{1;1}^2 + \frac{1}{n-1}\left(\sum_{j=2}^{n} V_{j;j}\right)^2 + \frac{1}{N-n}S(V)^2. \tag{6.3.16}$$

On the other hand, by using $n\sum_{k=1}^{n} a_k^2 \ge (\sum_{k=1}^{n} a_k)^2$, we have

$$\|\hat{D}V\|_{\hat{g}}^2 = V_{1;1}^2 + 2\sum_{j=2}^{n} V_{1;j}^2 + \sum_{i,j=2}^{n} V_{i;j}^2 \ge V_{1;1}^2 + \frac{1}{n-1}\left(\sum_{j=2}^{n} V_{j;j}\right)^2.$$

$$\tag{6.3.17}$$

Combining (6.3.16) with (6.3.17) gives (6.3.15). □

Based on Lemma 6.3.2, we prove the Laplacian comparison theorem, which was first obtained by Ohta-Sturm in [OS1] in a different way.

Theorem 6.3.5 (Laplacian). *Let (M, F, m) be an n-dimensional Finsler measure space and $r = d_F(p, \cdot)$ the distance function from a fixed point $p \in M$. Assume that $\mathrm{Ric}_N \ge K$ for some $N \in [n, \infty)$ and $K \in \mathbb{R}$. Then*

$$\mathbf{\Delta} r \le (N-1)\mathrm{ct}_{K/(N-1)}(r) \tag{6.3.18}$$

*pointwise on $M\backslash(\mathcal{C}_p \cup \{p\})$ and in the sense of distributions on $M\backslash\{p\}$,
where $\mathfrak{ct}_{K/(N-1)}$ is a function defined by (6.3.7) in which $c = K/(N-1)$.*

Proof. For any $x \in D_p = M\backslash(\mathcal{C}_p \cup \{p\})$, let $\gamma : [0,\ell] \to M$ be a normal minimal geodesic on (M, F) from $p = \gamma(0)$ to $q = \gamma(\ell)$. Then $t = d_F(p, \gamma(t)) = r(\gamma(t))$ and $\dot{\gamma}(t) = \nabla r(\gamma(t))$. We choose local $g_{\nabla r}$-orthonormal parallel frame fields $e_1, \ldots, e_n = \nabla r$ along γ. Then

$$\frac{d}{dr}\left(\nabla^2 r(e_i, e_j)\right) = \frac{d}{dr}\left(g_{\nabla r}\left(D_{e_i}^{\nabla r}\nabla r, e_j\right)\right) = g_{\nabla r}\left(D_{\nabla r}^{\nabla r}D_{e_i}^{\nabla r}\nabla r, e_j\right)$$

$$= g_{\nabla r}\left(R_{\nabla r}(\nabla r, e_i)\nabla r, e_j\right) + g_{\nabla r}\left(D_{[\nabla r, e_i]}^{\nabla r}\nabla r, e_j\right)$$

$$= -g_{\nabla r}\left(R_{\nabla r}(e_i, \nabla r)\nabla r, e_j\right) - \sum_k g_{\nabla r}\left(D_{e_i}^{\nabla r}\nabla r, e_k\right) \cdot g_{\nabla r}\left(D_{e_k}^{\nabla r}\nabla r, e_j\right)$$

$$= -g_{\nabla r}\left(R_{\nabla r}(e_i, \nabla r)\nabla r, e_j\right) - \sum_k \nabla^2 r(e_i, e_k) \cdot \nabla^2 r(e_k, e_j). \quad (6.3.19)$$

Since ∇r is a geodesic field, we have $\mathrm{Ric}(\nabla r) = \hat{\mathrm{Ric}}(\nabla r)$ (the Ricci curvature of $g_{\nabla r}$). Taking the trace on the both sides of (6.3.19) with respect to $g_{\nabla r}$ yields

$$\frac{d}{dr}(\mathrm{tr}_{\nabla r}\nabla^2 r) = -\mathrm{Ric}(\nabla r) - \|\nabla^2 r\|_{g_{\nabla r}}^2,$$

which is equivalent to

$$d(\mathbf{\Delta} r)(\nabla r) + \mathrm{Ric}_\infty(\nabla r) + \|\nabla^2 r\|_{g_{\nabla r}}^2 = 0, \quad (6.3.20)$$

where we used that $d(S(\nabla r))(\nabla r) = \nabla r(S(\nabla r)) = S_{|i}\nabla^i r = \dot{S}(\nabla r)$. Note that $\nabla^2 r$ is symmetric and $F(\nabla r) = 1$. Plugging (6.3.15) with $V = \nabla r$ into (6.3.20) yields

$$d(\mathbf{\Delta} r)(\nabla r) + \mathrm{Ric}_N(\nabla r) + \frac{(\mathbf{\Delta} r)^2}{N-1} \leq 0, \quad (6.3.21)$$

where $N \in (n, \infty)$. The case of $N = n$ is derived as the limit. Put $\varphi(t) = \mathbf{\Delta} r(\gamma(t))$ for $t \in (0, \ell]$. From (6.3.21), one obtains

$$-\dot{\varphi} \geq K + \frac{\varphi^2}{N-1}. \quad (6.3.22)$$

Moreover, along γ, we decompose the measure m as $dm = \sigma(t)dx$, where

$$\sigma(t) = e^{-\Psi(\gamma(t))}\sqrt{\det(g_{ij}(\dot{\gamma}))}.$$

Define a smooth Riemannian metric $\hat{g}$ inside D_p by

$$\hat{g}|_{\gamma(t)} = g_{\dot{\gamma}(t)} = (g_{ij}(\dot{\gamma})).$$

Let $\tilde{g} := (\exp_p)^*\hat{g}$. In general $\tilde{g}$ is singular at the origin, unless F is Riemannian. Regard T_pM as the cone $C(I_p)$ over the inditrix I_p of F and $\exp_x$ as a map $C(I_p) \to M$. By the Gauss lemma, we have $\tilde{g} = dt^2 + h_t$, where h_t is a family of Riemannian metrics on I_p. Then, by Lemma 3.1 in [Sh3],

$$\frac{1}{t^2}h_t \to \dot{g} \quad \text{as} \quad t \to 0,$$

where $\dot{g}$ is the induced Riemannian metric on I_p from $g_{\dot{\gamma}(0)}$. This means that the volume form

$$dm = e^{-\Psi(\gamma(t))}\sqrt{\det(g_{ij}(\dot{\gamma}))}\,dx \simeq e^{-\Psi(t,\cdot)}t^{n-1}dt \wedge \dot{\eta}, \quad \text{as} \quad t \to 0,$$

where $\dot{\eta}$ is the volume form of $\dot{g}$. Thus, by (6.1.5), we have

$$\mathbf{\Delta}\rho(\gamma(t)) = (\log \sigma(t))' \simeq -\Psi'(t) + \frac{n-1}{t}, \quad \text{as} \quad t \to 0,$$

which implies that $\lim_{t \to 0^+} t\mathbf{\Delta}\rho(\gamma(t)) = n - 1 \leq N - 1$. Note that the solution of the Ricatti equation

$$-\dot{\psi} = K + \frac{\psi^2}{N-1}, \quad \text{with} \quad \lim_{t \to 0^+} t\psi(t) = N - 1$$

is given by $\psi(t) = (N-1)\mathsf{ct}_{K/(N-1)}(t)$. Comparison principle for the ODE implies that $\varphi(t) \leq (N-1)\mathsf{ct}_{K/(N-1)}(t)$. The extension to a distributional inequality, valid also on the cut locus, follows by the well-known Calabi argument. $\square$

Corollary 6.3.2 ([OS1]). *Under the same assumptions as in Theorem 6.3.5, assume that $u(x) := f(r(x))$ for some nondecreasing smooth function $f : (0, \infty) \to \mathbb{R}$. Then*

$$\mathbf{\Delta}u(x) \leq f''(r) + (N-1)f'(r)\mathsf{ct}_{K/(N-1)}(r)$$

on $M \backslash (C_p \cup \{p\})$. Similarly, if $u(x) := f(\overleftarrow{r}(x))$ for some nonincreasing smooth function $f : (0, \infty) \to \mathbb{R}$, then

$$\mathbf{\Delta}u(x) \leq f''(\overleftarrow{r}) + (N-1)f'(\overleftarrow{r})\mathsf{ct}_{K/(N-1)}(\overleftarrow{r})$$

on $M \backslash (C_p \cup \{p\})$, where $\overleftarrow{r}(x) := \overleftarrow{d}_F(p, x) = d_F(x, p)$. In both cases, the estimates extend to hold in the sense of distributions on all of $M \backslash \{p\}$.

Proof. The first claim follows from Theorem 6.3.5 by an application of chain rule:

$$\boldsymbol{\Delta} f(r) = \mathrm{div}(\nabla f(r)) = \mathrm{div}(f'(r)\nabla r) = f'(r)\boldsymbol{\Delta} r + f''(r)dr(\nabla r)$$

on $M\backslash(\mathcal{C}_p \cup \{p\})$ and the fact that $dr(\nabla r) = F^2(\nabla r) = 1$.

For the case when f is nonincreasing, we should be careful since F is nonreversible in general. In this case, $-f$ is nondecreasing. Then, by a similar argument with $v = \overleftarrow{r}$,

$$\boldsymbol{\Delta} f(v) = \mathrm{div}(\nabla f(v)) = \mathrm{div}(-f'(v)\nabla(-v))$$

$$= f'(v)(-\boldsymbol{\Delta}(-v)) + f''(v)d(-v)(\nabla(-v)).$$

Note that $d(-v)(\nabla(-v)) = dv(\overleftarrow{\nabla}v) = 1$ and $\overleftarrow{\boldsymbol{\Delta}}v = -\boldsymbol{\Delta}(-v)$. The claim follows since the bound of Ric_N for (M, F, m) implies the same bound of $\overleftarrow{\mathrm{Ric}}_N$ for $(M, \overleftarrow{F}, m)$. $\qquad\square$

6.4 Volume Comparison Theorem

The Bishop–Gromov volume comparison theorem in Finsler geometry was first obtained by Shen under the assumptions that the Ricci curvature is bounded from below and S-curvature is bounded from above ([Sh3]). Afterwards, Ohta developed this result and gave a weighted version of the volume comparison theorem in [Oh1]. In this section, we shall use the Laplacian comparison theorem to derive the relative volume comparison theorem, including Ohta's comparison theorem as a special case ([Xia8]). Further, we give some applications of the volume comparison theorem, including the volume growth estimates of geodesic balls, etc. We refer to [Xia6]–[Xia8] and [Wu2] for more details.

Let (M, F) be a forward (resp., backward) complete Finsler manifold. Now we define the *forward and backward geodesic balls* of radius R with center at $x \in M$ by

$$B_R^+(x) := \{z \in M | d_F(x, z) < R\}, \quad B_R^-(x) := \{x \in M | d_F(z, x) < R\}.$$

If $\exp_x$ is defined on all of $\mathbf{B}_R(x)$, then $\exp_x(\mathbf{B}_R(x)) \subseteq B_R^+(x)$. For any $z \in D_x = M\backslash(\mathcal{C}_x \cup \{x\})$, we choose the geodesic polar coordinates (r, θ) centered at x such that $r(z) = F(v)$ and $\theta^\alpha(z) = \theta^\alpha(\frac{v}{F(v)})$, where $r(z) = d_F(x, z)$ is the distance function from a fixed point $x \in M$ and $v = \exp^{-1}(z) \in T_xM\backslash\{0\}$. As we know, the distance function r starting from $x \in M$ is smooth on D_x and $F(\nabla r) = 1$. By the Gauss lemma, the unit radial coordinate vector $\frac{\partial}{\partial r}$ is orthogonal to coordinate vectors $\frac{\partial}{\partial \theta^\alpha}$ with respect to $g_{\nabla r}$ for $1 \le \alpha \le n-1$. Therefore, writing $dm|_{\exp_x(rv_0)} = \sigma(r, \theta)drd\theta$, where $v_0 = \frac{v}{F(v)} \in I_x = \{v \in T_xM | F(v) = 1\}$, we have, from (6.1.5),

$$\Delta r = \frac{\partial}{\partial r}(\log \sigma). \tag{6.4.1}$$

Set $\mathcal{D}_r(x) = \{v_0 \in I_x | rv_0 \in \exp^{-1}(D_x \cap B_R^+(x))\}$ and $\tau(r) := \int_{\mathcal{D}_r(x)} \sigma(r, \theta)d\theta$. It is easy to see that $\mathcal{D}_{r_1}(x) \subset \mathcal{D}_{r_2}(x)$ for $r_1 > r_2$. The volume of $B_R^+(x)$ with respect to m is given by

$$m(B_R^+(x)) = \int_{B_R^+(x)} dm = \int_{B_R^+(x) \cap D_x} dm$$

$$= \int_0^R dr \int_{\mathcal{D}_r(x)} \sigma(r, \theta)d\theta = \int_0^R \tau(r)dr. \tag{6.4.2}$$

Similarly, we may define $m(B_R^-(x))$. For real numbers c and ζ, let

$$V_{c,\zeta}(r) := c_{n-1} \int_0^r \mathfrak{s}_c(t)^{\zeta-1}dt, \tag{6.4.3}$$

where $\mathfrak{s}_c(t)$ is defined by (5.3.9) and c_{n-1} is the (Euclidean) area of the unit sphere $\mathbb{S}^{n-1}$ in $\mathbb{R}^n$. Moreover, we denote by $B_{R,R_1}^+(x) := B_R^+(x) \backslash B_{R_1}^+(x)$ and $B_{R,R_1}^-(x) := B_R^-(x) \backslash B_{R_1}^-(x)$.

Theorem 6.4.1 (Relative volume comparison). *Let (M, F, m) be an n-dimensional forward (resp., backward) complete Finsler measure space satisfying $Ric_N \ge K$ for some $K \in \mathbb{R}$ and $N \in [n, \infty)$. Then, for any $x \in M$ and $0 \le r_1 < r$, $0 \le R_1 < R$ such that $r_1 \le R_1$, $r \le R$, we have*

$$\max\left\{\frac{m(B_{R,R_1}^+(x))}{m(B_{r,r_1}^+(x))}, \frac{m(B_{R,R_1}^-(x))}{m(B_{r,r_1}^-(x))}\right\} \le \frac{V_{K/(N-1),N}(R) - V_{K/(N-1),N}(R_1)}{V_{K/(N-1),N}(r) - V_{K/(N-1),N}(r_1)},$$

where $R \le \pi\sqrt{(N-1)/K}$ if $K > 0$.

To prove this theorem, we need the following lemma.

Lemma 6.4.1. *Suppose that f and g are positive integrable functions of a real variable $t \in [0, \infty)$ for which f/g is decreasing in t. Then for any $0 \le r_1 < r$, $0 \le R_1 < R$ such that $r_1 \le R_1$, $r \le R$, we have*

$$\frac{\int_{R_1}^R f(t)dt}{\int_{r_1}^r f(t)dt} \le \frac{\int_{R_1}^R g(t)dt}{\int_{r_1}^r g(t)dt}. \tag{6.4.4}$$

In particular, the function $\int_0^r f(t)dt / \int_0^r g(t)dt$ is decreasing in r.

Proof. By the assumption, we have

$$\frac{f(t)}{g(t)} \geq \frac{f(\tau)}{g(\tau)}$$

for any $0 \leq t < \tau$. From this, one obtains

$$\int_\sigma^\tau f(t)dt = \int_\sigma^\tau \frac{f(t)}{g(t)} g(t)dt \geq \frac{f(\tau)}{g(\tau)} \int_\sigma^\tau g(t)dt,$$

which implies that

$$\frac{\int_\sigma^\tau f(t)dt}{\int_\sigma^\tau g(t)dt} \geq \frac{f(\tau)}{g(\tau)}.$$

Let

$$G(\sigma, \tau) := \frac{\int_\sigma^\tau f(t)dt}{\int_\sigma^\tau g(t)dt}.$$

Obviously, $G(\sigma, \tau) = G(\tau, \sigma)$. Consequently,

$$\frac{\partial G}{\partial \tau} = \frac{g(\tau)}{\int_\sigma^\tau g(t)dt} \left(\frac{f(\tau)}{g(\tau)} - \frac{\int_\sigma^\tau f(t)dt}{\int_\sigma^\tau g(t)dt} \right) \leq 0.$$

Also, we have $\frac{\partial G}{\partial \sigma} \leq 0$. Thus, we get

$$\frac{\int_{R_1}^R f(t)dt}{\int_{R_1}^R g(t)dt} \leq \frac{\int_{R_1}^r f(t)dt}{\int_{R_1}^r g(t)dt} \leq \frac{\int_{r_1}^r f(t)dt}{\int_{r_1}^r g(t)dt},$$

which implies (6.4.4). $\qquad\square$

Now, we are ready to prove Theorem 6.4.1.

Proof. Note that $\overleftarrow{F}$ is also a Finsler metric if F is a Finsler metric. Then the backward geodesic ball $B_R^-(x_0)$ of F is exactly the forward geodesic ball of $\overleftarrow{F}$. Moreover, $\overleftarrow{\mathrm{Ric}}_N$ and Ric_N have the same lower bound. Hence it suffices to consider the volume comparison theorem for forward geodesic balls.

It follows from (6.4.1) and Theorem 6.3.5 that

$$\frac{\partial}{\partial r}(\log \sigma) \leq (N-1)\mathfrak{ct}_{K/(N-1)}(r) = \frac{d}{dr}\Big(\log \mathfrak{s}_{K/(N-1)}(r)^{N-1} \Big),$$

which implies that the function $\sigma(r,\theta)/\mathfrak{s}_{K/(N-1)}(r)^{N-1}$ is decreasing in r. Therefore, for any $0 < r < R$, we have

$$\frac{\int_{\mathcal{D}_r(x_0)} \sigma(r,\theta)d\theta}{\mathfrak{s}_{K/(N-1)}(r)^{N-1}} = \int_{\mathcal{D}_r(x_0)} \frac{\sigma(r,\theta)}{\mathfrak{s}_{K/(N-1)}(r)^{N-1}}d\theta \geq \int_{\mathcal{D}_R(x_0)} \frac{\sigma(r,\theta)}{\mathfrak{s}_{K/(N-1)}(r)^{N-1}}d\theta$$

$$\geq \int_{\mathcal{D}_R(x_0)} \frac{\sigma(R,\theta)}{\mathfrak{s}_{K/(N-1)}(R)^{N-1}}d\theta = \frac{\int_{\mathcal{D}_R(x_0)} \sigma(R,\theta)d\theta}{\mathfrak{s}_{K/(N-1)}(R)^{N-1}},$$

which means that

$$\frac{\int_{\mathcal{D}_r(x_0)} \sigma(r,\theta)d\theta}{\mathfrak{s}_{K/(N-1)}(r)^{N-1}}$$

is decreasing in r. Thus, by Lemma 6.4.1, we have

$$\frac{\int_{R_1}^{R} \int_{\mathcal{D}_t(x_0)} \sigma(t,\theta)d\theta dt}{\int_{r_1}^{r} \int_{\mathcal{D}_t(x_0)} \sigma(t,\theta)d\theta dt} \leq \frac{\int_{R_1}^{R} \mathfrak{s}_{K/(N-1)}(t)^{N-1}dt}{\int_{r_1}^{r} \mathfrak{s}_{K/(N-1)}(t)^{N-1}dt}, \tag{6.4.5}$$

which implies the conclusion. $\qquad\square$

Let $r_1 = R_1 = 0$ in Theorem 6.4.1 and its proof, we get the Bishop–Gromov's volume comparison theorem, which was first obtained by Ohta in [Oh1] in a different way.

Corollary 6.4.1 (Bishop–Gromov). *Let (M, F, m) be an n-dimensional forward or backward complete Finsler measure space satisfying $Ric_N \geq K$ for some $K \in \mathbb{R}$ and $N \in [n, \infty)$. Then $\frac{m(B_r^+(x))}{V_{K/(N-1),N}(r)}$ is decreasing in r, and*

$$\max\left\{ \frac{m(B_R^+(x))}{m(B_r^+(x))}, \frac{m(B_R^-(x))}{m(B_r^-(x))} \right\} \leq \frac{V_{K/(N-1),N}(R)}{V_{K/(N-1),N}(r)}$$

for any $x \in M$ and $0 < r < R$, where $R \leq \pi\sqrt{(N-1)/K}$ if $K > 0$.

In the following, we give some applications of the volume comparison theorem. The first result was due to Shen ([Sh3]).

Lemma 6.4.2. *For the Busemann–Hausdorff volume form $dm_{BH} = \sigma_{BH}(x)dx$, we have*

$$\lim_{r \to 0} \frac{m_{BH}(B_r^\pm(x))}{V_{K/(n-1),n}(r)} = 1$$

for any $x \in M$, where $m_{BH}(B_r^\pm(x))$ means the volume of geodesic ball $B_r^\pm(x)$ with respect to dm_{BH}.

Proof. Observe that

$$V_{K/(n-1),n}(r) = c_{n-1} \int_0^r \mathfrak{s}_{K/(n-1)}(t)^{n-1} dt = \text{Vol}(\mathbb{B}^n(r))(1 + o(r))$$

as $r \to 0+$. On the other hand, we remark that the forward geodesic ball $B_r^+(x) \subset M$ is mapped onto $\mathbf{B}_r(x) \subset \mathbb{R}^n$ by $\exp_x^{-1}$ for any sufficiently small $r > 0$. Clearly, $x' \to \text{Vol}(\mathbf{B}_r(x'))$ is continuous. Thus, there is an $x_0 \in B_r^+(x)$ such that, for any $x' \in B_r^+(x)$,

$$m_{BH}(B_r^+(x)) = \int_{B_r^+(x)} \frac{\text{Vol}(\mathbb{B}^n(1))}{\text{Vol}(\mathbf{B}_1(x'))} dx' = \frac{\text{Vol}(\mathbb{B}^n(r))}{\text{Vol}(\mathbf{B}_r(x_0))} \int_{B_r^+(x)} dx'$$

$$= \frac{\text{Vol}(B_r^+(x))}{\text{Vol}(\mathbf{B}_r(x_0))} \text{Vol}(\mathbb{B}^n(r)) \to \text{Vol}(\mathbb{B}^n(r)) \text{ (as } r \to 0).$$

The assertion follows. $\qquad\square$

The following corollary follows from Corollary 6.4.1 and Lemma 6.4.2.

Corollary 6.4.2. *Let $(M, F, d\mu_{BH})$ be an n-dimensional forward complete Finsler manifold equipped with the Busemann–Hausdorff measure m_{BH}. Assume that $Ric_n \geq K$ for some $K \in \mathbb{R}$. Then*

$$m_{BH}(B_r^+(x)) \leq V_{K/(n-1),n}(r)$$

for any $r > 0$. In particular, when $K = 0$, $m_{BH}(B_r^+(x)) \leq c(n)r^n$, where $c(n)$ is a positive constant only depending on n.

The classical integration arguments in the Laplace comparison theorem lead to the volume comparison of forward geodesic balls with different radius.

Proposition 6.4.1 ([Xia6]). *Let (M, F, m) be an n-dimensional Finsler measure space satisfying $Ric_N \geq K$ for $N \in [n, \infty)$ and $K \in \mathbb{R}$. Then, for any $0 < r_1 < r_2 < R$ ($\leq \pi\sqrt{(N-1)/K}$ if $K > 0$) and $x \in M$, it holds*

$$\max\left\{ \frac{m(B_{r_2}^+(x))}{m(B_{r_1}^+(x))}, \frac{m(B_{r_2}^-(x))}{m(B_{r_1}^-(x))} \right\} \leq \left(\frac{r_2}{r_1} \right)^N e^{r_2\sqrt{(N-1)|K|}}. \tag{6.4.6}$$

Proof. As in the proof of Theorem 6.4.1, we have

$$\frac{\partial}{\partial r}(\log \sigma) \leq (N-1)\mathfrak{ct}_{K/(N-1)}(r)$$

pointwise on D_x or on $M \backslash \{x\}$ (in the weak sense). Observe that the right-hand side of the above equality is less that $\frac{N-1}{r}\left(1 + r\sqrt{\frac{|K|}{N-1}}\right)$ by an element argument. Consequently, one obtains

$$\frac{\partial}{\partial r} \log \sigma \leq \frac{N-1}{r}\left(1 + r\sqrt{\frac{|K|}{N-1}}\right) \quad \text{on } D_x.$$

Integrating on both sides from r_1 to r_2 yields

$$\log \frac{\sigma(r_2, \theta)}{\sigma(r_1, \theta)} \leq (N-1) \log \frac{r_2}{r_1} + r_2\sqrt{(N-1)|K|},$$

which implies that

$$\frac{\sigma(r_2, \theta)}{\sigma(r_1, \theta)} \leq \left(\frac{r_2}{r_1}\right)^{N-1} e^{r_2\sqrt{(N-1)|K|}}.$$

From this, one obtains

$$s^{N-1}\sigma(t, \theta) \leq t^{N-1}\sigma(s, \theta) e^{r_2\sqrt{(N-1)|K|}}$$

for any $0 < s < r_1 < t < r_2 < R$. Integrating in t from r_1 to r_2 and then in s from 0 to r_1, we get

$$\frac{m(B_{r_2}^+(x)) - m(B_{r_1}^+(x))}{m(B_{r_1}^+(x))} \leq \frac{r_2^N - r_1^N}{r_1^N} e^{r_2\sqrt{(N-1)|K|}}.$$

Hence,

$$\frac{m(B_{r_2}^+(x))}{m(B_{r_1}^+(x))} \leq 1 + \left(\frac{r_2}{r_1}\right)^N e^{r_2\sqrt{(N-1)|K|}} - e^{r_2\sqrt{(N-1)|K|}}$$

$$\leq \left(\frac{r_2}{r_1}\right)^N e^{r_2\sqrt{(N-1)|K|}}.$$

The above arguments are also valid for the reverse metric $\overleftarrow{F}$. This finishes the proof. $\square$

Remark 6.4.1. The volume comparison (6.4.6) implies the volume doubling property, namely, there is a uniform positive constant c_0 such that $m(B_{2r}^+(x')) \leq c_0 m(B_r^+(x'))$ for any $x' \in B_R^+(x)$ and $0 < r < R/2$. This also follows from Corollary 6.4.1.

Letting $r_1 = 1$ and $r_2 = r$ in Proposition 6.4.1 yields the volume growth estimates of forward or backward geodesic balls.

Corollary 6.4.3. *Under the same assumptions as in Proposition 6.4.1, there is a positive constant $c = m(B_1^+(x))$ such that*

$$\max\left\{m(B_r^+(x)), m(B_r^-(x))\right\} \le cr^N e^r \sqrt{(N-1)|K|}. \tag{6.4.7}$$

for any $r \ge 1$.

Further, we have a more refined volume growth estimate of a forward geodesic ball $B_r^+(x)$ than (6.4.7) by means of Lemma 6.3.2.

Proposition 6.4.2 ([Xia7]). *Let (M, F, m) be a forward complete Finsler measure space satisfying $Ric_N \ge -K$ for some $N \in [n, +\infty)$ and $K > 0$. Then there exists a positive constant $C = C(K, N, m(B_1^+(x)))$ depending on $N, K, m(B_1^+(x))$ such that*

$$m(B_r^+(x)) \le Ce^r \sqrt{(N-1)K}$$

for any $x \in M$ and $r \ge 1$.

Proof. Let $\gamma : [0, r(z)] \to M$ be the minimizing geodesic from x to z with $\gamma(0) = x$ and $\gamma(r(z)) = z$, where $r = d_F(x, \cdot)$ is the distance function from x. In a geodesic polar coordinates (r, θ), we have

$$\mathbf{\Delta}r = \frac{\partial}{\partial r} \log \sigma(x, r, \theta) \tag{6.4.8}$$

from (6.4.1). In the following, we will omit the dependence of the quantities on θ. On the other hand, plugging (6.3.15) with $V = \nabla r$ into (6.3.20) yields

$$d(\mathbf{\Delta}r)(\nabla r) + Ric_N(\nabla r) + \frac{(\mathbf{\Delta}r)^2}{N-1} \le 0,$$

which implies that

$$\frac{\partial^2}{\partial r^2}(\log \sigma) + \frac{1}{N-1}\left(\frac{\partial}{\partial r}\log \sigma\right)^2 \le K.$$

Integrating this inequality from 1 to r and letting $u(r) := \frac{\partial \log \sigma}{\partial r}$ give

$$u(r) + \frac{1}{N-1}\int_1^r u^2(t)dt \le Kr + C_0$$

for some constant $C_0 > 0$. The Cauchy–Schwarz inequality implies that

$$u(r) + \frac{1}{(N-1)r}\left(\int_1^r u(t)dt\right)^2 \le Kr + C_0. \tag{6.4.9}$$

Now, we estimate $\int_1^r u(t)dt$. Consider the function

$$v(r) := -\int_1^r u(t)dt + r\sqrt{(N-1)K} + C_0\sqrt{(N-1)/K}. \quad (6.4.10)$$

Obviously, $v(1) > 0$. Assume that $R(> 1)$ is the first number such that $v(R) = 0$, namely,

$$\int_1^R u(t)dt = \left(R\sqrt{(N-1)K} + C_0\sqrt{(N-1)/K}\right).$$

Then

$$\frac{1}{(N-1)R}\left(\int_1^R u(t)dt\right)^2 = \frac{1}{(N-1)R}\left(R\sqrt{(N-1)K} + C_0\sqrt{(N-1)/K}\right)^2$$

$$\geq KR + 2C_0.$$

Consequently, $u(R) \leq -C_0 < 0$ by (6.4.9), which implies that $v'(R) = -u(R) + \sqrt{(N-1)K} > 0$. Thus, there is a number $\varepsilon > 0$ small enough such that $v(R - \varepsilon) < 0$. This contradicts the choice of R. Hence, $v(r) > 0$ for all $r \geq 1$. From this and (6.4.10), we have

$$\log \sigma(r) - \log \sigma|_{r=1} \leq r\sqrt{(N-1)K} + C_0\sqrt{(N-1)/K}.$$

This implies that $m(B_r^+(x)) \leq Ce^{r\sqrt{(N-1)K}}$ for some positive constant C depending on N, K and $m(B_1^+(x))$. $\qquad \square$

The following volume growth speed estimate of a geodesic ball ([Wu2]) is a generalization of the corresponding result on Riemannian manifolds with nonnegative Ricci curvature ([SY]).

Theorem 6.4.2. *Let (M, F, m) be an n-dimensional complete noncompact Finsler measure space with $Ric_N \geq 0$ for some $N \in [n, \infty)$. Then M must have infinite volume. Further, if M has finite reversibility Λ, then the volume $m(B_R^+(x))$ (resp., $m(B_R^-(x))$ of the forward (resp., backward) geodesic ball $B_R^+(x)$ (resp., $B_R^-(x)$) has at least linear growth for any $x \in M$, that is,*

$$\min\left\{m(B_R^+(x)), m(B_R^-(x))\right\} \geq CR,$$

where $C = C(N, \Lambda, m(B_1^+(x) \cap B_1^-(x)))$ is a positive constant.

Proof. Since M is complete, there is a geodesic $\gamma : [0, \infty) \to M$ such that $\gamma(0) = x$ and $d_F(x, \gamma(t)) = t$. By the triangle inequality, it is easy to check that

$$B_1^+(x) \cap B_1^-(x) \subset B_{t+1}^-(\gamma(t)) \backslash B_{t-1}^-(\gamma(t)), \quad \forall t > 1. \qquad (6.4.11)$$

Since $\mathrm{Ric}_N \geq 0$, one obtains that from Theorem 6.4.1

$$\frac{m(B_R^-(\gamma(t)))}{m(B_r^-(\gamma(t)))} \leq \frac{R^N}{r^N}$$

for any $0 < r < R$, which means that

$$m(B_r^-(\gamma(t))) \geq \frac{r^N}{R^N - r^N} \left(m(B_R^-(\gamma(t))) - m(B_r^-(\gamma(t))) \right). \qquad (6.4.12)$$

Thus, from (6.4.11)–(6.4.12), we get

$$m(B_{t-1}^-(\gamma(t))) \geq \frac{(t-1)^N}{(t+1)^N - (t-1)^N} \left(m(B_{t+1}^-(\gamma(t))) - m(B_{t-1}^-(\gamma(t))) \right)$$

$$\geq \frac{(t-1)^N}{(t+1)^N - (t-1)^N} m \left(B_1^+(x) \cap B_1^-(x) \right).$$

Since

$$\lim_{t \to +\infty} \frac{(t-1)^N}{t\left[(t+1)^N - (t-1)^N\right]} = \frac{1}{2N},$$

there is a constant $c = c(N) > 0$ such that

$$\frac{(t-1)^N}{(t+1)^N - (t-1)^N} \geq ct$$

for $t > 1$. Hence,

$$m(B_{t-1}^-(\gamma(t))) \geq ct \cdot m \left(B_1^+(x) \cap B_1^-(x) \right), \qquad (6.4.13)$$

which implies that M has infinite volume.

Further, if M has finite reversibility Λ, then

$$B_{t-1}^-(\gamma(t)) \subset B_{\Lambda(t-1)}^+(\gamma(t)) \subset B_{2\Lambda t}^+(x) \subset B_{2\Lambda^2 t}^-(x).$$

Combining this with (6.4.13) yields

$$m(B_{2\Lambda^2 t}^-(x)) \geq ct \cdot m \left(B_1^+(x) \cap B_1^-(x) \right),$$

which implies that

$$m(B_R^-(x)) \geq CR, \quad C := \frac{c}{2\Lambda^2} m \left(B_1^+(x) \cap B_1^-(x) \right) > 0.$$

Similarly, one can obtain that $m(B_R^+(x)) \geq CR$ via taking $\gamma : (-\infty, 0] \to M$ and $B_\bullet^+(\gamma(-t))$ instead of $\gamma : [0, \infty) \to M$ and $B_\bullet^-(\gamma(t))$, respectively, in previous arguments. $\qquad \square$

Chapter 7

Finsler Harmonic Functions

In this chapter, we shall introduce the basic theory of harmonic functions on Finsler manifolds, including the existence, the regularity and the Liouville properties for (super, sub) harmonic functions. For this, we need establish some necessary study tools, such as, the Bochner–Weitzenböck formula (inequality) and the L^p mean value inequality, etc. The (pointwise or integrated) Bochner–Weitzenböck formula was first established by Sturm–Ohta ([OS2]). We shall consider the integrated Bochner–Weitzenböck formula on Finsler measure spaces with boundary.

7.1 Bochner–Weitzenböck Formula

7.1.1 *Pointwise Bochner–Weitzenböck Formula*

Let (M, F, m) be an n-dimensional Finsler measure space. We define the linearization of Finsler Laplacian Δ on M_u by

$$L_u(\eta) = \frac{d}{dt}\Big|_{t=0}\Delta(u + t\eta) = \mathrm{div}\left[\frac{d}{dt}\Big|_{t=0}(\nabla(u + t\eta))\right] \qquad (7.1.1)$$

for any $u \in C^2(M_u)$ and $\eta \in C^2(M)$. Since $du \neq 0$ on M_u, $d(u + t\eta) \neq 0$ if t is small enough. Thus, $\nabla(u + t\eta) = g^{ij}(\nabla(u + t\eta))\frac{\partial(u+t\eta)}{\partial x^j}\frac{\partial}{\partial x^i}$. From this and $C_y(y, \cdot, \cdot) = 0$ for any $y \in TM$, we have

$$\frac{d}{dt}\Big|_{t=0}\nabla(u + t\eta) = \nabla^{\nabla u}\eta. \qquad (7.1.2)$$

Inserting (7.1.2) into (7.1.1) gives rise to $L_u(\eta) = \Delta^{\nabla u}\eta$. In particular, $L_u(u) = \Delta u$.

Fix a smooth vector field V on a Finsler manifold (M, F) and $x \in M_V := \{x \in M | V(x) \neq 0\}$. For any $t \in (0, \varepsilon)$ with sufficiently small $\varepsilon > 0$, we define the map γ_t generated by V and the vector field $\mathcal{V}_t$ in a neighborhood $\mathcal{U}$ of x by

$$\gamma_t(z) = \exp_z(tV_z), \quad \mathcal{V}_t(\gamma_t(z)) := \frac{d}{dt}(\gamma_t(z))$$

for any $z \in \mathcal{U}$. Obviously, $\mathcal{V}_t(\gamma_t(z))|_{t=0} = V_z$. In local coordinates $(x^i)_{i=1}^n$, write $\mathcal{V}_t = \mathcal{V}_t^i \frac{\partial}{\partial x^i}$ and $\frac{\partial \mathcal{V}_t}{\partial t} = \frac{\partial \mathcal{V}_t^i}{\partial t} \frac{\partial}{\partial x^i}$. As $\gamma(t) := \gamma_t(z)$ is a geodesic for each fixed $z \in \mathcal{U}$, we have $D_{\dot{\gamma}}^{\gamma} \dot{\gamma}(t) = 0$. On the other hand,

$$D_{\dot{\gamma}}^{\dot{\gamma}} \dot{\gamma}(t) = \left\{ \ddot{\gamma} + \Gamma_{jk}^i(\dot{\gamma})\dot{\gamma}^j \dot{\gamma}^k \right\} \frac{\partial}{\partial x^i}$$

$$= \left\{ \mathcal{V}_t^j \frac{\partial \mathcal{V}_t^i}{\partial x^j} + \frac{\partial \mathcal{V}_t^i}{\partial t} + \Gamma_{jk}^i(\mathcal{V}_t)\mathcal{V}_t^j \mathcal{V}_t^k \right\} (\gamma(t)) \frac{\partial}{\partial x^i}.$$

Consequently,

$$\frac{\partial \mathcal{V}_t}{\partial t} + D_{\mathcal{V}_t}^{\mathcal{V}_t} \mathcal{V}_t = 0. \tag{7.1.3}$$

Now, we put $\eta(t) := \exp_x(tV_x)$ and take an orthonormal basis $\{e_i\}_{i=1}^n$ of $(T_x M, g_V)$ such that $e_n = \dot{\eta}(0)/F(\dot{\eta}(0))$. Consider the geodesic variations $\alpha_i(t,s)$ of η defined by $\alpha_i(t,s) = \gamma_t(\exp_x(se_i))$ for $1 \leq i \leq n$. Their variation fields are given by

$$E_i(t) = \frac{\partial}{\partial s}\Big|_{s=0} \alpha_i(t,s) = \frac{\partial}{\partial s}\Big|_{s=0} \gamma_t(\exp_x(se_i)) = (d\gamma_t)_x(e_i),$$

which are linearly independent Jacobi fields along $\eta(t)$. We set $E_i'(t) := D_{\dot{\eta}}^{\dot{\eta}} E_i(t)$ for simplicity. Define the $n \times n$ matrix-valued function $B(t) = (b_{ij}(t))$ by $E_i'(t) = b_{ij}(t)E_j(t)$. Then we have the following Riccati type equation.

Lemma 7.1.1. *For η and B as above, we have*

$$\frac{d(\text{tr } B)}{dt} + \text{tr}(B(t)^2) + \text{Ric}(\dot{\eta}) = 0. \tag{7.1.4}$$

Proof. Assume that $A(t) = (a_{ij}(t))$ with $a_{ij}(t) = g_{\dot{\eta}}(E_i(t), E_j(t))$. Then

$$a_{ij}' = \dot{\eta} g_{\dot{\eta}}(E_i, E_j) = g_{\dot{\eta}}(D_{\dot{\eta}}^{\dot{\eta}} E_i, E_j) + g_{\dot{\eta}}(E_i, D_{\dot{\eta}}^{\dot{\eta}} E_j) = b_{ik} a_{kj} + a_{ik} b_{jk},$$

equivalently,

$$A' = BA + AB^T, \tag{7.1.5}$$

where B^T means the transpose of B. On the other hand, since $E_i(t)$ are Jacobi fields, i.e., $D^{\dot\eta}_{\dot\eta} D^{\dot\eta}_{\dot\eta} E_i + R_{\dot\eta}(E_i) = 0$, we have

$$\frac{d}{dt}\left(g_{\dot\eta}(E_i', E_j) - g_{\dot\eta}(E_i, E_j')\right) = -g_{\dot\eta}(R_{\dot\eta}(E_i), E_j) + g_{\dot\eta}(E_i, R_{\dot\eta}(E_j))$$

$$= -g_{\dot\eta}(R^k_{\ i}(\dot\eta)E_k, E_j) + g_{\dot\eta}(E_i, R^k_{\ j}(\dot\eta)E_k)$$

$$= -a_{kj}R^k_{\ i} + a_{ik}R^k_{\ j} = -R_{ji} + R_{ij} = 0.$$

Hence, $g_{\dot\eta}(E_i', E_j) - g_{\dot\eta}(E_i, E_j')$ are constants, equivalently,

$$BA - AB^T = B(0) - B(0)^T. \tag{7.1.6}$$

Combining this with (7.1.5) yields

$$A' = 2BA - B(0) + B(0)^T.$$

From this and (7.1.5), one obtains

$$A'' = 2B'A + 2B(BA + AB^T). \tag{7.1.7}$$

Also, we deduce from the Jacobi field equation of E_i that

$$a_{ij}'' = g_{\dot\eta}(D^{\dot\eta}_{\dot\eta} D^{\dot\eta}_{\dot\eta} E_i, E_j) + g_{\dot\eta}(E_i, D^{\dot\eta}_{\dot\eta} D^{\dot\eta}_{\dot\eta} E_j) + 2g_{\dot\eta}(D^{\dot\eta}_{\dot\eta} E_i, D^{\dot\eta}_{\dot\eta} E_j)$$

$$= -2a_{ik}R^k_{\ j} + 2b_{ik}a_{kl}b_{jl},$$

that is, $A'' = -2AR_{\dot\eta} + 2BAB^T$. Combining this with (7.1.7) yields $B' = -B^2 - AR_{\dot\eta}A^{-1}$. Taking the trace with respect to $g_{\dot\eta}$ yields (7.1.4). □

For the vector field V as above and $x \in M_V$, we define $\nabla V \in \Gamma(T_x^*M \otimes T_xM)$ by

$$\nabla V(w) := D^V_w V(x) \in T_xM, \quad w \in T_xM.$$

With this notation, $\nabla^2 u(X) := D^{\nabla u}_X \nabla u$ can be identified with $\nabla^2 u(X, \cdot)$ in the sense of (6.1.6) for any C^2 function u and vector field $X \in \Gamma(TM)$.

Lemma 7.1.2. (i) $B(t) = \nabla V_t(\eta(t))$ in the sense that $\nabla V_t(E_i(t)) = E_i'(t) = b_{ij}(t)E_j(t)$ for each $1 \le i \le n$.

(ii) *Given a smooth measure m on M, it holds that* $\operatorname{tr}(B(t)) = \operatorname{div}_m V_t(\eta) + d\Psi(\dot\eta)$, *where $dm = e^{-\Psi}\operatorname{Vol}_{\dot\eta}$ along η such that $\operatorname{Vol}_{\dot\eta}$ denotes the Riemannian volume measure of $g_{\dot\eta}$.*

Proof. (i) In local coordinates $(x^i)_{i=1}^n$, we write $\eta = \eta^i(t)\frac{\partial}{\partial x^i}$ and $E_i(t) = E_i^k(t)\frac{\partial}{\partial x^k}$. Then

$$E_i'(t) = D_{\dot\eta}^{\dot\eta} E_i(t) = \frac{dE_i^k}{dt}\frac{\partial}{\partial x^k} + \Gamma_{jk}^l(\dot\eta)\dot\eta^j E_i^k \frac{\partial}{\partial x^l}.$$

On the other hand,

$$\nabla \mathcal{V}_t(E_i(t)) = D_{E_i}^{\mathcal{V}_t}\mathcal{V}_t = d\mathcal{V}_t{}^k(E_i)\frac{\partial}{\partial x^k} + \Gamma_{kj}^l(\mathcal{V}_t)\mathcal{V}_t{}^j E_i^k \frac{\partial}{\partial x^l}.$$

Observe that $\dot\eta(t) = \mathcal{V}_t(\eta(t))$ and

$$\frac{dE_i^k}{dt} = \frac{d}{dt}\left\{\frac{d}{ds}\Big|_{s=0}(\gamma_t(\exp_x(se_i)))^k\right\}$$

$$= \frac{d}{ds}\Big|_{s=0}\left\{\frac{d}{dt}(\gamma_t(\exp_x(se_i)))^k\right\}$$

$$= \frac{d}{ds}\Big|_{s=0}\left\{\mathcal{V}_t^k(\gamma_t(\exp_x(se_i)))\right\}$$

$$= d\mathcal{V}_t^k(E_i).$$

Consequently, $E_i'(t) = D_{E_i}^{\dot\eta}\mathcal{V}_t = \nabla\mathcal{V}_t(E_i(t))$ and completes the proof of (i).

(ii) Choose a local coordinate system $(x^i)_{i=1}^n$ around $\eta(t)$ such that $\{\partial/\partial x^i|_{\eta(t)}\}$ is an orthonormal basis of $(T_{\eta(t)}M, g_{\dot\eta})$. We will suppress evaluations at $\eta(t)$. Recall first that

$$\nabla\mathcal{V}_t\left(\frac{\partial}{\partial x^i}\right) = D_{\partial/\partial x^i}^{\mathcal{V}_t}\mathcal{V}_t = \left(\frac{\partial \mathcal{V}_t^k}{\partial x^i} + \sum_{j=1}^n \Gamma_{ij}^k(\mathcal{V}_t)\mathcal{V}_t^j\right)\frac{\partial}{\partial x^k}.$$

Thus, we have

$$\mathrm{tr}(B(t)) = \mathrm{tr}(\nabla\mathcal{V}_t) = \frac{\partial \mathcal{V}_t^i}{\partial x^i} + \Gamma_{ij}^i(\mathcal{V}_t)\mathcal{V}_t^j. \tag{7.1.8}$$

On the other hand, $dm = e^{-\Psi}\sqrt{\det g_{\dot\eta}}dx$ along η with respect to m. Then

$$\mathrm{div}_m \mathcal{V}_t(\eta(t)) + d\Psi(\dot\eta(t)) = \frac{\partial \mathcal{V}_t^i}{\partial x^i} + \mathcal{V}_t^i\frac{\partial(\log\sqrt{\det g_{\dot\eta}})}{\partial x^i}. \tag{7.1.9}$$

By (3.1.2) and $\dot\eta(t) = \mathcal{V}_t(\eta(t))$, the second term on RHS of (7.1.9) is equal to

$$\frac{1}{2}g^{jk}(\mathcal{V}_t)\mathcal{V}_t^i\frac{\partial(g_{jk}(\mathcal{V}_t))}{\partial x^i} = \frac{1}{2}g^{jk}(\mathcal{V}_t)\mathcal{V}_t^i\left\{\frac{\partial g_{jk}}{\partial x^i} + 2C_{jkl}\frac{\partial \mathcal{V}_t^l}{\partial x^i}\right\}(\mathcal{V}_t)$$

$$= \frac{1}{2} g^{jk}(\mathcal{V}_t)\mathcal{V}_t^i \left\{ g_{jl}\Gamma_{ik}^l + g_{lk}\Gamma_{ij}^l + 2C_{jkl}N^l{}_i + 2C_{jkl}\frac{\partial \mathcal{V}_t^l}{\partial x^i} \right\}(\mathcal{V}_t)$$

$$= \mathcal{V}_t^i \Gamma_{ij}^j + g^{jk}(\mathcal{V}_t)C_{jkl}(\mathcal{V}_t)(2G^l(\mathcal{V}_t) + \ddot{\eta}) = \mathcal{V}_t^i \Gamma_{ij}^j. \tag{7.1.10}$$

Plugging (7.1.10) into (7.1.9) and togethering with (7.1.8) yield (ii). □

Based on Lemmas 7.1.1–7.1.2, Ohta–Sturm established the following point-wise Bochner–Weitzenböck formula for L_u (Theorem 3.3, [OS2]). It is fundamental to study the analysis and topology on Finsler manifolds.

Theorem 7.1.1 (Pointwise Bochner–Weitzenböck Formula and Inequality, [OS2]). *Let (M, F, m) be an n-dimensional Finsler measure space. Given $u \in C^\infty(M)$, we have*

$$\frac{1}{2}L_u \left(F^2(\nabla u) \right) = d(\boldsymbol{\Delta} u)(\nabla u) + Ric_\infty(\nabla u) + \|\nabla^2 u\|_{HS(\nabla u)}^2 \tag{7.1.11}$$

as well as

$$\frac{1}{2}L_u \left(F^2(\nabla u) \right) \geq d(\boldsymbol{\Delta} u)(\nabla u) + Ric_N(\nabla u) + \frac{(\boldsymbol{\Delta} u)^2}{N} \tag{7.1.12}$$

for $N \in [n, \infty]$ point-wise on M_u, where $\| \cdot \|_{HS(\nabla u)}$ stands for the Hilbert–Schmidt norm with respect to $g_{\nabla u}$.

Proof. Substituting Lemma 7.1.2 (ii) into (7.1.4) yields

$$\frac{d}{dt}\bigg|_{t=0+} (\mathrm{div}_m \mathcal{V}_t(\eta(t))) + \mathrm{tr}\left(B(0)^2 \right) + \mathrm{Ric}_\infty(\dot{\eta}(0)) = 0.$$

Thanks to (7.1.3), we have

$$\frac{d}{dt}\bigg|_{t=0+} (\mathrm{div}_m \mathcal{V}_t(\eta(t))) = d\left(\mathrm{div}_m V\right)(\dot{\eta}(0)) + \mathrm{div}_m \left(\frac{\partial \mathcal{V}_t}{\partial t}\bigg|_{t=0+}\right)(x)$$

$$= d\left(\mathrm{div}_m V\right)(\dot{\eta}(0)) - \mathrm{div}_m \left(D_V^V V\right)(x).$$

Combining these, one obtains at x

$$\mathrm{div}_m \left(D_V^V V\right) - d\left(\mathrm{div}_m V\right)(V) = \mathrm{Ric}_\infty(V) + \mathrm{tr}\left(B(0)^2\right). \tag{7.1.13}$$

Now, we take $u \in C^\infty(M)$ and $x \in M_u$, and apply (7.1.13) to $V = \nabla u$. Note that Lemma 7.1.2 (i) implies that $\mathrm{tr}(B(0)^2) = \|\nabla^2 u\|_{HS(\nabla u)}^2$, and Lemma 6.3.2(i) implies that

$$D_{\nabla u}^{\nabla u}(\nabla u) = \nabla^{\nabla u}\left(\frac{F(\nabla u)^2}{2}\right) \tag{7.1.14}$$

on M_u. Plugging these into (7.1.13) yields (7.1.11).

(7.1.12) is clear for $N = \infty$. For $N \in (n, \infty)$, taking $V = \nabla u$ in Lemma 6.3.2 yields

$$\|\nabla^2 u\|^2_{HS(\nabla u)} \geq \frac{(\Delta u)^2}{N} - \frac{S(\nabla u)^2}{N - n}$$

on M_u. Plugging this into (7.1.11) yields (7.1.12). □

Observe that

$$\nabla u \left(\frac{F^2(\nabla u)}{2} \right) = g_{\nabla u} \left(D^{\nabla u}_{\nabla u}(\nabla u), \nabla u \right) = \nabla^2 u(\nabla u, \nabla u). \quad (7.1.15)$$

Let

$$H_u := \frac{\nabla^2 u(\nabla u, \nabla u)}{F^2(\nabla u)}.$$

Then, from (7.1.11), Lemma 6.3.2 with $V = \nabla u$ and (7.1.15), we obtain the following more refined inequality.

Corollary 7.1.1. *Under the same assumptions as in Theorem 7.1.1, we have*

$$\frac{1}{2} L_u \left(F^2(\nabla u) \right) \geq d(\Delta u)(\nabla u) + Ric_N(\nabla u) + \frac{(\Delta u)^2}{N}$$

$$+ \frac{N}{N - 1} \left(\frac{\Delta u}{N} - H_u \right)^2 \quad (7.1.16)$$

for $N \in [n, \infty]$ point-wise on M_u.

7.1.2   *Integrated Bochner–Weitzenböck Formula and Inequality*

To extend the pointwise Bochner–Weitzenböck formula to a global one (in the weak sense), we need to be careful because some quantities are undefined on $M \setminus M_u$. Moreover, if $\partial M \neq \emptyset$, the normal vector of ∂M is not determined uniquely. We need an auxiliary lemma to overcome these difficulties.

Given a smooth measure m on a Finsler manifold (M, F), we define

$$W^{1,p}(M) := \left\{ u \in L^p(M) \Big| \int_M [F^*(du)]^p dm + \int_M [F^*(-du)]^p dm < \infty \right\}$$

for $p \geq 1$. It is a Sobolev space with respect to the norm

$$\|u\|_{1,p} := \|u\|_{L^p} + \|F^*(du)\|_{L^p} + \|\overleftarrow{F}^*(du)\|_{L^p}. \quad (7.1.17)$$

In fact, $W^{1,p}(M)$ is the completion of the space $C^\infty(M)$ consisting of smooth functions on M with respect to $\|\cdot\|_{1,p}$. Similarly, we define $W_0^{1,p}(M)$ as the completion of the space $C_0^\infty(M)$ consisting of smooth functions on M with compact support with respect to $\|\cdot\|_{1,p}$. Hence $(W_0^{1,p}(M), \|u\|_{1,p})$ is also a Sobolev space (cf. Chapter 10). In particular, if M is compact, then the metrics F and $\overleftarrow{F}$ are equivalent. In this case, $W^{1,p}(M)$ is isomorphic to the space

$$\tilde{W}^{1,p}(M) := \left\{ u \in L^p(M) \Big| \int_M [F^*(du)]^p dm < \infty \right\}$$

for $p \geq 1$, and the Sobolev norm $\|\cdot\|_{1,p}$ is equivalent to the norm

$$\|u\|_{1,p} := \|u\|_{L^p} + \|F^*(du)\|_{L^p}. \tag{7.1.18}$$

In general, $\tilde{W}^{1,p}(M)$ is not a linear space over $\mathbb{R}$ because of the non-reversibility of F ([KR]).

The following fact will play an important role to overcome the ill-posedness of ∇u on $M \backslash M_u$ ([OS2], [BH]).

Lemma 7.1.3. *The following statements hold.*

(1) *For each $h \in W_0^{1,2}(M)$, we have $dh = 0$ almost everywhere on $h^{-1}(0)$.*
(2) *If $h \in W_0^{1,2}(M) \cap L^\infty(M)$, then $d\left(h^2/2\right) = hdh = 0$ almost everywhere on $h^{-1}(0)$.*
(3) *The assertions (1) and (2) also hold true if h merely lies locally in the respective spaces.*

Lemma 7.1.3 implies that $\Delta u = 0$ almost everywhere on $M \backslash M_u$.

Theorem 7.1.2. *Let (M, F, m) be an n-dimensional compact Finsler measure space with smooth boundary (possibly $\partial M = \emptyset$). Given $u \in W^{2,2}(M) \cap C^1(M)$ with $\Delta u \in W^{1,2}(M)$, we have*

$$\int_M d\phi \left(\nabla^{\nabla u} \left(\frac{F^2(\nabla u)}{2} \right) \right) dm$$

$$+ \int_M \phi \left\{ d(\Delta u)(\nabla u) + Ric_\infty(\nabla u) + \|\nabla^2 u\|_{HS(\nabla u)}^2 \right\} dm$$

$$= \int_{\partial M} \phi g_\nu \left(\nu, D_{\nabla u}^{\nabla u} \nabla u \right) dm_\nu \tag{7.1.19}$$

and

$$\int_M d\phi \left(\nabla^{\nabla u} \left(\frac{F^2(\nabla u)}{2} \right) \right) dm$$

$$+ \int_M \phi \left\{ d(\mathbf{\Delta} u)(\nabla u) + Ric_N(\nabla u) + \frac{(\mathbf{\Delta} u)^2}{N} \right\} dm$$

$$\leq \int_{\partial M} \phi g_\nu \left(\nu, D^{\nabla u}_{\nabla u} \nabla u \right) dm_\nu \qquad (7.1.20)$$

for $N \in [n, \infty]$ and all nonnegative functions $\phi \in W^{1,2}(M) \cap L^\infty(M)$, where ν is the outward normal vector field of ∂M and m_ν is the measure on ∂M induced by m with respect to ν.

Proof. The proof follows the line of Theorem 3.6 in [OS2]. Since the proofs are similar, we only consider (7.1.20) with $N < \infty$.

Let us first treat the case of $u \in C^\infty(M)$. If $\phi \in W^{1,2}(M_u) \cap L^\infty(M)$, then, by (7.1.12) and the divergence lemma, we have

$$\int_M d\phi \left(\nabla^{\nabla u} \left(\frac{F^2(\nabla u)}{2} \right) \right) dm$$

$$+ \int_M \phi \left\{ d(\mathbf{\Delta} u)(\nabla u) + Ric_N(\nabla u) + \frac{(\mathbf{\Delta} u)^2}{N} \right\} dm$$

$$\leq \int_{\partial M} \phi g_\nu \left(\nu, \nabla^{\nabla u} \left(\frac{F^2(\nabla u)}{2} \right) \right) dm_\nu,$$

which implies (7.1.20), where we used (7.1.14). If $\phi \in W^{1,2}(M) \cap L^\infty(M)$, set

$$\phi_k := \min\{\phi, k^2 F^2(\nabla u)\}, \quad k \in \mathbb{N},$$

then $\phi_k \in W^{1,2}(M_u) \cap L^\infty(M)$ and $\lim_{k \to \infty} \phi_k(x) = \phi(x)$ for $x \in M_u$. Therefore, we have

$$\int_M d\phi_k \left(\nabla^{\nabla u} \left(\frac{F^2(\nabla u)}{2} \right) \right) dm$$

$$+ \int_M \phi_k \left\{ d(\mathbf{\Delta} u)(\nabla u) + Ric_N(\nabla u) + \frac{(\mathbf{\Delta} u)^2}{N} \right\} dm$$

$$\leq \int_{\partial M} \phi_k g_\nu \left(\nu, D^{\nabla u}_{\nabla u} \nabla u \right) dm_\nu. \qquad (7.1.21)$$

When k goes to infinity, the second term of the LHS of (7.1.21) converges to

$$\int_M \phi \left\{ d(\mathbf{\Delta} u)(\nabla u) + Ric_N(\nabla u) + \frac{(\mathbf{\Delta} u)^2}{N} \right\} dm, \qquad (7.1.22)$$

where we used that $\mathbf{\Delta} u = 0$ almost everywhere on $M \backslash M_u$. In the same way, when $k \to \infty$, the RHS of (7.1.21) converges to

$$\int_{\partial M \cap M_u} \phi g_\nu \left(\nu, D^{\nabla u}_{\nabla u} \nabla u \right) dm_\nu.$$

Note that $2D^{\nabla u}_{\nabla u} \nabla u = \nabla^{\nabla u} F^2(\nabla u)$ on M_u and vanishes a.e. on $\partial M \backslash M_u$ by Lemmas 6.3.2 and 7.1.3. Consequently, the integrand in the above integral is actually integrated on ∂M.

To see the limit of the first term on the left-hand side of (7.1.21), let

$$\Omega_k := \left\{ x \in M_u | \phi(x) > k^2 F^2(\nabla u(x)) \right\} = \left\{ x \in M_u | \phi(x) \neq \phi_k(x) \right\}$$

we find

$$\left| \int_M d(\phi - \phi_k) \left(\nabla^{\nabla u} \left(F^2(\nabla u) \right) \right) dm \right|$$

$$\leq \left| \int_{\Omega_k} d\phi \left(\nabla^{\nabla u} \left(F^2(\nabla u) \right) \right) dm \right|$$

$$+ k^2 \int_{\Omega_k} d \left(F^2(\nabla u) \right) \left(\nabla^{\nabla u} \left(F^2(\nabla u) \right) \right) dm.$$

The first term of the right-hand side tends to zero since Ω_k decreases to a null set as k goes to infinity. For the second term, note that

$$d \left(F^2(\nabla u) \right) \left(\nabla^{\nabla u} \left(F^2(\nabla u) \right) \right) = 4F^2(\nabla u) \cdot d(F(\nabla u)) \left(\nabla^{\nabla u} (F(\nabla u)) \right)$$

on M_u. Therefore, by the choice of Ω_k, one obtains

$$k^2 \int_{\Omega_k} d \left(F^2(\nabla u) \right) \left(\nabla^{\nabla u} \left(F^2(\nabla u) \right) \right) dm$$

$$\leq 4 \int_{\Omega_k} \phi \cdot d(F(\nabla u)) \left(\nabla^{\nabla u} (F(\nabla u)) \right) dm$$

$$\to 0 \quad (k \to \infty).$$

Hence, as k goes to infinity, the first term on LHS of (7.1.21) converges to

$$\int_M d\phi \left(\nabla^{\nabla u} \left(\frac{F^2(\nabla u)}{2} \right) \right) dm.$$

This proves (7.1.20) for $u \in C^\infty(M)$.

Next, we consider the general case when $u \in W^{2,2}(M) \cap C^1(M)$ with $\Delta u \in W^{1,2}(M)$. Since $u \in W^{2,2}(M)$, we can choose $\{u_k \in C^\infty(M)\}$ such that $u_k \to u$ in the $\|\cdot\|_{1,2}$-norm as $k \to \infty$. Since (7.1.20) holds with respect to u_k, by taking a limit, we obtain (7.1.20) for general u. $\qquad\square$

We denote by $W_c^{1,2}(M)$ the set of functions in $W^{1,2}(M)$ with compact support. The following corollary follows from Theorem 7.1.2.

Corollary 7.1.2 ([OS2]). *Let* (M, F, m) *be an n-dimensional Finsler measure space. Given* $u \in W_{loc}^{2,2}(M) \cap C^1(M) \cap W^{1,2}(M)$ *with* $\Delta u \in W_{loc}^{1,2}(M)$, *we have*

$$-\int_M d\phi \left(\nabla^{\nabla u} \left(\frac{F^2(\nabla u)}{2} \right) \right) dm$$

$$= \int_M \phi \left\{ d(\Delta u)(\nabla u) + Ric_\infty(\nabla u) + \|\nabla^2 u\|_{HS(\nabla u)}^2 \right\} dm$$

and

$$-\int_M d\phi \left(\nabla^{\nabla u} \left(\frac{F^2(\nabla u)}{2} \right) \right) dm$$

$$\geq \int_M \phi \left\{ d(\Delta u)(\nabla u) + Ric_N(\nabla u) + \frac{(\Delta u)^2}{N} \right\} dm$$

for $N \in [n, \infty]$ and all nonnegative functions $\phi \in W_c^{1,2}(M) \cap L^\infty(M)$.

Remark 7.1.1. In Corollary 7.1.2, the test function $\phi \in W_c^{1,2}(M) \cap L^\infty(M)$ can be extended to $\phi \in W_0^{1,2}(M) \cap L^\infty(M)$. In fact, for any $\phi \in W_0^{1,2}(M)$, there exist a sequence of functions $\{\phi_k \in C_0^\infty(M)\}$ such that $\{\phi_k\}$ converges to ϕ in $W^{1,2}(M)$ as $k \to \infty$. Note that $C^\infty(M)$ is dense in $W^{1,2}(M)$. Then $\phi_k \in W^{1,2}(M)$ with compact support, i.e., $\phi_k \in W_c^{1,2}(M)$. Since the two inequalities in Corollary 7.1.2 hold for any nonnegative $\phi_k \in W_c^{1,2}(M)$, they also hold for any nonnegative $\phi \in W_0^{1,2}(M)$ by an approximating argument.

When (M, F) is compact, we have $1 \in W^{1,2}(M)$. The following Reilly type formula and inequality follow from Theorem 7.1.2 by letting $\phi = 1$.

Corollary 7.1.3 (Reilly formula and inequality). *Let* (M, F, m) *be an* *n-dimensional compact Finsler measure space with boundary* ∂M *(possibly* $\partial M = \emptyset$*). For any function* $u \in W^{2,2}(M) \cap C^1(M) \cap W^{1,2}(M)$ *with* $\Delta u \in$ $W^{1,2}(M)$*, then we have*

$$\int_M \left\{ d(\Delta u)(\nabla u) + Ric_\infty(\nabla u) + \|\nabla^2 u\|^2_{HS(\nabla u)} \right\} dm$$

$$= \int_{\partial M} g_\nu \left(\nu, D^{\nabla u}_{\nabla u} \nabla u \right) dm_\nu \qquad (7.1.23)$$

and

$$\int_M \left\{ d(\Delta u)(\nabla u) + Ric_N(\nabla u) + \frac{(\Delta u)^2}{N} \right\} dm \leq \int_{\partial M} g_\nu \left(\nu, D^{\nabla u}_{\nabla u} \nabla u \right) dm_\nu,$$

$$(7.1.24)$$

where ν *is the outward normal vector field of* ∂M*.*

In particular, when $\partial M = \emptyset$, one obtains the following inequalities.

Corollary 7.1.4. *Assume that* (M, F, m) *is a compact Finsler measure space without boundary. Assume that* $Ric_N \geq K$ *for some* $K \in \mathbb{R}$ *and* $N \in$ $[n, \infty]$*. For any* $u \in W^{2,2}(M) \cap C^1(M) \cap W^{1,2}(M)$ *with* $\Delta u \in W^{1,2}(M)$*, then we have*

$$\int_M \left\{ d(\Delta u)(\nabla u) + K F^2(\nabla u) \right\} dm \leq 0, \quad \text{if} \quad N = \infty, \qquad (7.1.25)$$

and

$$\int_M \left\{ d(\Delta u)(\nabla u) + K F^2(\nabla u) + \frac{(\Delta u)^2}{N} \right\} dm \leq 0, \quad \text{if} \quad N < \infty. \quad (7.1.26)$$

In particular, when $K > 0$*, we have*

$$\int_M F^2(\nabla u) dm \leq \frac{N-1}{KN} \int_M (\Delta u)^2 dm. \qquad (7.1.27)$$

If $N = \infty$*, then the coefficient in the RHS of* (7.2.11) *should be read as* $1/K$*.*

Proof. (7.1.25)–(7.1.26) directly follow from (7.1.24). Moreover, by (7.1.26) and the integration by parts, we have

$$\int_M \left\{ K F^2(\nabla u) + \frac{(\Delta u)^2}{N} \right\} dm \leq - \int_M d(\Delta u)(\nabla u) dm = \int_M (\Delta u)^2 dm,$$

which implies (7.1.27) if $K > 0$. $\qquad \square$

Remark 7.1.2. By Theorem 6.3.4, the inequality (7.1.27) is also true on a forward (or backward) complete Finsler measure space with $\mathrm{Ric}_N \geq K > 0$. In particular, when $N \in [n, \infty]$, Corollary 7.1.4 is exactly Proposition 3.1 in [Oh3].

Remark 7.1.3. Theorems 7.1.1–7.1.2 and Corollaries 7.1.2–7.1.3 have been generalized to the p-Bochner–Weitzenböck formula (inequality) and the p-Reilly formula (inequality) for Finsler p-Laplacian. These inequalities are important to estimate the lower bound for the first p-eigenvalue ([Xia6]).

7.2 Mean Value Inequality

Given a smooth measure m and an open set Ω in M, let $H^1_{\mathrm{loc}}(M) := W^{1,2}_{\mathrm{loc}}(M)$ be the space of weakly differentiable functions u on Ω such that both u and $F^*(du)$ belong to $L^2_{\mathrm{loc}}(\Omega)$. We remark that $H^1_{\mathrm{loc}}(M)$ is defined only in terms of the differentiable structure of M. A function $u \in H^1_{\mathrm{loc}}(M)$ is said to be *subsolution (resp., supersolution)* of $\Delta u = w$ on M if

$$-\int_M d\varphi(\nabla u)dm \geq (resp., \leq) \int_M \varphi w\, dm \qquad (7.2.1)$$

for all nonnegative $\varphi \in C_0^\infty(M)$ (equivalently, for all nonnegative $\varphi \in H^1_{\mathrm{loc}}(M)$). A function $u \in H^1_{\mathrm{loc}}(M)$ is said to be a *weak solution* of $\Delta u = w$ if it is both a subsolution and a supersolution of $\Delta u = w$, i.e., (7.2.1) holds true with equality for all $\varphi \in C_0^\infty(M)$ (equivalently, $\varphi \in H^1_{\mathrm{loc}}(M)$). In particular, a function $u \in H^1_{\mathrm{loc}}(M)$ is said to be a Finsler *harmonic (resp., subharmonic, superharmonic)* on M if it is a weak solution (resp., subsolution, supersolution) of $\Delta u = 0$ in M.

Similar to the Riemannian case, the Finsler harmonic functions arise from the variation of the energy functional $E(u) = \int_M F^2(x, \nabla u)dm$. Moreover, there exists a unique solution u of $\Delta u = 0$ (in the weak sense) with Dirichlet or Neumann boundary condition (i.e., $\nabla u \in T(\partial M)$) if $\partial M \neq \emptyset$. Further, $u \in C^{1,\alpha}(M) \cap W^{2,2}_{loc}(M) \cap L^\infty(M)$ for some exponent $0 < \alpha < 1$ and u is smooth on the set M_u (cf. [GS], also Theorem 8.1.1 below). These results are nontrivial and different from those in Riemannian case.

Example 7.2.1 ([OS1]). Let $(\mathbb{R}^n, F, m)$ be a Minkowski space equipped with a Minkowski norm $F(x, \cdot) = \|\cdot\|$ for all $x \in \mathbb{R}^n$ and the Lebesgue measure m. Put $u(x) = f(\|x - y\|)$ for some nondecreasing C^2 function $f : \mathbb{R}^+ \to \mathbb{R}$ and some fixed point $y \in \mathbb{R}^n$. Then we have for any $x \neq y$,

$$\Delta u(x) = f''(\|x - y\|) + \frac{n-1}{\|x - y\|}f'(\|x - y\|). \qquad (7.2.2)$$

In fact,

$$du(x) = f'(\|x - y\|)d(\|x - y\|) = \frac{f'(\|x - y\|)}{\|x - y\|}(x - y).$$

For nondecreasing f, we find

$$\boldsymbol{\Delta} u(x) = \operatorname{div}\left(\frac{f'(\|x - y\|)}{\|x - y\|}(x - y)\right) = f''(\|x - y\|) + \frac{n - 1}{\|x - y\|}f'(\|x - y\|),$$

where we used

$$\sum_{i=1}^{n} \frac{\partial \|x - y\|}{\partial x^i}(x^i - y^i) = \|x - y\|$$

by Euler's Lemma. If f satisfies

$$f''(r) + \frac{n - 1}{r}f'(r) = 0, \quad \text{and} \quad f'(r) \geq 0, \tag{7.2.3}$$

then u is a Finsler harmonic function. In particular,

$$u(x) = \begin{cases} \|x - y\|^{-n+2}, & n \geq 3, \\ \log \|x - y\|, & n = 2, \end{cases}$$

is a Finsler harmonic function on $\mathbb{R}^n$. Similarly, since $\boldsymbol{\Delta}(\|x - y\|^2) = 2n$, $u(x) = \|x - y\|^2$ is a Finsler subharmonic function. If f is nonincreasing, then an analogous result holds for $v(x) = f(\|y - x\|)$, namely,

$$\boldsymbol{\Delta} v(x) = f''(\|y - x\|) + \frac{n - 1}{\|y - x\|}f'(\|y - x\|). \tag{7.2.4}$$

This is because, for nonincreasing f, the RHS of (7.2.2) coincides with $\overleftarrow{\boldsymbol{\Delta}} u(x) = -\boldsymbol{\Delta}(-u)(x)$, where $\overleftarrow{\boldsymbol{\Delta}}$ stands for the Finsler Laplacian of reverse Finsler metric $\overleftarrow{F}(x, \xi) = \| - \xi\|$. Similarly, for nondecreasing f, the RHS of (7.2.4) coincides with $\overleftarrow{\boldsymbol{\Delta}} v(x)$.

To establish the L^p mean value inequality, let us recall the local Sobolev inequality, which was due to Xia ([Xc]). It is worth mentioning that we only need assume the finiteness of the reversibility instead of the uniform smoothness and convexity for Finsler metrics from the proof of this inequality in [Xc].

Lemma 7.2.1 ([Xc]). *Let (M, F, m) be a forward complete Finsler measure space with finite reversibility Λ satisfying $Ric_N \geq -K$ for some*

$K > 0$. *Then there exist constants $\nu = \nu(N) > 2$ and $c_0 = c_0(N, \Lambda) > 0$
depending on N and Λ, such that*

$$\left(\int_{B_R} |u - \bar{u}|^{\frac{2\nu}{\nu-2}} dm \right)^{\frac{\nu-2}{\nu}} \le e^{c_0(1+\sqrt{K}R)} R^2 m\left(B_R\right)^{-\frac{2}{\nu}} \int_{B_R} F^{*2}(x, du) dm$$

$$(7.2.5)$$

*for any forward geodesic ball $B_R := B_R^+(x_0) \subset M$ and $u \in W_{loc}^{1,2}(M)$, where
$\bar{u} = \frac{1}{m(B_R)} \int_{B_R} u \, dm$. Consequently,*

$$\left(\int_{B_R} |u|^{\frac{2\nu}{\nu-2}} dm \right)^{\frac{\nu-2}{\nu}} \le e^{c_0(1+\sqrt{K}R)} R^2 m(B_R)^{-\frac{2}{\nu}}$$

$$\times \int_{B_R} \left\{ F^{*2}(x, du) + R^{-2} u^2 \right\} dm. \qquad (7.2.6)$$

With this, we prove the local L^2 mean value inequality for nonnegative
subharmonic functions as follows.

Proposition 7.2.1. *Let (M, F, m) be an n-dimensional forward complete
Finsler measure space with finite reversibility Λ. Assume that (7.2.6) holds.
If u is a nonnegative subharmonic function defined on a forward geodesic
ball $B_R := B_R^+(x_0)$, then there are constants $\nu > 2$, and $c = c(N, \nu, \Lambda) > 0$
such that for any $\delta \in [1/2, 1)$ we have*

$$\sup_{B_{\delta R}} u^2 \le \frac{e^{c(1+\sqrt{K}R)}}{(1-\delta)^\nu m(B_R)} \int_{B_R} u^2 dm. \qquad (7.2.7)$$

*In particular, when $Ric_N \ge -K$ for some $N \in [n, \infty)$ and $K > 0$, (7.2.7)
holds.*

Proof. Since u is a nonnegative subharmonic function on B_R, we have

$$\int_{B_R} d\varphi(\nabla u) dm \le 0 \qquad (7.2.8)$$

for any nonnegative function $\varphi \in C_0^\infty(B_R)$. In fact, this is also true for any
nonnegative $\varphi \in W_0^{1,2}(B_R)$. For any $\frac{1}{2} \le \delta < \delta' \le 1$, we choose a cut-off
function $\varphi(x)$ defined by

$$\varphi(x) = \begin{cases} 1 & \text{on } B_{\delta R}, \\ \frac{\delta' R - d_F(x_0, x)}{(\delta' - \delta) R} & \text{on } B_{\delta' R} \backslash B_{\delta R}, \\ 0 & \text{on } M \backslash B_{\delta' R}. \end{cases} \qquad (7.2.9)$$

Then $F^*(-d\varphi) \leq \frac{1}{(\delta'-\delta)R}$ and hence $F^*(d\varphi) \leq \frac{\Lambda}{(\delta'-\delta)R}$ a.e. on $B_{\delta'R}$. Replacing ϕ with $u\varphi^2$ in (7.2.8) and using $-d\varphi(\nabla u) \leq F^*(-d\varphi)F(\nabla u)$ yield

$$\int_{B_R} \varphi^2 F^2(\nabla u)dm \leq -2\int_{B_R} \varphi u d\varphi(\nabla u)dm \leq 2\int_{B_R} \varphi u F^*(-d\varphi)F(\nabla u)dm$$

$$\leq \frac{1}{2}\int_{B_R} \varphi^2 F^2(\nabla u)dm + 2\int_{B_R} u^2 F^{*2}(-d\varphi)dm,$$

that is,

$$\int_{B_R} \varphi^2 F^2(\nabla u)dm \leq 4\int_{B_R} u^2 F^{*2}(-d\varphi)dm,$$

which implies that

$$\int_{B_R} \varphi^2 F^{*2}(du)dm = \int_{B_R} \varphi^2 F^2(\nabla u)dm \leq \frac{4}{((\delta'-\delta)R)^2}\int_{B_{\delta'R}} u^2 dm.$$

$$(7.2.10)$$

On the other hand, by Hölder's inequality, (7.2.6), (1.1.4) and (7.2.9)–(7.2.10), we have

$$\int_{B_{\delta R}} u^{2(1+\frac{2}{\nu})}dm$$

$$= \int_{B_R} (u\varphi)^{2(1+\frac{2}{\nu})}dm \leq \left(\int_{B_R} (u\varphi)^{\frac{2\nu}{\nu-2}}dm\right)^{\frac{\nu-2}{\nu}} \cdot \left(\int_{B_R} (u\varphi)^2 dm\right)^{\frac{2}{\nu}}$$

$$\leq \left(A\int_{B_R} \{F^{*2}(d(u\varphi)) + R^{-2}u^2\varphi^2\}dm\right) \cdot \left(\int_{B_R} u^2\varphi^2 dm\right)^{\frac{2}{\nu}}$$

$$\leq A\int_{B_R} \{2\varphi^2 F^{*2}(du) + 2u^2 F^{*2}(d\varphi) + R^{-2}\varphi^2 u^2\}dm \cdot \left(\int_{B_{\delta'R}} u^2 dm\right)^{\frac{2}{\nu}}$$

$$\leq \frac{11A\Lambda^2}{((\delta'-\delta)R)^2}\left(\int_{B_{\delta'R}} u^2 dm\right)^{1+\frac{2}{\nu}},$$

where $A := e^{c_0(1+\sqrt{K}R)}R^2 m(B_R)^{-2/\nu}$, here ν and c_0 are chosen as in Lemma 7.2.1. For any $\tau \geq 1$, it is easy to see that u^τ is also a nonnegative subharmonic function. Let $\chi := 1 + \frac{2}{\nu}$. The above inequality implies that

$$\int_{B_{\delta R}} u^{2\chi\tau} \, dm \leq \frac{11A\Lambda^2}{((\delta' - \delta)R)^2} \left(\int_{B_{\delta' R}} u^{2\tau} \, dm \right)^{\chi}. \qquad (7.2.11)$$

For any $\delta \in [1/2, 1)$, let $\delta_0 = 1$ and $\delta_{i+1} = \delta_i - \frac{1-\delta}{2^{i+1}}$ on $B_{\delta_i R}$ for $i = 0, 1, \ldots$. Applying (7.2.11) for $\delta' = \delta_i$, $\delta = \delta_{i+1}$ and $\tau = \chi^i$, we have

$$\int_{B_{\delta_{i+1} R}} u^{2\chi^{i+1}} \, dm \leq \frac{4^{i+1}(11A\Lambda^2)}{((1-\delta)R)^2} \left(\int_{B_{\delta_i R}} u^{2\chi^i} \, dm \right)^{\chi}.$$

By iteration, one obtains

$$\left(\int_{B_{\delta_{i+1} R}} u^{2\chi^{i+1}} \, dm \right)^{\frac{1}{\chi^{i+1}}}$$

$$\leq 4^{\sum j\chi^{-j}} (11A\Lambda^2)^{\sum \chi^{-j}} [(1-\delta)R]^{-2\sum \chi^{-j}} \cdot \int_{B_R} u^2 \, dm, \qquad (7.2.12)$$

in which $\sum$ denotes the summation on j from 1 to $i+1$. Since $\sum_{j=1}^{\infty} \chi^{-j} = \frac{\nu}{2}$ and $\sum_{j=1}^{\infty} j\chi^{-j}$ converges, (7.2.12) implies (7.2.7) by taking $i \to \infty$. This finishes the proof. $\qquad \square$

From Proposition 7.2.1, one obtains the local $L^p (0 < p \leq 2)$ mean value inequality.

Theorem 7.2.1 ([Xia8]). *Let (M, F, m) be an n-dimensional forward complete Finsler measure space with finite reversibility Λ. Assume that $Ric_N \geq -K$ for some $N \in [n, \infty)$ and $K > 0$. If u is a nonnegative subharmonic function defined on a forward geodesic ball $B_R := B_R^+(x_0)$, then, for any $0 < p \leq 2$ and $\delta \in [1/2, 1)$, there are constants $\nu > 2$ and $c = c(N, \nu, p, \Lambda) > 0$ such that*

$$\sup_{B_{\delta R}} u^p \leq \frac{e^{c(1+\sqrt{K}R)}}{(1-\delta)^{\nu} m(B_R)} \int_{B_R} u^p \, dm. \qquad (7.2.13)$$

Proof. We remark that the case when $p = 2$ has been proved in Proposition 7.2.1. It suffices to prove (7.2.13) for $0 < p < 2$.

For any $\delta \in [1/2, 1)$, choose $\varepsilon \in (0, 1)$ with $1/2 < \delta + \varepsilon \leq 1$. It follows from the proof of Proposition 7.2.1 that there are positive constants $\nu > 2$

and $\tilde{c} = \tilde{c}(N, \nu, \Lambda)$ such that

$$\sup_{B_{\delta R}} u^2 \leq e^{\tilde{c}(1+\sqrt{K}R)} \varepsilon^{-\nu} m(B_{(\delta+\varepsilon)R})^{-1} \int_{B_{(\delta+\varepsilon)R}} u^2 dm$$

$$\leq e^{\tilde{c}(1+\sqrt{K}R)} \varepsilon^{-\nu} m(B_{2^{-1}R})^{-1} \int_{B_{(\delta+\varepsilon)R}} u^2 dm.$$

On the other hand,

$$\int_{B_{(\delta+\varepsilon)R}} u^2 dm \leq \sup_{B_{(\delta+\varepsilon)R}} u^{2-p} \int_{B_{(\delta+\varepsilon)R}} u^p dm \leq (\sup_{B_{(\delta+\varepsilon)R}} u^2)^{1-p/2} \int_{B_R} u^p dm.$$

Thus,

$$\sup_{B_{\delta R}} u^2 \leq e^{\tilde{c}(1+\sqrt{K}R)} \varepsilon^{-\nu} m(B_{2^{-1}R})^{-1} (\sup_{B_{(\delta+\varepsilon)R}} u^2)^{1-p/2} \int_{B_R} u^p dm.$$

Let $\mu = 1 - p/2 > 0$ and

$$\mathcal{M}(\delta) := \sup_{B_{\delta R}} u^2, \quad \mathcal{R} = e^{\tilde{c}(1+\sqrt{K}R)} m(B_{2^{-1}R})^{-1} \int_{B_R} u^p dm.$$

Choose $\delta_0 = \delta$ and $\delta_i = \delta_{i-1} + \frac{1-\delta}{2^i}$ for $i = 1, 2, \ldots$. Then

$$\mathcal{M}(\delta_{i-1}) \leq \mathcal{R} 2^{i\nu} (1-\delta)^{-\nu} \mathcal{M}(\delta_i)^\mu.$$

By iterating, we get

$$\mathcal{M}(\delta_0) \leq \mathcal{R}^{\sum \mu^{i-1}} 2^{\nu \sum i\mu^{i-1}} (1-\delta)^{-\nu \sum \mu^{i-1}} \mathcal{M}(\delta_j)^{\mu^j}, \qquad (7.2.14)$$

in which $\sum$ denotes the summation on i from 1 to j. Obviously, we have $\lim_{j\to\infty} \delta_j = 1$ and $\lim_{j\to\infty} \mu^j = 0$. Moreover, $\sum_{i=1}^\infty \mu^{i-1} = 2/p$ and $\sum_{i=1}^\infty i\mu^{i-1}$ converges. Letting $j \to \infty$ on the both sides of (7.2.14), there is a positive constant $\bar{c} = \bar{c}(N, \nu, p)$ such that

$$\sup_{B_{\delta R}} u^p \leq \mathcal{M}(\delta_0)^{p/2} \leq e^{\bar{c}(1+\sqrt{K}R)} (1-\delta)^{-\nu} m(B_{2^{-1}R})^{-1} \int_{B_R} u^p dm.$$

Therefore, (7.2.13) follows from (6.4.6), i.e.,

$$m(B_R) \leq 2^N e^{R\sqrt{(N-1)K}} m(B_{2^{-1}R}).$$

The proof is finished. $\qquad\qquad\qquad\qquad\qquad\qquad\qquad\qquad\qquad\qquad\square$

The following corollary directly follows from Theorem 7.2.1.

Corollary 7.2.1. *Let (M, F, m) be a forward complete and noncompact Finsler measure space with finite reversibility Λ and with $Ric_N \geq 0$ for* *some $N \in [n, \infty)$. If u is a nonnegative subharmonic function on M, then there are constants $\nu = \nu(N) > 2$ and $C = C(N, \nu, p, \Lambda) > 0$ such that for any $\delta \in [1/2, 1)$ and $0 < p \le 2$, we have*

$$\sup_{B_{\delta R}^+(x)} u^p \le \frac{C}{(1 - \delta)^\nu m(B_R^+(x))} \int_{B_R^+(x)} u^p dm, \quad \forall x \in M.$$

In particular, it holds on M

$$u^p(x) \le \sup_{B_{R/2}^+(x)} u^p \le \frac{C}{m(B_R^+(x))} \int_{B_R^+(x)} u^p dm.$$

Remark 7.2.1. For the case when $p > 2$, if $u \in L_{loc}^p(M)$, then Theorem 7.2.1 and Corollary 7.2.1 also hold. In fact, we may assume that $u > 0$. Otherwise, we replace u by $\tilde{u} := u + \epsilon > 0$ for some sufficiently small positive number ϵ. If $\tilde{u}$ satisfies (7.2.13), so does u by letting $\epsilon \to 0$. Since $u \in L_{loc}^p(M)$, we have $u^{p/2} \in L_{loc}^2(M)$ with $F^*(du^{p/2}) \in L_{loc}^2(M)$.

On the other hand, $u^{p/2}(p > 2)$ is also a subharmonic function on B_R, which can be seen from

$$\int_{B_R} d\phi(\nabla u^{p/2}) dm = \frac{p}{2} \int_{B_R} u^{p/2-1} d\phi(\nabla u) dm$$

$$= \frac{p}{2} \int_{B_R} d\left(\phi u^{p/2-1}\right)(\nabla u) dm$$

$$- \left(\frac{p}{2} - 1\right) \int_{B_R} \phi u^{p/2-2} F^2(\nabla u) dm \le 0$$

for any nonnegative $\phi \in C_0^\infty(B_R)$. Applying (7.2.7) to $u^{p/2}$ yields (7.2.13).

7.3 Some L^p Liouville Theorems

In this section, we study the Liouville properties for $L^p(p > 0)$ subharmonic functions based on Sections 7.1–7.2. We argue this according to two cases when $p > 1$ and $0 < p \le 1$.

7.3.1 $L^p(p > 1)$ Liouville Theorems

A celebrated theorem of S.T.Yau says that every nonnegative $L^p(p > 1)$ subharmonic function on a complete Riemannian manifold M must be a constant ([Yau]). This is also true in Finsler geometry.

Theorem 7.3.1. *Let (M, F, m) be an n-dimensional forward complete and noncompact Finsler manifold and u be a positive subharmonic function defined on a forward geodesic ball $B_{2R}^+(x)$ for any $x \in M$. Then for any $p > 1$, there is a constant $C = C(p) > 0$ such that*

$$\int_{B_R^+(x)} u^{p-2} F^2(\nabla u) dm \le \frac{C(p)}{R^2} \int_{B_{2R}^+(x)} u^p dm.$$

In particular, there does not exist a nonconstant positive $L^p(p > 1)$ subharmonic function.

Proof. We simply denote $B_R := B_R^+(x)$. For any $\varphi \in C_0^\infty(B_{2R})$, we have

$$\int_{B_{2R}} d\varphi(\nabla u) dm \le 0.$$

Let $\varphi = \phi^2 u^{p-1}$, where $\phi = \phi(x)$ is a cut-off function with $0 \le \phi \le 1$ such that it is 1 on B_R and 0 outside B_{2R} and $F(-d\phi) \le \frac{1}{R}$. By Cauchy–Schwarz's inequality, we get

$$\frac{p-1}{2} \int_{B_R} u^{p-2} F^2(\nabla u) dm \le \frac{2}{(p-1)R^2} \int_{B_{2R}} u^p dm,$$

which implies the conclusion. $\square$

In the following, we introduce a more general result than Theorem 7.3.1, which is from [ZX]. Fix a point $x_0 \in M$, denote the volume of a forward geodesic ball $B_r^+(x_0)$ with respect to the measure m by $m(r)$. Let

$$m_p(r) = \int_{B_r^+(x_0)} u^p dm$$

for $p \in \mathbb{R}$. Note that $m_p(r) = m(r)$ if $p = 0$.

Theorem 7.3.2 ([ZX]). *Let (M, F, m) be an n-dimensional forward complete and noncompact Finsler measure space with finite reversibility Λ. Assume that*

$$\int_1^\infty \frac{r}{m_p(r)} dr = \infty. \tag{7.3.1}$$

(i) *If $p \in (-\infty, 1)$ and u is a nonnegative superharmonic function on M, then u is a constant.*

(ii) *If $p \in (1, \infty)$ and u is a nonnegative subharmonic function on M, then u is a constant.*

Proof. We prove this according to the cases when $p > 0$ $(p \neq 1)$, $p < 0$ and $p = 0$, respectively.

Case 1. $p > 0$ and $p \neq 1$. By the assumption, u is a nonnegative subharmonic function if $p > 1$ and u is a nonnegative superharmonic function if $0 < p < 1$. Without loss of generality, we always assume that $u > 0$ in this case. Otherwise, we replace u by $\tilde{u} := u + \epsilon > 0$ for a sufficiently small positive number ϵ and then take a limit.

Let $x_0 \in M$ be a fixed point and $\gamma : [0, r] \to M$ a unit speed minimizing geodesic from x_0 to x for any $x \in M$. Thus, $r(x) := d(x_0, x) = r$. Set

$$\varphi(x) := \min\{(R - r(x))^+, R - r_0\}, \tag{7.3.2}$$

where $r_0, R \in \mathbb{R}^+$ such that $0 < r_0 < R$ and $(R - r(x))^+ = \max\{R - r(x), 0\}$. Let $\Omega = \bar{B}_R^+(x_0) \backslash B_{r_0}^+(x_0)$. For the sake of simplicity, denote $B_r := B_r^+(x_0)$. Then from (7.3.2), we get

$$\varphi(x) = \begin{cases} R - r_0, & x \in \bar{B}_{r_0}, \\ R - r(x), & x \in \Omega, \\ 0, & x \in M \backslash B_R. \end{cases}$$

It is obvious that $d\varphi = -dr$ on Ω and $d\varphi = 0$ on $M \backslash \Omega$. It follows from Lemma 2.1.1, (2.1.1) and $F^*(dr) = F(\partial/\partial r) = 1$ that

$$F^{*2}(d\varphi) = F^{*2}(-dr) \leq \Lambda^2 F^{*2}(dr) = \Lambda^2 \tag{7.3.3}$$

on Ω and $F^*(d\varphi) = 0$ on $M \backslash \Omega$. Set $v := u^{\frac{p}{2}}$. Then

$$\nabla v = \frac{p}{2} u^{\frac{p}{2}-1} \nabla u, \quad F(\nabla v) = \frac{p}{2} u^{\frac{p}{2}-1} F(\nabla u) \tag{7.3.4}$$

on $M_u = M_v$. (7.3.4) actually holds on M since $M \backslash M_u = M \backslash M_v$. Thus, by the assumption, we get

$$0 \geq (p-1) \int_{B_R} d(u^{p-1}\varphi^2)(\nabla u)dm$$

$$= (p-1)^2 \int_{B_R} u^{p-2}\varphi^2 F^2(\nabla u)dm + 2(p-1) \int_{\Omega} u^{p-1}\varphi d\varphi(\nabla u)dm$$

$$= 4\left(1 - p^{-1}\right)^2 \int_{B_R} \varphi^2 F^2(\nabla v)dm + 4\left(1 - p^{-1}\right) \int_{\Omega} v\varphi d\varphi(\nabla v)dm.$$

$$\tag{7.3.5}$$

Note that $|d\varphi(\nabla v)| \leq \Lambda F^*(d\varphi)F(\nabla v)$ by (1.1.5). One obtains from (7.3.5) that

$$(1 - p^{-1})^2 \int_{B_R} \varphi^2 F^2(\nabla v)dm \leq |1 - p^{-1}|\Lambda \int_{\Omega} v\varphi F^*(d\varphi)F(\nabla v)dm.$$

$$\tag{7.3.6}$$

By Hölder's inequality and (7.3.6), we get

$$\left(1-p^{-1}\right)^2 \int_{B_R} \varphi^2 F^2(\nabla v)dm$$

$$\leq \Lambda \left(\int_\Omega v^2 F^{*2}(d\varphi)dm\right)^{\frac{1}{2}} \left(\left(1-p^{-1}\right)^2 \int_\Omega \varphi^2 F^2(\nabla v)dm\right)^{\frac{1}{2}},$$

that is,

$$\left((1-p^{-1})^2 \int_{B_R} \varphi^2 F^2(\nabla v)dm\right)^2$$

$$\leq \Lambda^2 \left(\int_\Omega v^2 F^{*2}(d\varphi)dm\right) \left((1-p^{-1})^2 \int_\Omega \varphi^2 F^2(\nabla v)dm\right). \quad (7.3.7)$$

Let

$$G(r) := (1-p^{-1})^2 \int_{B_r} F^2(\nabla v)dm, \quad H := (1-p^{-1})^2 \int_\Omega \varphi^2 F^2(\nabla v)dm.$$

Assume that u is nonconstant. Let $\rho_0 > 0$ be so large that u is nonconstant in B_{ρ_0}. Fix $R > r_0 > \rho_0$. Then all terms in (7.3.7) are positive. Hence,

$$\int_\Omega v^2 F^*(d\varphi)^2 dm \geq \frac{1}{\Lambda^2} \frac{\left((1-p^{-1})^2 \int_{B_R} \varphi^2 F^2(\nabla v)dm\right)^2}{(1-p^{-1})^2 \int_\Omega \varphi^2 F^2(\nabla v)dm}$$

$$= \frac{1}{\Lambda^2} \frac{\left(\begin{array}{c}(1-p^{-1})^2 \int_{B_{r_0}} (R-r_0)^2 F^2(\nabla v)dm \\ +(1-p^{-1})^2 \int_\Omega \varphi^2 F^2(\nabla v)dm\end{array}\right)^2}{H}$$

$$= \frac{1}{\Lambda^2} \frac{\left((R-r_0)^2 G(r_0) + H\right)^2}{H}$$

$$\geq \frac{1}{\Lambda^2} (R-r_0)^2 G(r_0) \left\{\frac{(R-r_0)^2 G(r_0)}{H} + 1\right\}. \quad (7.3.8)$$

Since

$$H \leq (1-p^{-1})^2 (R-r_0)^2 \int_{\bar{B}_R \backslash B_{r_0}} F^2(\nabla v)dm$$

$$= (G(R) - G(r_0))(R-r_0)^2,$$

(7.3.8) implies that

$$\int_\Omega v^2 F^{*2}(d\varphi)dm \geq \frac{1}{\Lambda^2} (R-r_0)^2 G(r_0) \left\{\frac{G(r_0)}{G(R)-G(r_0)} + 1\right\}$$

$$= \frac{1}{\Lambda^2} (R-r_0)^2 \frac{G(r_0)G(R)}{G(R)-G(r_0)}. \quad (7.3.9)$$

On the other hand, from (7.3.3), we have

$$\int_\Omega v^2 F^{*2}(d\varphi)dm \le \Lambda^2 \int_\Omega u^p dm = \Lambda^2(m_p(R) - m_p(r_0)). \qquad (7.3.10)$$

By (7.3.9)–(7.3.10), we get

$$\Lambda^2\left(m_p(R) - m_p(r_0)\right) \ge \frac{1}{\Lambda^2}(R - r_0)^2 \frac{G(r_0)G(R)}{G(R) - G(r_0)},$$

that is,

$$\frac{1}{G(r_0)} - \frac{1}{G(R)} \ge \frac{1}{\Lambda^4} \frac{(R - r_0)^2}{m_p(R) - m_p(r_0)}. \qquad (7.3.11)$$

For fixed r_0, take $R_k = 2^k r_0$, $k \in N^+$, that is,

$$r_0 = R_0 < R_1 < \cdots < R_n.$$

Then from (7.3.11), we obtain

$$\frac{1}{G(r_0)} \ge \frac{1}{G(R_n)} + \frac{1}{\Lambda^4}\sum_{k=1}^n \frac{(R_k - R_{k-1})^2}{m_p(R_k) - m_p(R_{k-1})}$$

$$\ge \frac{1}{4\Lambda^4}\sum_{k=1}^n \frac{(R_k)^2}{m_p(R_k)}. \qquad (7.3.12)$$

Observe that

$$\int_{2r_0}^\infty \frac{r}{m_p(r)}dr = \sum_{k=1}^\infty \int_{2^k r_0}^{2^{k+1} r_0} \frac{r}{m_p(r)}dr \le \sum_{k=1}^\infty \left(\frac{2^{k+1} r_0}{m_p(2^k r_0)} \times 2^k r_0\right)$$

$$= \sum_{k=1}^\infty \frac{2R_k^2}{m_p(R_k)},$$

which implies that $\sum_{k=1}^\infty \frac{(R_k)^2}{m_p(R_k)} = \infty$ from (7.3.1). Hence, let $n \to \infty$ in (7.3.12), we get $\int_{B_{r_0}} F^2(\nabla v)dm \le 0$, which implies that $F^2(\nabla v) = 0$ on $B_{r_0}^+(x_0)$. Since r_0 is arbitrary, we obtain that $F^2(\nabla v) = 0$ on M, which is impossible by the assumption. Thus, u is a constant.

Case 2. $p < 0$. In this case, u is a nonnegative superharmonic function by the assumption. We also assume that $u > 0$ similar to Case 1. Set $v := u^{\frac{p}{2}}$. Then $dv = -|\frac{p}{2}|u^{\frac{p}{2}-1}du$ and $F^2(\nabla v) = F^{*2}(dv) \le \frac{p^2}{4}\Lambda^2 u^{p-2}F^2(\nabla u)$

by Lemma 2.1.1. Thus, we get

$$u^{p-2}F^2(\nabla u) \geq \frac{4}{p^2}\frac{1}{\Lambda^2}F^2(\nabla v).$$

Similarly, from (2.1.1), we have

$$u^{p-2}F^2(\nabla u) \leq \frac{4}{p^2}\Lambda^2 F^2(\nabla v).$$

Let φ be a function as in Case 1. Similar to Case 1, we obtain

$$0 \leq -(p-1)\int_{B_R} d(u^{p-1}\varphi^2)(\nabla u)dm$$

$$= -(p-1)^2\int_{B_R} u^{p-2}\varphi^2 F^2(\nabla u)dm - 2(p-1)\int_{\Omega} u^{p-1}\varphi d\varphi(\nabla u)dm$$

$$\leq -4(1-p^{-1})^2\frac{1}{\Lambda^2}\int_{B_R}\varphi^2 F^2(\nabla v)dm$$

$$+ 4\Lambda\left(\int_{\Omega} v^2 F^{*2}(d\varphi)dm\right)^{\frac{1}{2}}\left((1-p^{-1})^2\int_{\Omega}\Lambda^2\varphi^2 F^2(\nabla v)dm\right)^{\frac{1}{2}},$$

$$(7.3.13)$$

which implies that

$$\left((1-p^{-1})^2\frac{1}{\Lambda^2}\int_{B_R}\varphi^2 F^2(\nabla v)dm\right)^2$$

$$\leq \Lambda^4\left(\int_{\Omega} v^2 F^{*2}(d\varphi)dm\right)\left((1-p^{-1})^2\int_{\Omega}\varphi^2 F^2(\nabla v)dm\right).$$

From this and (7.3.3), we have

$$\left((1-p^{-1})^2\int_{B_R}\varphi^2 F^2(\nabla v)dm\right)^2$$

$$\leq \Lambda^{10}\left(\int_{\Omega} v^2 dm\right)\left((1-p^{-1})^2\int_{\Omega}\varphi^2 F^2(\nabla v)dm\right).$$

The rest follows the same arguments as in Case 1. Consequently, u is a constant.

Case 3. $p = 0$. Let u be a nonnegative superharmonic function and let $m(r)$ satisfy (7.3.1), where $m(r)$ is the volume of B_r. Set $u_k = \min\{u, k\}$,

$k \in \mathbb{R}^+$. For every $k \in \mathbb{R}^+$,

$$u_k = \begin{cases} k, & u \geq k, \\ u, & u < k. \end{cases}$$

Obviously, u_k is differentiable on $\{u > k\}$ and $\{u < k\}$, respectively, and

$$du_k = 0, \quad u > k,$$
$$du_k = du, \quad u < k.$$

Let $E = \{x \in M | u(x) = k\}$, $D^+ = \{x \in E | du \neq 0\}$ and $D^- = \{x \in E | du = 0\}$. Then we claim that $m(D^+) = 0$, where $m(D^+)$ is the measure of D^+ with respect to the measure m. Given a point $x' \in D^+ \subset M$, there is a local coordinate system $(U; x^i)$ in M, such that $x' \in U$. Since $du \neq 0$ on D^+, we can assume $\frac{\partial u}{\partial x^n} \neq 0$ on D^+. Take $z^1 = x^1, z^2 = x^2, \ldots, z^n = u(x) - k$ on U. Then we have

$$\left| \frac{\partial(z^1, z^2, \ldots, z^n)}{\partial(x^1, x^2, \ldots, x^n)} \right| = \frac{\partial u}{\partial x^n} \neq 0.$$

Therefore, $(z^1, z^2, \ldots, z^n)$ can be regarded as a local coordinate system in U and $(z^1, z^2, \ldots, z^n)|_{D^+} = (z^1, z^2, \ldots, z^{n-1}, 0)$. Thus, $\dim(D^+) \leq n - 1$, that is, $m(D^+) = 0$.

Moreover, we claim that $du_k = 0$ on D^-. Take a C^∞ curve $\gamma(t) \subset M$ such that $u(\gamma(0)) = k$. Then

$$\lim_{t \to 0} \left| \frac{\min\{u(\gamma(t)), k\} - \min\{u(\gamma(0)), k\}}{t - 0} \right|$$

$$= \lim_{t \to 0} \left| \frac{\min\{u(\gamma(t)) - k, 0\}}{t} \right|$$

$$= \lim_{t \to 0} \left| \min\left\{ \frac{u(\gamma(t)) - k}{t}, 0 \right\} \right| = 0,$$

where the last equality follows from $du = 0$ on D^-. Thus, we get $du_k = 0$ a.e. on $\{u = k\}$. Therefore, we have $\nabla u_k = \mathcal{L}^{-1}(du_k) = 0$ a.e. on $\{u \geq k\}$ and $\nabla u_k = \mathcal{L}^{-1}(du_k) = \nabla u$ on $\{u < k\}$. Hence, we obtain

$$\int_M d\varphi(\nabla u_k) dm = \int_{\{u<k\}} d\varphi(\nabla u) dm. \tag{7.3.14}$$

Similar to the previous arguments, we get $\nabla u = 0$ a.e. on $\{u = k\}$. Since $\Delta u = 0$ a.e. on $\{M \backslash M_u\}$ from Lemma 7.1.3 and $\Delta u \leq 0$ on M_u, we obtain

$$\int_{\{u<k\}} d\varphi(\nabla u) dm = \int_{\{u=k\}} \varphi \nabla u \, dm - \int_{\{u<k\}} \varphi \Delta u \, dm$$

$$= -\int_{\{u<k\} \cap M_u} \varphi \Delta u \, dm \geq 0. \qquad (7.3.15)$$

Thus, from (7.3.14)–(7.3.15), we have $\int_M d\varphi(\nabla u_k) \geq 0$, that is, u_k is a nonnegative superharmonic function. For any $q > 0$ and $q \neq 1$, by the same arguments as in Case 1, one obtains (7.3.7) for u_k, where p is replaced by q and v is replaced by $v_k := (u_k)^{q/2}$. Note that

$$\int_{B_R} u_k^q dm \leq \int_{B_R} k^q dm = k^q m(R),$$

which implies that (7.3.1) holds for u_k. By Case 1, u_k is a constant, which means that u is a constant since k is an arbitrary nonnegative constant. The proof is finished. □

Remark 7.3.1. In Theorem 7.3.2, we assume that the reversibility Λ is finite for the sake of convenience. In fact, we need not the assumption when $p > 1$. This can be seen from the proof of Theorem 7.3.2. Since $p > 1$, i.e., $1 - p^{-1} > 0$, and $F^*(-d\varphi) = F^*(dr) = 1$, we can use the inequality $-d\varphi(\nabla v) \leq F^*(-d\varphi) F(\nabla v) = F(\nabla v)$ in the second term on the RHS of (7.3.5). Following the same arguments after (7.3.5), we still have $F(\nabla v) = 0$ on $B_{r_0}^+(x_0)$, which implies the conclusion.

Note that the assumption (7.3.1) is satisfied if $\int_M u^p dm < \infty$ or

$$m_p(r) \leq Cr^2 \quad \text{or} \quad m_p(r) \leq Cr^2 \log r,$$

where C is a positive constant. Thus, from Theorem 7.3.2 and Remark 7.3.1, one obtains the following corollary which extends the corresponding results in Riemannian geometry ([SY], [CY]).

Corollary 7.3.1. *Let* (M, F, m) *be an* n*-dimensional forward complete and noncompact Finsler measure space.*

(i) *If* $p \in (-\infty, 1)$ *and* $\Lambda < \infty$, *then every nonnegative superharmonic function with* $\int_M u^p dm < \infty$ *on* M *is a constant. In particular, every nonnegative superharmonic function on* M *with* $m(M) < \infty$ *is a constant.*

(ii) *If $p \in (1, \infty)$, then every nonnegative L^p subharmonic function is a constant.*

In general, Theorem 7.3.2 and Corollary 7.3.1 are not valid for non-negative subharmonic functions with $\int_M u^p dm < \infty$ when $p \leq 1$ or for nonnegative L^p superharmonic functions when $p \geq 1$ without additional assumptions, even in Riemannian case (cf. [LS]). Next we further discuss the Liouville property for nonnegative $L^p (p > 0)$ subharmonic functions as an example. Because of (ii) in Corollary 7.3.1, it suffices to consider the case when $0 < p \leq 1$.

7.3.2 $L^p (0 < p \leq 1)$ *Liouville Theorems*

It is known that $\overleftarrow{\Delta} u = -\Delta(-u)$. Then u is a subharmonic (resp., superharmonic) function of Δ if and only if $-u$ is a superharmonic (resp., subharmonic) function of $\overleftarrow{\Delta}$. However, the "nonnegativity" is not preserved. In the following, we only focus on the Liouville properties for nonnegative L^p subharmonic functions. In particular, we give several different geometric conditions to ensure that the $L^p (0 < p \leq 1)$ Liouville theorem holds ([Xia8]).

Theorem 7.3.3. *Let (M, F, m) be an n-dimensional forward complete Finsler measure space with finite reversibility.*

(i) *If M is noncompact and $Ric_N \geq 0$ for some $N \in [n, \infty)$, then M has infinite volume and every nonnegative $L^p (0 < p \leq 1)$ subharmonic function on M is identically zero.*

(ii) *If $Ric_N \geq -K$ for some constant $K > 0$ and $N \in [n, \infty)$, and the volume of every unit geodesic ball in M has a uniform lower bound by a positive constant, then every nonnegative $L^p (0 < p \leq 1)$ subharmonic function on M is a constant.*

Proof. (i) Since $Ric_N \geq 0$, M has infinite volume by Theorem 6.4.2. Further, the volume of a geodesic ball grows linearly with respect to its radius if F has finite reversibility. From this and Corollary 7.2.1, we get the conclusion by taking the limit $R \to \infty$.

(ii) For any $x \in M$, Theorem 7.2.1 implies that

$$u^p(x) \leq \sup_{B_{1/2}(x)} u^p(x) \leq C \cdot m(B_1(x))^{-1} \int_{B_1(x)} u^p dm,$$

where C is a positive constant independent of x. From this and the assumption, there is a positive constant $\tilde{C}$ such that $\sup_M u \leq \tilde{C}$. This means that

$$\int_M u^2 dm \leq \tilde{C}^{2-p} \int_M u^p dm < \infty,$$

namely, $u \in L^2(M)$. Hence, u is a constant by Corollary 7.3.1. $\quad\square$

Remark 7.3.2. By Theorem 7.2.1 and Remark 7.2.1, Theorem 7.3.3(i) holds for any $p > 0$.

In fact, a weaker hypothesis guarantees the Liouville property of nonnegative $L^p(0 < p \leq 1)$ subharmonic functions, although not every complete manifold enjoys this property.

Theorem 7.3.4. *Let (M, F, m) be a forward complete and noncompact Finsler measure space with finite reversibility Λ. For any $N \in [n, \infty)$, there is a sufficiently small positive constant $\delta = \delta(N, \Lambda)$ depending only on N and Λ such that if*

$$Ric_N \geq -\delta(N, \Lambda)r^{-2}(x) \tag{7.3.16}$$

whenever the distance function $r(x) := d_F(x_0, x)$ from some point x_0 is sufficiently large, then M has infinite volume and every nonnegative $L^p(0 < p \leq 1)$ subharmonic function on (M, F) is identically zero.

Proof. Since u is a nonnegative $L^p(0 < p \leq 1)$ subharmonic function, we may choose $\delta = \frac{1}{2}$ in Theorem 7.2.1 such that

$$\sup_{B_{2^{-1}R}^+(x)} u^p \leq e^{c\left(1+R\sqrt{K(x,5R)}\right)} m(B_R^+(x))^{-1} \int_{B_R^+(x)} u^p dm \tag{7.3.17}$$

for some positive constant $c = c(N, p, \nu, \Lambda)$, where the term $-K(x, 5R)$ denotes the nonpositive lower bound of the weighted Ricci curvature Ric_N on $B_{5R}^+(x)$. Next, we shall utilize (7.3.17) to show that u vanishes at infinity. In the following, we always denote by $-K(x, \bullet)$ the nonpositive lower bound of Ric_N on $B_\bullet^+(x)$ for any $x \in M$.

For any $x \in M$, consider a unit speed minimal geodesic $\gamma : [-t, 0] \to M$ with $\gamma(-t) = x$ and $\gamma(0) = x_0$, where $t = d_F(x, x_0)$. Define a set of values $t_i \in [0, t](0 \leq i \leq k)$ by

$$t_0 = 0, \quad t_1 = 1 + \tau, \quad \ldots, \quad t_i = 2\sum_{j=0}^{i} \tau^j - 1 - \tau^i,$$

where $\tau > 1$ to be chosen later, and k is the largest number for which $t_k \leq t$. Denote $x_i = \gamma(-t_i)$. Then

$$d_F(x_{i+1}, x_i) = d_F(\gamma(-t_{i+1}), \gamma(-t_i)) = t_{i+1} - t_i = \tau^i + \tau^{i+1},$$

$$d_F(x, x_k) < \tau^k + \tau^{k+1}.$$

Obviously, the set of forward geodesic balls $B_{R_i}^+(x_i)$ with $R_i = \tau^i$ covers $\gamma([1 - 2\sum_{j=0}^{k}\tau^j, 0])$. By the triangle inequality and (2.1.2), it is easy to check that

$$B_{\Lambda^{-1}R_{i-1}}^+(x_{i-1}) \subset B_{R_{i-1}}^+(x_{i-1}) \cap B_{R_{i-1}}^-(x_{i-1}) \subset B_{R_i+2R_{i-1}}^+(x_i) \backslash B_{R_i}^+(x_i)$$

$$(7.3.18)$$

for each $0 \leq i \leq k$. We claim that for a fixed $\tau > 2\Lambda/[(e^{-1} + \Lambda^N)^{1/N} - \Lambda] > 1$, there is a positive constant C independent of k such that

$$m(B_{R_k}^+(x_k)) \geq C \left(e\varrho\Lambda^N\right)^{-k} m(B_1^+(x_0)), \qquad (7.3.19)$$

where $\varrho := \frac{(\tau+2)^N - \tau^N}{\tau^N}$. In fact, it follows from Theorem 6.4.1 that

$$m(B_{R_i}^+(x_i)) \geq T_i \left\{ m(B_{R_i+2R_{i-1}}^+(x_i)) - m(B_{R_i}^+(x_i)) \right\},$$

$$\geq T_i \cdot m\left(B_{\Lambda^{-1}R_{i-1}}^+(x_{i-1})\right), \qquad (7.3.20)$$

where the last inequality follows from (7.3.18) and we set

$$T_i = \frac{V_{-K(x_i, R_i+2R_{i-1})/(N-1), N}(R_i)}{\left(\begin{array}{c} V_{-K(x_i, R_i+2R_{i-1})/(N-1), N}(R_i + 2R_{i-1}) \\ -V_{-K(x_i, R_i+2R_{i-1})/(N-1), N}(R_i) \end{array}\right)}.$$

Further, by Proposition 6.4.1, we have

$$m(B_{\Lambda^{-1}R_{i-1}}^+(x_{i-1})) \geq \Lambda^{-N} e^{-R_{i-1}\sqrt{(N-1)K(x_{i-1}, R_{i-1})}} m(B_{R_{i-1}}^+(x_{i-1})).$$

Combining this with (7.3.20) yields

$$m(B_{R_i}^+(x_i)) \geq \Lambda^{-N} T_i \cdot e^{-R_{i-1}\sqrt{(N-1)K(x_{i-1}, R_{i-1})}} m(B_{R_{i-1}}^+(x_{i-1})).$$

Iterating this inequality, we conclude that

$$m(B_{R_k}^+(x_k)) \geq \Lambda^{-kN} \left(\Pi_{i=1}^{k} T_i\right) \cdot e^{-\sum_{i=1}^{k} R_{i-1}\sqrt{(N-1)K(x_{i-1}, R_{i-1})}} m(B_1^+(x_0)).$$

$$(7.3.21)$$

Since $d_F(x_i, x_0) = 2\sum\limits_{j=0}^{i}\tau^j - 1 - \tau^i$ and $R_i = \tau^i$, we have

$$r(x) \geq \Lambda^{-1}d_F(x, x_0) \geq \Lambda^{-1}\{d_F(x_i, x_0) - d_F(x_i, x)\} \geq \Lambda^{-1}\left(2\sum_{j=1}^{i-2}\tau^j - 1\right)$$

for any $x \in B^+_{R_i + 2R_{i-1}}(x_i)$. Thus, the assumption on Ric_N implies that

$$\sqrt{K(x_i, R_i + 2R_{i-1})} \leq \Lambda\delta^{1/2}(N, \Lambda)\left(2\sum_{j=0}^{i-2}\tau^j - 1\right)^{-1}$$

$$= \Lambda\delta^{1/2}(N, \Lambda)\frac{\tau - 1}{2\tau^{i-1} - \tau - 1}$$

for sufficiently large i. For a fixed $\tau > 2\Lambda/[(e^{-1} + \Lambda^N)^{1/N} - \Lambda] > 1$,

$$(R_i + 2R_{i-1})\sqrt{K(x_i, R_i + 2R_{i-1})}$$

$$\leq \Lambda\delta^{1/2}(N, \Lambda)\frac{\tau - 1}{2\tau^{i-1} - \tau - 1}(\tau^i + 2\tau^{i-1}), \qquad (7.3.22)$$

which can be made arbitrarily small by choosing sufficiently small $\delta = \delta(N, \Lambda, \tau)$, where $\tau = \tau(N, \Lambda)$ chosen as above. Hence, T_i has the following approximation:

$$T_i \sim \frac{R_i^N}{(R_i + 2R_{i-1})^N - R_i^N} = \frac{\tau^N}{(\tau + 2)^N - \tau^N} = \varrho^{-1}.$$

Note that $B^+_{R_{i-1}}(x_{i-1}) \subset B^+_{R_i + 2R_{i-1}}(x_i)$. Thus, by choosing a small $\delta(N, \Lambda)$ in (7.3.22), we have

$$R_{i-1}\sqrt{(N-1)K(x_{i-1}, R_{i-1})}$$

$$\leq (R_i + 2R_{i-1})\sqrt{(N-1)K(x_i, R_i + 2R_{i-1})} \leq 1.$$

Combining these with (7.3.21) gives (7.3.19). Now, we estimate $m(B^+_R(x))$ for some $R < R_k$. We argue this according to two cases.

Case 1. $d_F(x, x_k) \leq \frac{1}{10}R_k$. In this case, we have $B^+_{R_k/10}(x_k) \subset B^+_{R_k/5}(x)$ and hence

$$m(B^+_{R_k/5}(x)) \geq m(B^+_{R_k/10}(x_k)) \geq 10^{-N}e^{-R_k\sqrt{(N-1)K(x_k, R_k)}}m(B^+_{R_k}(x_k)).$$

$$(7.3.23)$$

Note that $R_k\sqrt{K(x_k, R_k)}$ is bounded from above as above. Thus, from (7.3.23) and (7.3.19), one obtains

$$m(B^+_{R_k/5}(x)) \geq c_1\left(e\varrho\Lambda^N\right)^{-k}m(B^+_1(x_0)), \qquad (7.3.24)$$

where $c_1 = c_1(p, \nu, N, \Lambda)$ is a positive constant. Applying (7.3.17) to $B^+_{R_k/10}(x)$ and using (7.3.24) yield

$$u^p(x) \leq \sup_{B^+_{R_k/10}(x)} u^p \leq c_2 \left(e\varrho\Lambda^N\right)^k m(B^+_1(x_0))^{-1} \int_{B^+_{R_k/5}(x_k)} u^p dm$$

for some positive constant $c_2 := c_2(N, \Lambda)$, which implies that $u \to 0$ as $x \to \infty$ (i.e., $k \to \infty$) since $e\varrho\Lambda^N < 1$ by the choice of τ.

Case 2. $d_F(x, x_k) \geq \frac{1}{10}R_k$. In this setting, we have

$$B^+_{R_k/(20\Lambda)}(x_k) \subset B^+_{R_k/20}(x_k) \cap B^-_{R_k/20}(x_k) \subset B^+_{R_{k+1}+21R_k/20}(x) \backslash B^+_{R_k/20}(x).$$

By arguing as above, one obtains

$$m(B^+_{R_k/20}(x)) \geq T\left\{m(B^+_{R_{k+1}+21R_k/20}(x)) - m(B^+_{R_k/20}(x))\right\}$$

$$\geq T \cdot (20\Lambda)^{-N} e^{-R_k\sqrt{(N-1)K(x_k,R_k)}} m(B^+_{R_k}(x_k)), \quad (7.3.25)$$

where

$$T := \frac{V_{-K(x,R_{k+1}+21R_k/20)/(N-1),N}(R_k/20)}{\left(\begin{array}{c} V_{-K(x,R_{k+1}+21R_k/20)/(N-1),N}(R_{k+1}+21R_k/20) \\ -V_{-K(x,R_{k+1}+21R_k/20)/(N-1),N}(R_k/20) \end{array}\right)}.$$

Since $(R_{k+1} + 21R_k/20)\sqrt{K(x, R_{k+1} + 21R_k/20)} \leq (R_{k+1} + 2R_k)$ $\sqrt{K(x, R_{k+1} + 2R_k)}$, which can be made sufficiently small by the choice of the above δ. Thus, we can approximate by

$$T \sim \frac{1}{(20\tau + 21)^N - 1},$$

which implies that T is bounded from below by a constant only depending on Λ, N. Combining (7.3.25) and (7.3.19) gives

$$m(B^+_{R_k/20}(x)) \geq c_3 \left(e\varrho\Lambda^N\right)^{-k} m(B^+_1(x_0)), \quad (7.3.26)$$

where $c_3 = c_3(N, \Lambda)$ is a positive constant. Applying (7.3.17) to $B^+_{R_k/40}(x)$ and using (7.3.26) yield

$$u^p(x) \leq \sup_{B^+_{R_k/40}(x)} u^p \leq c_4 \left(e\varrho\Lambda^N\right)^k m(B^+_1(x_0))^{-1} \int_{B^+_{R_k/20}(x_k)} u^p dm,$$

$$(7.3.27)$$

where $c_4 := c_4(p, \nu, N, \Lambda)$ is a positive constant. Thus, $u \to 0$ as $x \to \infty$ (i.e., $k \to \infty$) by the choice of τ. Consequently, u vanishes at infinity in any case. This means that $u \in L^p(M) \cap L^\infty(M)$ and hence $u \in L^2(M)$. So, u is a constant. Moreover, it is easy to see from (7.3.24) and (7.3.26) that the volume of M is infinite by taking $k \to \infty$, and $R\sqrt{K(x, 5R)}$ is bounded from above. Hence, $u \equiv 0$ by taking $R \to \infty$ in (7.3.17). □

7.4 Gradient Estimates

In this section, we shall derive the local and global estimates for harmonic functions on a complete Finsler measure space (M, F, m). As applications, one obtains the L^∞-Liouville and Harnack properties for Finsler harmonic functions.

7.4.1 *Compact Case*

First of all, we prove the following lemma as a preparation.

Lemma 7.4.1. *Let (M, F, m) be an $n(\geq 2)$-dimensional Finsler measure space. Then, for any C^2 harmonic function u and a real number $N \in [n, \infty)$, the following inequality:*

$$F(\nabla u)\boldsymbol{\Delta}^{\nabla u}(F(\nabla u)) \geq \frac{1}{N-1}g_{\nabla u}\left(\nabla^{\nabla u}F(\nabla u), \nabla^{\nabla u}F(\nabla u)\right) + Ric_N(\nabla u)$$

$$(7.4.1)$$

holds on M_u.

Proof. For any $u \in C^2(M)$, we have on M_u

$$\frac{1}{2}\boldsymbol{\Delta}^{\nabla u}(F(\nabla u)^2) = F(\nabla u)\boldsymbol{\Delta}^{\nabla u}F(\nabla u) + g_{\nabla u}(\nabla^{\nabla u}F(\nabla u), \nabla^{\nabla u}F(\nabla u)).$$

$$(7.4.2)$$

On the other hand, by Theorem 7.1.1 and $\boldsymbol{\Delta}u = 0$, one obtains

$$\frac{1}{2}\boldsymbol{\Delta}^{\nabla u}(F(\nabla u)^2) = Ric_\infty(\nabla u) + \|\nabla^2 u\|_{HS(\nabla u)}^2 \qquad (7.4.3)$$

on M_u. Combining (7.4.2) with (7.4.3) yields

$$F(\nabla u)\boldsymbol{\Delta}^{\nabla u}(F(\nabla u))$$

$$= \|\nabla^2 u\|_{HS(\nabla u)}^2 + Ric_\infty(\nabla u) - g_{\nabla u}(\nabla^{\nabla u}F(\nabla u), \nabla^{\nabla u}F(\nabla u)). \quad (7.4.4)$$

For any $x \in M_u$, choose a local orthonormal frame $\{e_i\}_{i=1}^n$ with respect to $g_{\nabla u}$ such that $\nabla u = u_1 e_1$, where $u_1 = F(\nabla u) > 0$.

Then $\nabla^{\nabla u} F(\nabla u) = \sum_{j=1}^{n} u_{1;j} e_j$, where ";" is the covariant derivative with respect to the Levi-Civita connection of $g_{\nabla u}$. Then,

$$\|\nabla^2 u\|_{HS(\nabla u)}^2 - g_{\nabla u}(\nabla^{\nabla u} F(\nabla u), \nabla^{\nabla u} F(\nabla u))$$

$$= \sum_{i,j=1}^{n} u_{i;j}^2 - \sum_{j=1}^{n} u_{1;j}^2 \geq \sum_{i=2}^{n} u_{1;i}^2 + \sum_{i=2}^{n} u_{i;i}^2$$

$$\geq \sum_{i=2}^{n} u_{1;i}^2 + \frac{1}{n-1} \left(\sum_{i=2}^{n} u_{i;i} \right)^2, \tag{7.4.5}$$

where we used $u_{i;j} = u_{j;i}$ $(1 \leq i,j \leq n)$ from the symmetry of $\nabla^2 u$. Moreover, by Lemma 6.1.1 and $\Delta u = 0$, we have $\sum_{i=2}^{n} u_{i;i} = -u_{1;1} + S(\nabla u)$. Plugging this into (7.4.5) and using the inequality $(a+b)^2 \geq \frac{a^2}{1+\delta} - \frac{b^2}{\delta}$ with $\delta = \frac{N-n}{n-1}$ yield

$$\|\nabla^2 u\|_{HS(\nabla u)}^2 - g_{\nabla u}(\nabla^{\nabla u} F(\nabla u), \nabla^{\nabla u} F(\nabla u))$$

$$\geq \frac{1}{N-1} \sum_{j=1}^{n} u_{1;j}^2 - \frac{S(\nabla u)^2}{N-n}. \tag{7.4.6}$$

Combining (7.4.4) with (7.4.6) yields (7.4.1).   □

Assume that (M, F) is a Finsler manifold with a smooth boundary ∂M. For any $x \in \partial M$, there exist exactly two normal vectors ν with $g_\nu(\nu, \nu) = 1$ such that

$$T_x(\partial M) = \{X \in T_x M | g_\nu(\nu, X) = 0\}$$

(see Section 2.4). Let γ be the geodesic of $F_{\partial M}$ on ∂M with $\gamma(0) = x$ and $\dot{\gamma}(0) = X \in T_x(\partial M)$. Define $\Lambda_\nu(X) := g_\nu(\nu, D_{\dot{\gamma}}^{\dot{\gamma}} \dot{\gamma}(0))$, which is called the *normal curvature* of ∂M at x in the direction X. ∂M is said to be *convex* if Λ_ν is nonpositive for any $(x, X) \in T(\partial M)$. We remark that the convexity of M means that $D_{\dot{\gamma}}^{\dot{\gamma}} \dot{\gamma}(0)$ lies at the same side of $T_x M$ for any $x \in M$. Hence the choice of the normal vector is not essential for the definition of convexity. Based on Lemma 7.4.1, one obtains

Theorem 7.4.1. *Let (M, F, m) be an n-dimensional compact Finsler measure space without boundary or with a convex boundary. Assume that $Ric_N \geq K$ for some real numbers K and $N \in [n, \infty)$ and u is a harmonic function on M with a Neumann boundary condition, i.e., $\nabla u \in T(\partial M)$ if*

$\partial M \neq \emptyset$. *Then $K \leq 0$ and*

$$F(x, \nabla u) \leq \sqrt{-(N-1)K}(u - \inf_M u). \qquad (7.4.7)$$

In particular, if $Ric_N \geq 0$, then u is a constant.

Proof. Note that Ric_N is always bounded on a compact Finsler manifold. Without loss of generality, assume that u is positive. Otherwise, we can replace u by $u - \inf_M u + \varepsilon$, where ε is a sufficient small positive number. Let

$$f(x) := F(\nabla \log u) = \frac{1}{u}F(\nabla u).$$

Since M is compact, f attains its maximum at some point x_0 in the closure of M.

Case 1. $x_0 \in \text{Int}(M)$. If $f(x_0) = 0$, we have $F(\nabla u) = f(x) = 0$ on M and (7.4.7) holds obviously. If $f(x_0) \neq 0$, then $F(\nabla u)(x_0) \neq 0$, and hence $x_0 \in M_u$. Thus, ∇u and f are smooth around x_0.

Applying the maximal principle to $f(x)$ on $(M, g_{\nabla u})$ yields

$$\nabla^{\nabla u} f(x_0) = 0, \quad \mathbf{\Delta}^{\nabla u} f(x_0) \leq 0.$$

For any $x \in M_u$, we have

$$\nabla^{\nabla u} f = u^{-1}\nabla^{\nabla u} F(\nabla u) - u^{-2}F(\nabla u)\nabla u$$

and

$$\mathbf{\Delta}^{\nabla u} f = u^{-1}\mathbf{\Delta}^{\nabla u} F(\nabla u) - 2u^{-2}g_{\nabla u}(\nabla^{\nabla u} F(\nabla u), \nabla u) - u^{-2}F(\nabla u)\mathbf{\Delta} u$$
$$+ 2u^{-3}F^3(\nabla u).$$

In particular, at x_0, we have

$$\nabla^{\nabla u} F(\nabla u) = f\nabla u \qquad (7.4.8)$$

and

$$\mathbf{\Delta}^{\nabla u} f = u^{-1}\mathbf{\Delta}^{\nabla u} F(\nabla u) - 2u^{-2}g_{\nabla u}(\nabla^{\nabla u} F(\nabla u), \nabla u) + 2f^3 \leq 0. \qquad (7.4.9)$$

Multiplying $F(\nabla u)$ on both sides of (7.4.9) and using Lemma 7.4.1 yield

$$[(N-1)u]^{-1} g_{\nabla u}(\nabla^{\nabla u} F(\nabla u), \nabla^{\nabla u} F(\nabla u)) + u^{-1}KF(\nabla u)^2$$
$$- 2u^{-2}g_{\nabla u}(\nabla^{\nabla u} F(\nabla u), \nabla u)F(\nabla u) + 2f^3 F(\nabla u) \leq 0. \qquad (7.4.10)$$

By (7.4.8) and (7.4.10), one obtains

$$\frac{1}{N-1}f^2(x_0) + K \leq 0. \tag{7.4.11}$$

From this, we can see that $K \leq 0$ and $f(x_0) \leq \sqrt{-(N-1)K}$, which implies that $F(\nabla u) \leq \sqrt{-(N-1)K}u$ for any $x \in M_u$. Finally, replacing u by $u - \inf_M u + \varepsilon$ and letting $\varepsilon \to 0$ yield (7.4.7).

Case 2. $x_0 \in \partial M$ if $\partial M \neq \emptyset$. We may assume that $x_0 \in M_u$ as in Case 1. Otherwise nothing needs to be done. Let ν be a normal vector field of ∂M that points outward. By Proposition 2.4.1, there is another outward normal vector field $\nu_{\nabla u}$ satisfying (2.4.4). Consequently, $df(\nu_{\nabla u}) \geq 0$ near x_0. On the other hand, the Neumann boundary condition implies that $g_{\nabla u}(\nu_{\nabla u}, \nabla u) = 0$, namely, $du(\nu_{\nabla u}) = 0$. Therefore, by (6.1.8), we have

$$df(\nu_{\nabla u}) = \frac{1}{2uF(\nabla u)}d(g_{\nabla u}(\nabla u, \nabla u))(\nu_{\nabla u})$$

$$= \frac{1}{uF(\nabla u)}g_{\nabla u}(\nabla u, D^{\nabla u}_{\nu_{\nabla u}}\nabla u)$$

$$= \frac{1}{uF(\nabla u)}g_{\nabla u}(\nu_{\nabla u}, D^{\nabla u}_{\nabla u}\nabla u).$$

Since ∂M is convex, we have $g_{\nu}\left(\nu, D^{\nabla u}_{\nabla u}\nabla u\right) \leq 0$, equivalently, $g_{\nabla u}\left(\nabla u, D^{\nabla u}_{\nabla u}\nabla u\right) \leq 0$ by Proposition 2.4.1. Thus, $df(\nu_{\nabla u}) \leq 0$ and hence $df(\nu_{\nabla u}) = 0$ at x_0. Obviously, df vanishes along the tangent vector of ∂M because of its maximality. So, $df(x_0) = 0$. It is obvious that $\mathbf{\Delta}^{\nabla u}f(x_0) \leq 0$. The rest proof follows that of Case 1. This finishes the proof. $\square$

7.4.2 *Complete and Noncompact Case*

For a forward complete and noncompact Finsler manifold (M, F), the maximum principle does not work because of the nonlinearity of Finsler Laplacian. We will use Moser's iteration to derive the local gradient estimate for Finsler harmonic functions. In this case, we need assume that F is uniformly smooth and uniformly convex, i.e., there exist two uniform positive constants κ^*, κ such that F satisfies (2.1.4), which implies that the reversibility $\Lambda < \infty$ and F^* is also uniformly smooth and convex by Lemma 2.1.2.

Theorem 7.4.2 ([Xc]). *Let (M, F, m) be an $n(\geq 2)$-dimensional forward complete and noncompact Finsler measure space equipped with a uniformly* *convex and uniformly smooth Finsler metric F. Assume that $Ric_N \geq -K$ for some $N \in [n, \infty)$ and $K \geq 0$. Let u be a positive harmonic function in a forward geodesic ball $B_{2R}^+(x_0)$. Then there exists a positive constant $C = C(N, \kappa, \kappa^*)$ depending on N, κ and κ^*, such that*

$$\sup_{x \in B_R^+(x_0)} \{F(x, \nabla \log u(x)), F(x, \nabla(-\log u(x)))\} \leq C\frac{1 + \sqrt{K}R}{R}. \quad (7.4.12)$$

In particular, $F(x, \nabla \log u(x))$ and $F(x, \nabla(-\log u(x)))$ are bounded on M.

Proof. Since $\Delta u = 0$ on $B_{2R} := B_{2R}^+(x_0)$, we have $u \in C^{1,\alpha}(B_{2R}) \cap W_{loc}^{2,2}(B_{2R})$ (see Section 7.2). Denote $v = \log u$. Then $\nabla v = \frac{1}{u}\nabla u$ and $\Delta v = -F^2(x, \nabla v)$ on $M_v = M_u$.

Let $f(x) := F^2(x, \nabla v)$. Then $f \in W_{loc}^{1,2}(B_{2R}) \cap C^\alpha(B_{2R})$. Moreover, f is smooth on $M_v \cap B_{2R}$. It follows from Theorem 7.1.2 that for $\phi \in W_0^{1,2}(B_{2R}) \cap L^\infty(B_{2R})$,

$$\int_M d\phi(\nabla^{\nabla v}f)dm \leq \int_M \phi\left(2df(\nabla v) - 2Ric_N(\nabla v) - \frac{2f^2}{N}\right)dm. \quad (7.4.13)$$

Note that, it follows from Lemma 7.1.3 that $\nabla^{\nabla v}f = 0$ a.e. on $f^{-1}(0) = M \backslash M_v$. Therefore the both sides of (7.4.13) is actually integrated over $M_v \cap B_{2R}$.

To extend the above integral inequality to B_{2R}, we consider the function $f_\varepsilon = (f - \varepsilon)^+$ for $\varepsilon > 0$ and the nonnegative function $\eta \in C_0^\infty(B_{2R}) \cap M_v$ with $0 \leq \eta \leq 1$. Note that $f_\varepsilon = f - \varepsilon, df_\varepsilon = df$ a.e. in $\{f > \varepsilon\}$ and $f_\varepsilon = 0, df_\varepsilon = 0$ a.e. in $\{f \leq \varepsilon\}$. Choose $\phi = f_\varepsilon^t \eta^2$ as the test function in (7.4.13), where $t > 1$ is to be determined later. Then we have

$$\int_{B_{2R} \cap \{f > \varepsilon\}} \{t\eta^2 f_\varepsilon^{t-1}df(\nabla^{\nabla v}f)dm + 2\eta f_\varepsilon^t d\eta(\nabla^{\nabla v}f)\} dm$$

$$\leq \int_{B_{2R} \cap \{f > \varepsilon\}} \eta^2 f_\varepsilon^t \left\{2df(\nabla v) + 2Kf - \frac{2f^2}{N}\right\} dm. \quad (7.4.14)$$

Since F satisfies (2.1.4), one obtains from (2.1.6) that

$$\tilde{\kappa}^* F^2(x, \nabla f) \leq df(\nabla^{\nabla v}f) = g^{ij}(x, \nabla v)f_i f_j \leq \tilde{\kappa}F^2(x, \nabla f). \quad (7.4.15)$$

From this and the Cauchy–Schwarz inequality, we get

$$|d\eta(\nabla^{\nabla v}f)| = |df(\nabla^{\nabla v}\eta)| \leq \tilde{\kappa}F(x, \nabla\eta)F(x, \nabla f). \quad (7.4.16)$$

Plugging (7.4.15)–(7.4.16) into (7.4.14) and passing ε to 0 yield

$$\tilde{\kappa}^* t \int_{B_{2R}} \eta^2 f^{t-1} F^2(\nabla f) dm - 2\tilde{\kappa} \int_{B_{2R}} \eta f^t F(\nabla f) F(\nabla \eta) dm$$

$$\leq 2 \int_{B_{2R}} \eta^2 f^{t+1/2} F(\nabla f) dm + 2K \int_{B_{2R}} \eta^2 f^{t+1} dm - \frac{2}{N} \int_{B_{2R}} \eta^2 f^{t+2} dm,$$

(7.4.17)

where we used Cauchy–Schwarz's inequality $df(\nabla v) \leq f^{1/2} F(\nabla f)$. Note that

$$2\tilde{\kappa} \int_{B_{2R}} \eta f^t F(\nabla f) F(\nabla \eta) dm$$

$$\leq \frac{\tilde{\kappa}^* t}{2} \int_{B_{2R}} \eta^2 f^{t-1} F^2(\nabla f) dm + \frac{2\tilde{\kappa}^2}{\tilde{\kappa}^* t} \int_{B_{2R}} f^{t+1} F^2(\nabla \eta) dm \quad (7.4.18)$$

and the first term of the RHS of (7.4.17) is less than or equal to

$$\frac{\tilde{\kappa}^* t}{4} \int_{B_{2R}} \eta^2 f^{t-1} F^2(\nabla f) dm + \frac{4}{\tilde{\kappa}^* t} \int_{B_{2R}} \eta^2 f^{t+2} dm$$

$$\leq \frac{\tilde{\kappa}^* t}{4} \int_{B_{2R}} \eta^2 f^{t-1} F^2(\nabla f) dm + \frac{1}{N} \int_{B_{2R}} \eta^2 f^{t+2} dm \quad (7.4.19)$$

if we take $t \geq \max\{1, 4N/\tilde{\kappa}^*\}$. Hence, for $t > \max\{1, 4N/\tilde{\kappa}^*\} \geq 1$, it follows from (7.4.17)–(7.4.19) that

$$\frac{1}{4} \tilde{\kappa}^* t \int_{B_{2R}} \eta^2 f^{t-1} F^2(\nabla f) dm$$

$$\leq \frac{2\tilde{\kappa}^2}{\tilde{\kappa}^* t} \int_{B_{2R}} f^{t+1} F^2(\nabla \eta) dm + 2K \int_{B_{2R}} \eta^2 f^{t+1} dm$$

$$- \frac{1}{N} \int_{B_{2R}} \eta^2 f^{t+2} dm. \quad (7.4.20)$$

Recall that $F(\nabla f) = F^*(df)$ and $F^*(\xi + \eta) \leq F^*(\xi) + F^*(\eta)$ (see Lemma 1.1.2). By (7.4.20), there exist positive constants $c_i = c_i(\tilde{\kappa}, \tilde{\kappa}^*) > 1$ for $i = 1, 2$ and $c_3 = c_3(\tilde{\kappa}, \tilde{\kappa}^*, N)$ depending on $\tilde{\kappa}, \tilde{\kappa}^*, N$ such that for $t > \max\{1, 4N/\tilde{\kappa}^*\}$ large enough,

$$\int_{B_{2R}} F^{*2} \left(d(\eta f^{(t+1)/2}) \right) dm$$

$$\leq \int_{B_{2R}} f^{t+1} F^{*2}(d\eta) dm + \frac{(t+1)^2}{4} \int_{B_{2R}} \eta^2 f^{t-1} F^{*2}(df) dm$$

$$\leq c_1 \int_{B_{2R}} f^{t+1} F^{*2}(d\eta) dm + c_2 Kt \int_{B_{2R}} \eta^2 f^{t+1} dm$$

$$- c_3 t \int_{B_{2R}} \eta^2 f^{t+2} dm. \tag{7.4.21}$$

Next, we shall estimate the local L^t norm of f based on this inequality.

Let $\chi := \frac{\nu}{\nu-2}$. Taking $u = \eta f^{(t+1)/2}$ in (7.2.6) and using (7.4.21), one obtains

$$\left(\int_{B_{2R}} \eta^{2\chi} f^{\chi(t+1)} dm \right)^{\frac{1}{\chi}}$$

$$\leq 4 e^{c_0(1+2\sqrt{K}R)} R^2 m(B_{2R})^{-\frac{2}{\nu}} \left\{ \int_{B_{2R}} F^{*2} \left(d(\eta f^{(t+1)/2}) \right) dm \right.$$

$$\left. + \frac{1}{4} R^{-2} \int_{B_{2R}} \eta^2 f^{t+1} dm \right\}$$

$$\leq e^{2c_0(1+\sqrt{K}R)} m(B_{2R})^{-\frac{2}{\nu}} \left\{ c_1 R^2 \int_{B_{2R}} f^{t+1} F^{*2}(d\eta) dm - c_3 t R^2 \right.$$

$$\times \int_{B_{2R}} \eta^2 f^{t+2} dm + c_2 t(1+\sqrt{K}R)^2 \int_{B_{2R}} \eta^2 f^{t+1} dm \right\}. \tag{7.4.22}$$

Choosing $t = t_0 = c_0(1 + \sqrt{K}R)$ in (7.4.22), we have

$$\left(\int_{B_{2R}} \eta^{2\chi} f^{\chi(t_0+1)} dm \right)^{\frac{1}{\chi}}$$

$$\leq e^{2t_0} m(B_{2R})^{-\frac{2}{\nu}} \left\{ c_1 R^2 \int_{B_{2R}} f^{t_0+1} F^{*2}(d\eta) dm \right.$$

$$\left. - c_3 t R^2 \int_{B_{2R}} \eta^2 f^{t_0+2} dm + c_4 t_0^3 \int_{B_{2R}} \eta^2 f^{t_0+1} dm \right\},$$

$$\tag{7.4.23}$$

where $c_4 := c_2/c_0^2$. Note that all $c_i = c_i(\kappa, \kappa^*, N)(1 \leq i \leq 4)$ only depending on κ, κ^* and N by the definitions of $\tilde{\kappa}, \tilde{\kappa}^*$. In the following, we always denote the positive constants c_i as $c_i = c_i(\kappa, \kappa^*, N)(i \geq 1)$ depending on κ, κ^*, N unless otherwise specified. Now we decompose the region B_{2R} into two subregions, one is $\{f \geq \frac{2c_4}{c_3}(\frac{t_0}{R})^2\}$, another is the complement of the first subregion in B_{2R}. Thus,

$$c_4 t_0^3 \int_{B_{2R}} \eta^2 f^{t_0+1} dm$$

$$\leq \frac{1}{2} c_3 t_0 R^2 \int_{B_{2R}} \eta^2 f^{t_0+2} dm + c_5^{t_0+1} t_0^3 \left(\frac{t_0}{R}\right)^{2(t_0+1)} \cdot m(B_{2R}). \quad (7.4.24)$$

For the first term of the RHS in (7.4.23), we let $\eta = \psi^{t_0+2}$ with $\psi(z) = \tilde{\psi}(d_F(x,z)) \in C_0^\infty(B_{2R})$ satisfying

$$0 \leq \psi \leq 1, \quad \psi = 1 \quad \text{in} \quad [0, 3R/2), \quad |\psi'| \leq \frac{c_6}{R}.$$

Note that $F^*(d(d_F(x,\cdot))) = 1$ a.e. in B_{2R}. Thus, ψ satisfies $F^*(d\psi) \leq \frac{c_6 \Lambda}{R}$. Since F is uniformly convex and smooth, we have $1 \leq \Lambda \leq \min\{\sqrt{\kappa}, \sqrt{1/\kappa^*}\}$. Hence, $F^*(d\psi) \leq \frac{c_6'}{R}$, where $c_6' = c_6'(\kappa, \kappa^*)$, and $c_1 R^2 F^{*2}(d\eta) \leq c_7 t_0^2 \eta^{\frac{2(t_0+1)}{t_0+2}}$ for some positive constant $c_7 = c_7(\kappa, \kappa^*)$. By Hölder's and Young's inequalities, one obtains

$$c_1 R^2 \int_{B_{2R}} f^{t_0+1} F^{*2}(d\eta) dm$$

$$\leq c_7 t_0^2 \int_{B_{2R}} f^{t_0+1} \eta^{\frac{2(t_0+1)}{t_0+2}} dm$$

$$\leq c_7 t_0^2 \left(\int_{B_{2R}} \eta^2 f^{t_0+2} dm\right)^{\frac{t_0+1}{t_0+2}} \cdot m(B_{2R})^{\frac{1}{t_0+2}}$$

$$\leq \frac{1}{2} c_3 t_0 R^2 \int_{B_{2R}} \eta^2 f^{t_0+2} dm + c_8^{t_0+1} t_0^{t_0+3} R^{-(2(t_0+1))} m(B_{2R}).$$

Plugging the above inequality and (7.4.24) into (7.4.23) yields

$$\left(\int_{B_R} \eta^{2\chi} f^{\chi(t_0+1)} dm\right)^{\frac{1}{\chi}}$$

$$\leq e^{2t_0} m(B_{2R})^{1-\frac{2}{\nu}} \left\{ c_5^{t_0+1} t_0^3 \left(\frac{t_0}{R}\right)^{2(t_0+1)} + c_8^{t_0+1} t_0^{t_0+3} R^{-(2(t_0+1))} \right\}$$

$$\leq c_9^{t_0+1} e^{2t_0} t_0^3 \left(\frac{t_0}{R}\right)^{2(t_0+1)} m(B_{2R})^{1-\frac{2}{\nu}}.$$

Note that $t_0^{\frac{3}{t_0+1}} \le e^3$. Taking the $\frac{1}{t_0+1}$-th power on both sides of the above inequality and letting $t_1 = \chi(t_0 + 1)$ yield

$$\|f\|_{L^{t_1}(B_{\frac{3}{2}R})} \le c_{10} \left(\frac{t_0}{R}\right)^2 m(B_{2R})^{\frac{1}{t_1}},$$

which implies that $f \in L^{t_1}(B_{\frac{3}{2}R})$ with

$$\|f\|_{L^{t_1}(B_{\frac{3}{2}R})} \le \frac{c(1+\sqrt{K}R)^2}{R^2} m(B_{2R})^{\frac{1}{t_1}}, \qquad (7.4.25)$$

where $c := c(\kappa, \kappa^*, N) > 0$.

Finally, we derive our desired gradient estimate by Moser's iteration with (7.4.25). In fact, by (7.4.22), we have

$$\left(\int_{B_{2R}} \eta^{2\chi} f^{\chi(t+1)} dm\right)^{\frac{1}{\chi}} \le e^{2t_0} m(B_{2R})^{-\frac{2}{\nu}} \left\{c_1 R^2 \int_{B_{2R}} f^{t+1} F^{*2}(d\eta) dm\right.$$

$$\left. + c_4 t_0^2 t \int_{B_{2R}} \eta^2 f^{t+1} dm\right\}. \qquad (7.4.26)$$

Let t_0, t_1 be chosen as above and $t_{k+1} = \chi t_k$. Moreover, we choose $R_k = R + \frac{R}{2^k}$ and $\eta_k \in C_0^\infty(B_{R_k})$ satisfying

$$0 \le \eta_k \le 1, \quad \eta_k = 1 \quad \text{in} \quad B_{R_{k+1}}, \quad F^*(x, d\eta_k) \le \tilde{c}\frac{2^k}{R}, \quad k = 1, 2, \ldots,$$

where $\tilde{c}$ is a certain positive constant depending on the reversibility Λ on $B_{3R/2}$. Denote $c_{11} = \max\{c_1\tilde{c}^2, c_4\}$. Taking $t+1 = t_k, \eta = \eta_k$ in (7.4.26), one obtains

$$\|f\|_{L^{t_{k+1}}(B_{R_{k+1}})} \le \left(c_{11}e^{2t_0}\right)^{\frac{1}{t_k}} m(B_{2R})^{-\frac{2}{\nu t_k}} (4^k + t_0^2 t_k)^{\frac{1}{t_k}} \|f\|_{L^{t_k}(B_{R_k})}$$

$$= \left(c_{11}e^{2t_0}\right)^{\frac{1}{t_k}} m(B_{2R})^{-\frac{2}{\nu t_k}} \left(4^k + t_0^2 \chi^{k-1} t_1\right)^{\frac{1}{t_k}} \|f\|_{L^{t_k}(B_{R_k})}.$$

Note that $\sum_k \frac{1}{t_k} = \frac{\nu}{2t_1}$ and $\sum_k \frac{k}{t_k}$ converges. By the standard Moser's iteration and using (7.4.25), we get

$$\|F^2(x, \nabla v)\|_{L^\infty(B_R)}$$

$$= \|f\|_{L^\infty(B_R)} \le c_{12} \left(c_{11}e^{2t_0}\right)^{\sum_k \frac{1}{t_k}} m(B_{2R})^{-\frac{2}{\nu}\sum_k \frac{1}{t_k}} \left(t_0^3\right)^{\sum_k \frac{1}{t_k}} \|f\|_{L^{t_1}(B_{R_1})}$$

$$\le C\frac{(1+\sqrt{K}R)^2}{R^2},$$

which implies that

$$\|F(x, \nabla(\log u))\|_{L^\infty(B_R)} \leq C\frac{1 + \sqrt{K}R}{R}.$$

The previous arguments also work if we use $-v$ instead of v. Thus one obtains the same upper bound for $\|F(x, \nabla(-v))\|_{L^\infty(B_R)}$. □

Several standard applications of Theorem 7.4.2 yield the Harnack inequality and the Liouville properties.

Corollary 7.4.1. *Let* (M, F, m), Ric_N *be as in Theorem 7.4.2 and* u *be a positive harmonic function in a forward geodesic ball* $B_{2R}^+(x_0) \subset M$ *for any* $x_0 \in M$. *Then there exists a positive constant* $C = C(N, \kappa, \kappa^*)$ *such that*

$$\sup_{x \in B_R^+(x_0)} u(x) \leq e^{C(1+\sqrt{K}R)} \inf_{x \in B_R^+(x_0)} u(x).$$

If $Ric_N \geq 0$ *in* $B_{2R}^+(x_0)$, *then there exists a uniform positive constant* $c = c(N, \kappa, \kappa^*)$ *such that*

$$\sup_{x \in B_R^+(x_0)} u(x) \leq c \inf_{x \in B_R^+(x_0)} u(x).$$

Proof. Let $x_1, x_2 \in B_R^+(x_0)$ such that $u(x_1) = \min_{B_R^+(x_0)} u$ and $u(x_2) = \max_{B_R^+(x_0)} u$, and $\gamma : [0, 1] \to M$ be a minimizing geodesic from x_1 to x_2. Then, by Theorem 7.4.2, we have

$$\log u(x_2) - \log u(x_1) = \int_0^1 \frac{d}{dt} \log u(\gamma(t))dt = \int_0^1 \frac{1}{u}du(\dot\gamma)dt$$

$$\leq \int_0^1 F(\nabla \log u)F(\dot\gamma)dt \leq C(1 + \sqrt{K}R),$$

which implies the conclusion. □

Corollary 7.4.2. *Let* (M, F, m) *be as in Theorem 7.4.2. Assume that* $\partial M = \emptyset$ *and* $Ric_N \geq 0$. *Then*

(i) *any harmonic function bounded from below on* (M, F, m) *must be a constant. In particular, any positive harmonic function on* (M, F, m) *must be a constant.*

(ii) *if* u *be a sublinear growth harmonic function, i.e.,* $|u(x)| = o(r(x))$, *where* $r(x) = d_F(x_0, x)$ *for some* $x_0 \in M$, *then* u *is a constant.*

Proof. Let $u(x)$ be a harmonic function and $u(x) > k$ for some constant k. Then $\tilde{u} = u - k$ is a positive harmonic function. Thus, (i) follows directly from Theorem 7.4.2 by letting $R \to \infty$.

(ii) For any $R > 0$, let $v := u + \sup_{B_{2R}^+(x_0)} |u| > 0$. Then $\nabla v = \nabla u$ and v is a harmonic function (in the weak sense) in $B_{2R}^+(x_0)$. Hence, Theorem 7.4.2 implies that

$$\sup_{x \in B_R^+(x_0)} \left\{ F(x, \nabla \log v(x)), F(x, \nabla(-\log v(x))) \right\} \leq CR^{-1}.$$

In particular,

$$\sup_{B_R^+(x_0)} F(\nabla u) = \sup_{B_R^+(x_0)} F(\nabla v) \leq CR^{-1} \sup_{B_R^+(x_0)} v \leq 2CR^{-1} \sup_{B_{2R}^+(x_0)} |u|$$

which tends to zero as $R \to \infty$ by the assumption on u. Hence, u must be a constant. $\quad\square$

Remark 7.4.1. (1) Theorem 7.4.2 does not coincide with the local gradient estimate on a weighted Riemannian manifold $(M, g_{\nabla u}, m)$ since Ric_N and the weighted Ricci curvature $\hat{\mathrm{Ric}}_N$ (also denoted by $\mathrm{Ric}_N^{\nabla u}$ in some references) of $(M, g_{\nabla u}, m)$ are different. In fact, the weighted Ricci curvature Ric_N depends on the Finsler structure F and the measure m, while the weighted Ricci curvature $\hat{\mathrm{Ric}}_N$ only depends on u. A simple example is $(\mathbb{R}^n, \| \cdot \|, m_{BH})$, where $\| \cdot \|$ is a Minkowski norm and m_{BH} denotes the Busemann–Hausdorff measure. In this case, Ric_n vanishes but $\hat{\mathrm{Ric}}_n$ does not.

(2) The global gradient estimate for harmonic functions on a complete and noncompact Finsler measure space will be given in Corollary 8.5.3. More generally, we obtained the local and global gradient estimates for $p(> 1)$-harmonic functions on complete Finsler measure spaces and their applications in [Xia7].

Chapter 8

The Eigenvalue Problem

One of the fundamental problems in geometric analysis is to study the eigenvalue problem for the Laplacian. The eigenvalue problem of Riemannian Laplacian has been well developed ([SY] and the references therein). As its generalization, the eigenvalue problem of Finsler Laplacian was first introduced by Shen ([Sh1]). In this chapter, we shall introduce some recent progress on the eigenvalue problem in Finsler geometry.

8.1 Eigenvalue and Eigenfunction

For any nonzero function $u \in W^{1,p}(M)\backslash\{0\}(p > 1)$, we define the Rayleigh quotient of u by

$$E(u) := \frac{\int_M [F^*(x, du)]^p dm}{\int_M |u|^p dm}.$$

Note that $E(u)$ is C^1 on $W^{1,p}(M)\backslash\{0\}$. Then $d_u E = 0$ if and only if u satisfies $\mathbf{\Delta}_p u = -\lambda |u|^{p-2} u$ in the weak sense, i.e.,

$$\int_M [F^*(x, du)]^{p-2} d\varphi(\nabla u) dm = \lambda \int_M |u|^{p-2} u \varphi dm, \qquad (8.1.1)$$

where $\varphi \in C_0^\infty(M)$ and $\lambda = E(u)$. In fact, φ can be chosen as $\varphi \in W_0^{1,p}(M)$. If there are a constant λ and a function $u \in W^{1,p}(M)(p > 1)$ satisfying (8.1.1), then we say that λ is a *p-eigenvalue* of $\mathbf{\Delta}_p$ and u is a *p-eigenfunction* on M associated to λ.

Proposition 8.1.1. *Let (M, F, m) be an n-dimensional compact Finsler measure space (with possibly non-empty boundary). Assume u is a p-eigenfunction corresponding to the p-eigenvalue λ. Then $u \in L^\infty(M)$.*

Proof. Sobolev's embedding theorem (i.e., Theorem 10.0.4 in Appendix) implies that it suffices to consider the case when $1 < p \leq n$. Let $\bar{u} = u^+ = \max\{0, u\}$. For some $k > 0$, define

$$\bar{u}_k := \begin{cases} \bar{u}, & \text{if } u \leq k, \\ k, & \text{if } u \geq k. \end{cases}$$

Then $0 \leq \bar{u}_k \leq \bar{u}$, $\bar{u}_k \in W^{1,p}(M) \cap L^\infty(M)$ and

$$d\bar{u}_k = \begin{cases} d\bar{u}, & \text{if } 0 < u < k, \\ 0, & \text{otherwise} \end{cases}$$

(cf. Lemma 7.6 in [GT]). By the definition of Legendre transformation, we have $\nabla \bar{u}_k = \nabla \bar{u} = \nabla u$ if $0 < u < k$ and zero otherwise.

Let $\varphi = \bar{u}_k^{\ell+1}$ for some constant $\ell > 0$. Then $\varphi \in W^{1,p}(M)$ since $\bar{u}_k$ has a upper bound k and $u \in W^{1,p}(M)$. In particular, when $\partial M \neq \emptyset$ and $u|_{\partial M} = 0$, we have $\bar{u}_k|_{\partial M} = 0$ and hence $\varphi \in W_0^{1,p}(M)$. Obviously, $\varphi = \bar{u}_k = 0$ if $u \leq 0$. Thus, we only need to consider the case when $u > 0$. In this case, $\bar{u} = u$ and

$$d\varphi(\nabla u) = (1 + \ell)\bar{u}_k^\ell d\bar{u}_k(\nabla u) = (1 + \ell)\bar{u}_k^\ell F^{*2}(d\bar{u}_k). \tag{8.1.2}$$

Plugging (8.1.2) into (8.1.1) yields

$$(1 + \ell) \int_M [F^*(du)]^{p-2}\bar{u}_k^\ell F^{*2}(d\bar{u}_k)dm = \lambda \int_M (|u|^{p-2}u)\bar{u}_k^{\ell+1}dm$$

$$\leq \lambda \int_M \bar{u}^{p+\ell}dm. \tag{8.1.3}$$

Let $v_k := \bar{u}_k^{t+1}$, where $t := \ell/p$. Then $[F^*(dv_k)]^p = (1 + t)^p \bar{u}_k^\ell [F^*(d\bar{u}_k)]^p$. From this and (8.1.3), one obtains

$$\frac{1 + tp}{(1 + t)^p} \int_M [F^*(dv_k)]^p dm \leq \lambda \int_M \bar{u}^{p(1+t)}dm,$$

which implies that

$$\|v_k\|_{1,p} \leq c_1 \|\bar{u}\|_{L^{p(1+t)}}^{1+t},$$

where $c_1 := 1 + \frac{\lambda^{1/p}(1+t)}{(1+tp)^{1/p}}$ is a positive constant depending on p and ℓ. Since $p > 1$ and $t > 0$, we have $c_1 \leq (1 + \lambda^{1/p})(1 + t)^{\frac{p-1}{p}}$. It is easy to see that there exists a positive constant c_3 independent of t such that the right side of this inequality is less than $c_3^{\sqrt{1+t}}$. Hence, $c_1 \leq c_3^{\sqrt{1+t}}$. On the other hand, by Theorem 10.0.4 in Appendix, there is a positive constant c_2 such that

$$\|v_k\|_{L^{p^*}} \leq c_2 \|v_k\|_{1,p},$$

here we take $p^* = \frac{np}{n-p}$ if $1 < p < n$ and $p^* = 2p$ if $p = n$. Thus, we have

$$\|v_k\|_{L^{p^*}} \leq c_2 c_3^{\sqrt{1+t}} \|\bar{u}\|_{L^{p(1+t)}}^{1+t}.$$

Letting $k \to \infty$, Fatou's lemma implies that

$$\|\bar{u}\|_{L^{(1+t)p^*}} \leq c_2^{\frac{1}{1+t}} c_3^{\frac{1}{\sqrt{1+t}}} \|\bar{u}\|_{L^{p(1+t)}}. \tag{8.1.4}$$

Choose t, denoted by t_1, such that $(t_1 + 1)p = p^*$ (i.e., take $\ell := p^* - p > 0$). Then (8.1.4) becomes

$$\|\bar{u}\|_{L^{(1+t_1)p^*}} \leq c_2^{\frac{1}{1+t_1}} c_3^{\frac{1}{\sqrt{1+t_1}}} \|\bar{u}\|_{L^{p^*}}.$$

Next, choose t, denoted by t_2, such that $(t_2 + 1)p = (t_1 + 1)p^*$ in (8.1.4) yields

$$\|\bar{u}\|_{L^{(1+t_2)p^*}} \leq c_2^{\frac{1}{1+t_2}} c_3^{\frac{1}{\sqrt{1+t_2}}} \|\bar{u}\|_{L^{(1+t_1)p^*}}.$$

By induction on t, we obtain

$$\|\bar{u}\|_{L^{(1+t_i)p^*}} \leq c_2^{\frac{1}{1+t_i}} c_3^{\frac{1}{\sqrt{1+t_i}}} \|\bar{u}\|_{L^{(1+t_{i-1})p^*}},$$

where the sequence $\{t_i\}$ is chosen such that $(1 + t_i)p = (1 + t_{i-1})p^*$. It is easy to see that $1 + t_i = (\frac{p^*}{p})^i > 1$. Hence,

$$\|\bar{u}\|_{L^{(1+t_i)p^*}} \leq c_2^{\sum_i (\frac{p}{p^*})^i} c_3^{\sum_i (\frac{p}{p^*})^{i/2}} \|\bar{u}\|_{L^{p^*}}.$$

Letting $i \to \infty$, there is a positive constant c only depending on p and n such that

$$\|\bar{u}\|_{L^\infty} \leq c\|\bar{u}\|_{L^{p^*}} \leq c\|\bar{u}\|_{L^p} \leq c\|u\|_{L^p},$$

which means that $u^+ \in L^\infty(M)$. In the same way, we have $u^- \in L^\infty(M)$. Therefore, $u = u^+ + u^- \in L^\infty(M)$. □

For the sake of simplicity, we only consider the case when $p = 2$ from now on. For the arguments on Finsler p-eigenvalue problem, we refer to

[Xia5]–[Xia7] and [YH1]–[YH2] for more details. We say that u is an *eigenfunction* on M associated to the *eigenvalue* λ if $\Delta u = -\lambda u$ in the weak sense, i.e.,

$$\int_M d\varphi(\nabla u)dm = \lambda \int_M u\varphi dm$$

for any $\varphi \in C_0^\infty(M)$.

Let $\mathcal{H}$ be $W_0^{1,2}(M)$ or the set of the functions $u \in W^{1,2}(M)$ with $\int_M u\,dm = 0$. Then we have the following existence and regularity for eigenfunctions, which were first obtained by Ge-Shen ([GS]) in a different way. Here, we give another proof (cf.[Xia5]).

Theorem 8.1.1. *For any n-dimensional compact Finsler measure space* (M, F, m) *without boundary or with a smooth boundary, there exists a function* $u \in \mathcal{H}$ *with* $\|u\|_2 = 1$, *which minimizes* $E(u)$. *Further,* $u \in C^{1,\beta}(M) \cap W_{loc}^{2,2}(M) \cap L^\infty(M)$ *for some constant* β *with* $0 < \beta < 1$. *Moreover,* u *is smooth on the set* M_u.

Denote $\lambda_1 := \inf_{u \in \mathcal{H}\setminus\{0\}} E(u) > 0$. Then λ_1 is a critical value of $E(u)$, which is called the *first eigenvalue* of Δ, and u is a critical point of $E(u)$, which is called the *first eigenfunction* corresponding to λ_1.

To prove Theorem 8.1.1, we need some preparations. Let $(\mathcal{U}, \psi)$ be a local coordinate chart in M such that $\bar{\mathcal{U}}$ is compact and ψ maps $\mathcal{U}$ diffeomorphically onto an open ball of $\mathbb{R}^n$. It induces a standard local coordinate system (x^i, ξ_i) in T^*M. Set $J^i(x,\xi) := \frac{1}{2}[F^{*2}(x,\xi)]_{\xi_i}(x,\xi)$ and

$$g^{*ij}(x,\xi) = \frac{\partial J^i}{\partial \xi_j}(x,\xi) = \frac{1}{2}[F^{*2}(x,\xi)]_{\xi_i\xi_j}(x,\xi). \tag{8.1.5}$$

Obviously, $J^i(x,\xi) = g^{*ij}(x,\xi)\xi_j$ and $J(x,\xi) = 0$ if $\xi = 0$. From this, we can see that $J = J^i(x,\xi)\frac{\partial}{\partial x^i}$ defines the inverse of the Legendre transformation $\mathcal{L} : TM\setminus\{0\} \to T^*M\setminus\{0\}$, that is, $\mathcal{L}^{-1}(x,\xi) = J(x,\xi)$. Thus, there is a constant $C > 1$ such that

$$C^{-1}|y| \leq F(x,y) \leq C|y|,$$

where we identify $y \in T_xM$ with $d\psi(y) \in \mathbb{R}^n$ for simplicity and $|\cdot|$ means the Euclidean norm in $\mathbb{R}^n$. From this, for any $x_1, x_2 \in U$, we have

$$C^{-1}|\psi(x_2) - \psi(x_1)| \leq d_F(x_1, x_2) \leq C|\psi(x_2) - \psi(x_1)|, \tag{8.1.6}$$

$$C^{-2}d_F(x_2, x_1) \leq d_F(x_1, x_2) \leq C^2 d_F(x_2, x_1). \tag{8.1.7}$$

Similarly, without loss of generality, we can take the same constant $C > 1$ as above such that

$$C^{-1}|\xi| \le F^*(x,\xi) \le C|\xi|, \tag{8.1.8}$$

$$|J^i(x,\xi)| \le C|\xi|, \tag{8.1.9}$$

$$\sum_{i,j=1}^{n} |\frac{\partial J^i}{\partial x^j}|(x,\xi) \le C|\xi|, \tag{8.1.10}$$

$$C^{-1}|\eta|^2 \le g^{*ij}(x,\xi)\eta_i\eta_j \le C|\eta|^2 \tag{8.1.11}$$

for all $\xi \in T_x^*M\backslash\{0\}$, $\eta \in T_zM$ and $x \in \overline{\mathcal{U}}$, where we identify $\xi \in T_x^*M$ with $(\psi^{-1})^*(\xi) \in \mathbb{R}^n$. Further, from (8.1.8)–(8.1.11), we have

$$C^{-2}F^*(x,\xi) \le F^*(x,-\xi) \le C^2F^*(x,\xi), \tag{8.1.12}$$

$$|J(x,\xi) - J(x',\xi)| \le C|\xi||x-x'|, \quad x' \in \mathcal{U} \tag{8.1.13}$$

$$|[J^i(x',\xi) - J^i(x,\xi)]\eta_i| \le C|\xi||\eta||x'-x|, \quad x' \in \mathcal{U}, \tag{8.1.14}$$

$$|[J^i(x,\xi) - J^i(x,\zeta)]\eta_i| \le C|\eta||\xi - \zeta|, \tag{8.1.15}$$

$$C^{-1}|\xi - \eta|^2 \le \{J^i(x,\xi) - J^i(x,\eta)\}(\xi_i - \eta_i) \le C|\xi - \eta|^2. \tag{8.1.16}$$

For the proofs of (8.1.5)–(8.1.16), we refer to Section 6.2 in [BCS] and Section 3 in [Xia5]. Consider the following degenerated divergence type equation:

$$\text{div}A(x,u,du) + B(x,u,du) = 0, \quad x \in \Omega, \tag{8.1.17}$$

where Ω is a bounded domain in $\mathbb{R}^n$ and $u \in W^{1,p}(\Omega) \cap L^\infty(\Omega)(1 < p < \infty)$. Let $A = (A^i)$ and $(a^{ij}) = (\partial A^i/\partial \xi_j)$ (here and in what follows, we use z and ξ instead of the variables u and du). The "degeneracy" of (8.1.17) means that the eigenvalues of the matrix (a^{ij}) become zero or infinite at $\xi = 0$. The structure assumptions on A and B are as follows:

(1) $A \in C^0(\Omega \times \mathbb{R} \times \mathbb{R}^n) \cap C^1(\Omega \times \mathbb{R} \times \mathbb{R}^n\backslash\{0\})$ and B is a Carathéodory function, i.e., B is measurable in x and continuous in z and ξ;

(2) $A(x,z,0) = 0$;

(3) $\sum\limits_{i,j=1}^{n} a^{ij}(x,z,\xi)\eta_i\eta_j \ge \gamma(c+|\xi|)^{p-2}|\eta|^2$;

(4) $\sum\limits_{i,j=1}^{n} |a^{ij}(x,z,\xi)| \le \Gamma(c+|\xi|)^{p-2}$;

(5) $\sum_{i,j=1}^{n} |\frac{\partial A^i}{\partial x_j}(x,z,\xi)| + |\frac{\partial A^i}{\partial z}(x,z,\xi)| \le \Gamma(c+|\xi|)^{p-2}|\xi|$;

(6) If Ω is a bounded domain with $C^{1,\beta}$ boundary, where $0 < \beta \leq 1$, then we need to assume that $|A(x,z,\xi) - A(y,w,\xi)| \leq \Gamma(1 + |\xi|)^{p-1} \left(|x-y|^\beta + |z-w|^\beta\right)$;

(7) $|B(x,z,\xi)| \leq \Gamma(1+|\xi|)^p$,

where $c \in [0,1]$ is a nonnegative constant and γ, Γ are positive constants.

Under suitable conditions on A and B, Tolksdorf and Lieberman respectively proved the interior regularity for the weak solutions u of (8.1.17) and the boundary regularity for the weak solutions of the Dirichlet or Neumann problem of (8.1.17) (see Theorem 1 in [To] and Theorems 1–2 in [Lieb]). With the help of these results, we can prove Theorem 8.1.1.

Proof. The proof of the existence is standard. In fact, it suffices to verify that the functional $E(u)$ satisfies the Palais–Smale condition. Namely, if $\{u_k\}$ is a sequence $\{u_k\} \subset \mathcal{H}$ satisfying

$$\|u_k\|_2 = 1, \quad E(u_k) \to C, \quad d_{u_k}E \to 0 \quad \text{in} \quad (W^{1,2})^{-1}(M), \qquad (8.1.18)$$

then $u_k \to u$ strongly in $\mathcal{H}$.

From (8.1.18), we have $E(u_k) \leq 1 + C$ and hence $\{u_k\}$ is bounded in $\mathcal{H}$. Since $W^{1,2}(M)$ is reflexive, passing to a subsequence if necessary, we may assume that $\{u_k\} \to u$ weakly in $W^{1,2}(M)$ for some $u \in W^{1,2}(M)$ ([GT], P86). By Theorem 10.0.4 in Appendix, $u_k \to u$ strongly in $L^2(M)$ and $u \in W_0^{1,2}(M)$ if $u_k \in W_0^{1,2}(M)$. By Minkowski's inequality (see Lemma 1.1.2), one obtains that $|\|u_k\|_2 - \|u\|_2| \leq \|u_k - u\|_2 \to 0$ as $k \to \infty$, which means that

$$\int_M |u|^2 dm = \lim_{k\to\infty} \int_M |u_k|^2 dm = 1. \qquad (8.1.19)$$

Note that $\int_M u_k dm = 0$. Since $\|u_k - u\|_1 \leq \text{Vol}(M)^{1/2} \|u_k - u\|_2 \to 0$ as $k \to \infty$, we have

$$\int_M u\, dm = \lim_{k\to\infty} \int_M u_k dm = 0.$$

Thus, $u \in \mathcal{H}$. By (8.1.18), we have

$$o(1) = \frac{1}{2} d_{u_k} E(u_k - u)$$

$$= \int_M J^i(du_k) \frac{\partial}{\partial x^i}(u_k - u) dm - E(u_k) \int_M u_k(u_k - u) dm$$

$$= \int_M \left\{ J^i(du_k) - J^i(du) \right\} \frac{\partial}{\partial x^i}(u_k - u)dm - E(u_k) \int_M u_k(u_k - u)dm$$

$$+ \int_M J^i(du) \frac{\partial}{\partial x^i}(u_k - u)dm. \tag{8.1.20}$$

Observe that

$$\left| \int_M u_k(u_k - u)dm \right| \leq \int_M |u_k(u_k - u)|dm \leq \|u_k\|_2 \|u_k - u\|_2 = o(1)$$

and

$$\left| \int_M J^i(du) \frac{\partial}{\partial x^i}(u_k - u)dm \right| = \left| \int_M \mathrm{div}(J(du))(u_k - u)dm \right|$$

$$\leq \tilde{C} \int_M F^*(du)|u_k - u|dm \leq \tilde{C} \|F^*(du)\|_2 \|u_k - u\|_2 = o(1)$$

by the divergence lemma, (8.1.8) and (8.1.10). It follows from these and (8.1.20) that

$$o(1) = \int_M \left\{ J^i(du_k) - J^i(du) \right\} \left(\frac{\partial u_k}{\partial x^i} - \frac{\partial u}{\partial x^i} \right) dm.$$

From this and (8.1.16), one obtains that $\|F^*(du_k - du)\|_2 = o(1)$ and hence $u_k \to u$ strongly in $\mathcal{H}$.

Next, we prove the regularity. For any point $x \in M$, we choose a local coordinate chart $(\mathcal{U}, \psi)$ as before. Then, $F^*(x, \xi)$ and $J(x, \xi) = J^i(x, \xi)\frac{\partial}{\partial x^i}$ satisfy (8.1.8)–(8.1.16). We write $dm = \sigma(x)dx$. By the assumption, u satisfies $\boldsymbol{\Delta} u = -\lambda u$ in the weak sense, i.e.,

$$\int_M \left(J^i(du) \frac{\partial \varphi}{\partial x^i} - \lambda \varphi u \right) \sigma(x)dx = 0, \tag{8.1.21}$$

where $\varphi \in C_0^\infty(M)$. We choose $\varphi \in C_0^\infty(M)$ such that $\mathrm{supp}(\varphi) \subset \mathcal{U}$. Thus, (8.1.21) can be regarded as the degenerated divergence type equation (8.1.17) defined on $\psi(\mathcal{U}) \subset \mathbb{R}^n$ with $A(x, u, du) = \sigma J(x, du)$ and $B(x, u, du) = \lambda \sigma u$. Note that $u \in L^\infty(M)$ by Proposition 8.1.1 and σ is bounded since M is compact. It is easy to see that $A(x, u, du)$ and $B(x, u, du)$ satisfy the structure assumptions (1)–(5) and (7) with $p = 2$ by (8.1.10)–(8.1.16) for all $x \in \psi(\mathcal{U})$, $\xi \in \mathbb{R}^n \backslash \{0\}$ and all $\eta \in \mathbb{R}^n$. Hence $u \in C^{1,\beta}(M) \cap W^{2,2}_{loc}(M)$ by Theorem 1 and Proposition 1 in [To].

Now, we consider the regularity of u near the boundary if $\partial M \neq \emptyset$. Let $(\mathcal{U}, \psi)$ be a local coordinate system at $x \in \partial M$ such that

$\psi(\mathcal{U}) = \mathbb{B}_1^{n,+} := \mathbb{B}_1^n \cap \mathbb{R}_+^n$ and $\psi(\mathcal{U} \cap \partial M) = \Gamma_1 := \mathbb{B}_1^n \cap \partial\mathbb{R}_+^n$, where $\mathbb{B}_1^n$ is the unit ball centered at the origin in $\mathbb{R}^n$. Since $u \in \mathcal{H}$, u satisfies either $u|_{\partial M} = 0$ or $\int_M u\,dm = 0$ when $\partial M \neq \emptyset$. If the first case occurs, then we have $u \circ \psi^{-1}|_{\Gamma_1} = 0$. By the same arguments as above, $A(x, u, du) = \sigma J(x, du)$ and $B = \lambda\sigma u$ satisfy the structure assumptions (3)–(4) and (6)–(7) with $p = 2$ from (8.1.10)–(8.1.16) for all $(x, z, \xi) \in \Gamma_1 \times [-M_0, M_0] \times \mathbb{R}^n$. From Theorem 1 in [Lieb], we have $u \in C^{1,\beta}(\bar{\mathcal{U}})$ in this case. If $\int_M u\,dm = 0$, then there is a weak solution $u \in W^{1,2}(M)$ satisfying $\mathbf{\Delta}u = -\lambda_1 u$ in M with $\nabla u \in T_x(\partial M)$, i.e., $g_\nu(\nu, \nabla u) = 0$, where ν is the outward normal vector of ∂M, which implies that

$$g_{\nabla u}(\nu_{\nabla u}, \nabla u)(x) = 0 \quad \text{on} \quad \partial M \Leftrightarrow Du(\nu_{\nabla u})(x) = 0 \quad \text{on} \quad \partial M,$$

by Proposition 2.4.1. Thus, $J(x, du) \cdot \nu_{\nabla u} = 0$ on $\mathcal{U} \cap \partial M$. Consequently, $u \in C^{1,\beta}(\bar{\mathcal{U}})$ from Theorem 2 in [Lieb]. Hence, we have $u \in C^{1,\beta}(M)$. Further, the elliptic regularity ensures that u is a smooth function on M_u. This finishes the proof. $\qquad\square$

Since the Finsler Laplacian is a nonlinear weakly differential operator, it is difficult to study the spectrum of Finsler Laplacian. On compact reversible Finsler measure spaces, Kristály–Shen–Yuan–Zhao recently describe the nonlinear spectrum for Finsler Laplacian by means of faithful dimension pairs and give some estimates for spectral gaps ([KSYZ]). It is not clear how to define the nonlinear spectrum on a nonreversible Finsler measure space up to now.

8.2 Upper Bound Estimates for λ_1

A fundamental result for the upper bound estimates of the first eigenvalue on complete Riemannian manifolds was due to Cheng in 1975 via establishing the first eigenvalue comparison theorem with the simply connected space forms ([Che]). However, this approach seems not work in Finsler geometry since we do not have good model spaces in Finsler geometry like "space forms" in Riemannin geometry. However, we can give an upper bound estimate of λ_1 by means of the oscillatory behavior for some ODE.

For a noncompact Finsler measure space (M, F, m) (it need not be complete), we define the *first eigenvalue* on M by

$$\lambda_1 := \inf_\Omega \lambda_1(\Omega) = \inf_\Omega \inf_{u|_{\partial\Omega}=0} \frac{\int_\Omega [F^*(x, du)]^2 dm}{\int_\Omega u^2 dm},$$

where Ω runs through all compact subdomains with C^1 boundary in M, and the associating eigenfunction is said to be the *first eigenfunction* on M. Obviously, for an exhaustion $\Omega_1, \Omega_2, \ldots$ of M such that $\bar{\Omega}_i \subset \Omega_{i+1}$ for all $i \geq 1$ and $M = \cup_{i=1}^{\infty} \Omega_i$, $\{\lambda_1(\Omega_i)\}$ is a decreasing sequence with respect to $\{\Omega_i\}$. Consequently, $\lambda_1 = \lim_{i \to \infty} \lambda_1(\Omega_i)$, which is independent of the choice of $\{\Omega_i\}$.

Proposition 8.2.1. *Let (M, F, m) be a forward complete Finsler measure space with finite reversibility Λ and infinite volume, i.e., $m(M) = +\infty$. Assume that $m(B_r^+(x)) \leq cr^k e^{ar}$ for any $r \geq r_0 > 0$ and $x \in M$, where $c > 0$, $k \geq 0$ and $a \geq 0$ are constants independent of r, and λ_1 is the first eigenvalue of Finsler Laplacian. (1) If $a = 0$, then $\lambda_1(M) = 0$. (2) If $a > 0$, then $\lambda_1(M) \leq \left(\frac{a\Lambda}{2}\right)^2$.*

To prove Proposition 8.2.1, we study the oscillatory behavior of the following ODE on $\psi = \psi(t)$:

$$(\psi' \nu)' + \lambda \psi \nu = 0, \quad t \geq t_0, \tag{8.2.1}$$

where $\nu := \nu(t)$ is a positive continuous function on $[t_0, \infty)$ and λ is a positive constant. Recall that equation (8.2.1) is said to be *oscillatory* if all solutions of (8.2.1) have arbitrary large zeroes on $[T, \infty)$.

Lemma 8.2.1. *Let $\int_{t_0}^{+\infty} \nu(\zeta) d\zeta = +\infty$ and $v(t) := \int_{t_0}^{t} \nu(\zeta) d\zeta \leq ct^k e^{at}$ for some nonnegative constants k, a and a positive constant c. Then (8.2.1) is oscillatory provided either (i) $a = 0$ or (ii) $\lambda > \left(\frac{a}{2}\right)^2$ when $a > 0$.*

Proof. We argue this by contradiction. Since (8.2.1) is invariant up to a sign, we may assume that there exists a solution ψ of (8.2.1) and a sufficiently large positive constant $T > t_0$ such that $\psi > 0$ on $[T, \infty)$. Set

$$\tilde{\psi} = -\frac{\psi' \nu}{\psi}, \quad t \in [T, \infty).$$

Then, by (8.2.1), one obtains

$$\tilde{\psi}' = \lambda \nu + \frac{1}{\nu} |\tilde{\psi}|^2 \geq 2\lambda^{1/2} |\tilde{\psi}|. \tag{8.2.2}$$

Obviously, $\tilde{\psi}(t)$ is increasing on $[T, \infty)$.

Case 1. $\int_T^\infty \frac{1}{\nu(\zeta)}d\zeta < +\infty.$

Note that $\tilde{\psi}' \geq \lambda\nu$, which means that $\tilde{\psi}(t) \geq \lambda(v(t) - v(T)) + \tilde{\psi}(T)$. We may assume that $\tilde{\psi}(t) > 0$ for $t \geq T$ since $\lim_{t\to+\infty} v(t) = +\infty$. Further, it follows from (8.2.2) that $\tilde{\psi}(t) \geq \tilde{\psi}(T)e^{2(t-T)\sqrt{\lambda}}$. On the other hand, (8.2.2) implies that $\tilde{\psi}' \geq \frac{1}{\nu}\tilde{\psi}^2$. Solving this inequality yields

$$\frac{1}{\tilde{\psi}(t)} \leq \frac{1}{\tilde{\psi}(t-1)} - \int_{t-1}^t \frac{1}{\nu}d\zeta. \tag{8.2.3}$$

Note that the assumption that $\int_T^\infty \frac{1}{\nu(\zeta)}d\zeta < +\infty$ ensures that the second term on the right-hand side of (8.2.3) is meaningful for any $t \geq T$. By Hölder's inequality, one obtains

$$1 = \int_{t-1}^t d\zeta \leq \left(\int_{t-1}^t \frac{1}{\nu}d\zeta\right)^{\frac{1}{2}} \cdot \left(\int_{t-1}^t \nu d\zeta\right)^{\frac{1}{2}},$$

which implies that

$$\int_{t-1}^t \frac{1}{\nu}d\zeta \geq \left(\int_{t-1}^t \nu d\zeta\right)^{-1} \geq \frac{1}{v(t)}.$$

Plugging the above inequality into (8.2.3) yields

$$0 < \frac{1}{\tilde{\psi}(t)} \leq \frac{1}{\tilde{\psi}(t-1)} - \frac{1}{v(t)}$$

$$\leq \frac{1}{\tilde{\psi}(T)e^{2(t-1-T)\sqrt{\lambda}}} - \frac{1}{ct^k e^{at}}$$

$$= \left(\frac{ct^k}{\tilde{\psi}(T)e^{[(2\sqrt{\lambda}-a)t-2\sqrt{\lambda}(1+T)]}} - 1\right)\frac{1}{ct^k e^{at}}.$$

Since $\lambda > 0$, the RHS of the last equality is less than or equal to zero when t is large enough if $a = 0$ or $\lambda > \left(\frac{a}{2}\right)^2$ when $a > 0$. We have a contradiction.

Case 2. $\int_T^\infty \frac{1}{\nu(\zeta)}d\zeta = +\infty.$ Let

$$s(t) = \int_T^t \frac{1}{\nu(\zeta)}d\zeta,$$

which is a non-degenerate transformation of parameters. We write $s = s(t)$ or $t = t(s)$. Then

$$\frac{d}{dt}(\psi'(t)\nu(t)) = \frac{d}{ds}(\psi'(s)) \cdot \frac{1}{\nu(s)},$$

where we used that $\psi'(s) = \frac{d}{ds}(\psi(t(s))) = \psi'(t)\nu(t)$. Thus, (8.2.1) becomes

$$\frac{d}{ds}(\psi'(s)) + \lambda\psi(s)\nu(s)^2 = 0. \qquad (8.2.4)$$

Let $\bar{\psi}(s) := -\frac{\psi'(s)}{\psi(s)}$. Then, by (8.2.4),

$$\bar{\psi}'(s) = \lambda\nu(s)^2 + \left(\frac{\psi'(s)}{\psi(s)}\right)^2 = \lambda\nu(s)^2 + (\bar{\psi}(s))^2 \geq 2\sqrt{\lambda}\nu(s)|\bar{\psi}(s)|. \qquad (8.2.5)$$

Obviously, $\bar{\psi}'(s) \geq \lambda\nu(s)^2$, which means that

$$\bar{\psi}(s) - \bar{\psi}(s(T)) \geq \lambda \int_{s(T)}^{s} \nu(s)^2 ds = \lambda \int_{T}^{t} \nu(\zeta)d\zeta \to +\infty, \quad \text{as } t \to +\infty.$$

So, we may assume that $\bar{\psi}(s) > 0$. Further, the inequality (8.2.5) implies that

$$\bar{\psi}(s) \geq \bar{\psi}(s(T))\exp\left(2\sqrt{\lambda}\int_{s(T)}^{s}\nu(s)ds\right) = \bar{\psi}(s(T))\exp(2\sqrt{\lambda}(t-T)). \qquad (8.2.6)$$

Note that $s = s(t)$ and $t = t(s)$ are increasing by the assumption, and hence $s \to +\infty$ if and only if $t \to +\infty$. By (8.2.6), we have $\lim_{s\to+\infty}\bar{\psi}(s) = +\infty$. On the other hand, (8.2.5) implies that $\bar{\psi}'(s) \geq \bar{\psi}^2$. By a similar argument to (8.2.3), we have

$$0 < \frac{1}{\bar{\psi}(s)} \leq \frac{1}{\bar{\psi}(s-1)} - 1 < 0$$

for a sufficiently large s. Thus, we get a contradiction. The proof is completed. $\qquad \square$

Proof of Proposition 8.2.1. By (6.4.2), we have

$$m(B_r^+(x)) = \int_0^r \tau(t)dt.$$

Since $m(M) = +\infty$, we have $\int_{r_0}^{+\infty} \tau(t)dt = +\infty$ for some $r_0 > 0$. For each $\lambda > 0$, by Lemma 8.2.1, if $a = 0$ or $\lambda > \left(\frac{a}{2}\right)^2$ when $a > 0$, there exists a nontrivial oscillatory solution ψ_λ of (8.2.1) on $[r_0, +\infty)$ in which $\nu(t)$ is replaced by $\tau(t)$. Consequently, there exist two numbers $r_1^\lambda, r_2^\lambda \in [r_0, +\infty)$ with $r_1^\lambda < r_2^\lambda$ such that $\psi_\lambda(r_1^\lambda) = \psi_\lambda(r_2^\lambda) = 0$ and $\psi_\lambda(r) \neq 0$ for any $r \in (r_1^\lambda, r_2^\lambda)$. This means that either $\psi_\lambda(r) > 0$ or $\psi_\lambda(r) < 0$ on $(r_1^\lambda, r_2^\lambda)$. Let $r(z) := d_F(x,z)$, $\Omega_\lambda := B_{r_2^\lambda}^+(x) \backslash B_{r_1^\lambda}^+(x) \subset M$ and $u_\lambda(z) := \psi_\lambda(r(z))$.

Then we have $F^*(du_\lambda) = F^*(\psi'_\lambda dr) \leq \Lambda|\psi'_\lambda|$ a.e. on Ω_λ. By the principle of the variation, we have

$$0 \leq \lambda_1(M) \leq \lambda_1(\Omega_\lambda)$$

$$= \inf_{u_\lambda} \frac{\int_{\Omega_\lambda} F^{*2}(du_\lambda)dm}{\int_{\Omega_\lambda} |u_\lambda|^2 dm} \leq \frac{\Lambda^2 \int_{\Omega_\lambda} (\psi'_\lambda(r))^2 dm}{\int_{\Omega_\lambda} (\psi_\lambda(r))^2 dm}$$

$$= \frac{\Lambda^2 \int_{r_1^\lambda}^{r_2^\lambda} (\psi'_\lambda(r))^2 \tau(r) dr}{\int_{r_1^\lambda}^{r_2^\lambda} (\psi_\lambda(r))^2 \tau(r) dr} = -\frac{\Lambda^2 \int_{r_1^\lambda}^{r_2^\lambda} (\psi'_\lambda \tau)' \psi_\lambda dr}{\int_{r_1^\lambda}^{r_2^\lambda} |\psi_\lambda|^2 \tau dr} = \lambda\Lambda^2.$$

In the case of $a = 0$, $\lambda_1(M) = 0$ since λ is an arbitrary positive constant. For the case when $a > 0$, $\lambda_1(M) \leq \left(\frac{a\Lambda}{2}\right)^2$ since λ is an arbitrary positive constant greater than $\left(\frac{a}{2}\right)^2$. This finishes the proof. □

Theorem 8.2.1. *Let (M, F, m) be an n-dimensional forward complete Finsler measure space with finite reversibility Λ and λ_1 be the first eigenvalue of Finsler Laplacian on M. Assume that $Ric_N \geq -K$ for some $N \in [n, \infty)$ and $K \geq 0$.*

(i) *If $m(M) < +\infty$, then $\lambda_1 = 0$.*
(ii) *If $m(M) = +\infty$ and $Ric_N \geq -K$ for some $K \geq 0$ and $N \in [n, +\infty)$,*
then $\lambda_1(M) \leq \left(\frac{\Lambda\sqrt{(N-1)K}}{2}\right)^2$. In particular, $\lambda_1 = 0$ when $K = 0$.

Proof. (i) The variational principle of $\lambda_1(M)$ asserts that

$$\lambda_1(M) \int_M |u|^2 dm \leq \int_M F^{*2}(du)dm \qquad (8.2.7)$$

for any $u \in W_0^{1,2}(M)$. In particular, we choose $u(x) = -\psi(x)$, where ψ is a cut-off function defined by

$$\psi(x) = \begin{cases} 1 & \text{on} \quad B_R^+(x_0), \\ \frac{2R - d_F(x_0, x)}{R} & \text{on} \quad B_{2R}^+(x_0) \backslash B_R^+(x_0), \\ 0 & \text{on} \quad M \backslash B_{2R}^+(x_0). \end{cases} \qquad (8.2.8)$$

Then $F^*(du) \leq \frac{1}{R}$ a.e. on $B_{2R}^+(x_0)$ and hence (8.2.7) implies that

$$\lambda_1 m(B_R^+(x_0)) \leq R^{-2} m(B_{2R}^+(x_0)),$$

which implies that $\lambda_1(M) = 0$ by taking $R \to \infty$. The assertion (ii) follows from Corollary 6.4.3 and Proposition 8.2.1. □

The above approach to estimate the upper bound of the first eigenvalue is also suitable for the case of Finsler p-Laplacian. All related results as above also hold for the Finsler p-Laplacian ([Xia7]). When F is Riemannian with Ric$\geq -(n-1)k(k \geq 0)$, then Theorem 8.2.1 implies that $\lambda \leq \frac{(n-1)^2 k}{4}$, which was first obtained by Cheng ([Che]) in a different way.

8.3 Lower Bound Estimates for λ_1

We shall study the lower bound estimates for the first eigenvalue λ_1 on a compact Finsler measure space (M, F, m) when the weighted Ricci curvature bounded from below. In particular, we shall focus on the sharp lower bound estimates for λ_1.

8.3.1 *General Lower Bound Estimates*

The following general lower bound estimates for λ_1 are based on the Reilly inequality and the Bochner–Weitzenböck inequality (see Section 7.1).

Theorem 8.3.1. *Let (M^n, F, m) be an n-dimensional compact Finsler space without boundary or with a convex boundary. Assume that $Ric_N \geq K (K \in \mathbb{R})$ for some $N \in [n, \infty]$, and λ_1 is the first (nonzero) eigenvalue of Finsler Laplacian, i.e., $\Delta u = -\lambda_1 u$ in M (in the weak sense) with a Neumann boundary condition $\nabla u \in T_x(\partial M)$ when $\partial M \neq \emptyset$.*

(1) *If $K > 0$, then*

$$\lambda_1 \geq \frac{NK}{N-1}, \tag{8.3.1}$$

where the RHS of the above inequality is read as K when $N = \infty$.
(2) *If $K \leq 0$, then*

$$\lambda_1 \geq \frac{4\exp[-(2 + d\sqrt{-KN})]}{(N+2)d^2}, \tag{8.3.2}$$

where d is the diameter of M.

Proof. First of all, $u \in C^{1,\beta}(M) \cap W_{loc}^{2,2}(M) \cap L^{\infty}(M)$ by Theorem 8.1.1.
(1) By the assumption, we have

$$\int_M d\varphi(\nabla u)dm = \lambda_1 \int_M \varphi u\, dm$$

for any function $\varphi \in W^{1,2}(M) \cap L^\infty(M)$. In particular, letting $\varphi = u$ in the above equality yields

$$\lambda_1 \int_M u^2 dm = \int_M F^2(\nabla u) dm. \qquad (8.3.3)$$

On the other hand, $\nabla u = 0$ and $\Delta u = 0$ a.e. on $M \backslash M_u$ by Lemma 7.1.3. For the first Neumann eigenvalue λ_1, $g_\nu(\nu, D^{\nabla u}_{\nabla u} \nabla u) \leq 0$ by the convexity of ∂M, where ν is an outward norm vector on ∂M. Thus, by (7.1.24) and the assumptions, we have

$$\frac{\lambda_1^2}{N} \int_M u^2 dm + K \int_M F^2(\nabla u) dm - \lambda_1 \int_M F^2(\nabla u) dm \leq 0.$$

From this and (8.3.3), one obtains

$$\left(-\frac{N-1}{N} \lambda_1^2 + K\lambda_1 \right) \int_M u^2 dm \leq 0,$$

which implies (8.3.1) when $N < \infty$. For the case when $N = \infty$, by (7.1.23) and (8.3.3), we get (8.3.1).

(2) Since $K \leq 0$, we write $\mathrm{Ric}_N \geq -K(K \geq 0)$ in the following. Since u is the eigenfunction of Δ corresponding to λ_1, u must change the sign. Without loss of generality, we can assume that $\min u = -1$ and $\max u \leq 1$ by rescaling.

Let

$$\bar{u} := \log(a + u)$$

for some constant $a > 1$. Then $\nabla \bar{u} = \frac{1}{a+u} \nabla u$. For any function $\varphi \in W^{1,2}(M) \cap L^\infty(M)$, we have

$$\int_M d\varphi(\nabla \bar{u}) dm = \int_M \frac{1}{a+u} d\varphi(\nabla u) dm$$

$$= \int_M d\left(\frac{\varphi}{a+u} \right)(\nabla u) dm + \int_M \varphi \frac{F^2(\nabla u)}{(a+u)^2} dm$$

$$= \int_M \varphi \left(F^2(\nabla \bar{u}) + \frac{u\lambda_1}{a+u} \right) dm.$$

This means that it holds on M in the weak sense

$$\Delta \bar{u} = -F^2(\nabla \bar{u}) - \frac{\lambda_1 u}{a+u}. \qquad (8.3.4)$$

Consider the function $G := F^2(\nabla \bar{u})$ defined on M. Since M is compact, there is a point $x_0 \in M$ such that G attains the maximum. We first consider the case when x_0 lies in the interior of M and then the case when x_0 is on the boundary of M if $\partial M \neq \emptyset$.

Case 1. G attains its maximum at $x_0 \in M$.

Obviously, $\nabla \bar{u}(x_0) \neq 0$, i.e., $\nabla u(x_0) \neq 0$. Otherwise, $G(x_0) = 0$, which implies that $u(x)$ is constant on M and hence $\lambda_1 = 0$. This is excluded. Hence, $x_0 \in M_u$ and (8.3.4) holds strongly around x_0. Denote $f(u) := \frac{1}{a+u}$. Then $\mathbf{\Delta} \bar{u} = -G - \lambda_1 f u$. From this, one obtains

$$d(\mathbf{\Delta} \bar{u})(\nabla \bar{u}) = -dG(\nabla \bar{u}) - a\lambda_1 fG. \tag{8.3.5}$$

Since $L_{\bar{u}}$ is a linear and strictly elliptic operator at the points that $\nabla \bar{u} \neq 0$, we have $dG(x_0) = 0$ and $L_{\bar{u}}(G)(x_0) \leq 0$. From this and (7.1.11), we have $\lambda_1 \geq 0$ if $N = \infty$. This is trivial. Now, we consider the case when $N \in [n, \infty)$. Applying (7.1.12) to the function $\frac{1}{2}G$ and using (8.3.4)–(8.3.5) yield

$$\frac{1}{2} L_{\bar{u}}(G) \geq d(\mathbf{\Delta} \bar{u})(\nabla \bar{u}) + Ric_N(\nabla \bar{u}) + \frac{(\mathbf{\Delta} \bar{u})^2}{N}$$

$$\geq \frac{1}{N} G^2 - KG + \left(\frac{2u}{N} - a \right) \lambda_1 fG - dG(\nabla \bar{u}). \tag{8.3.6}$$

At x_0, we have

$$G - NK + (2u - Na)\lambda_1 f \leq 0.$$

Note that $|u| < a$. So,

$$G(x) \leq G(x_0) \leq NK - (2u(x_0) - Na)\lambda_1 f(u(x_0))$$

$$\leq NK + (N+2)\left(\frac{a}{a-1} \right) \lambda_1.$$

Applying the inequality $(x+y)^{\frac{1}{2}} \leq x^{\frac{1}{2}} + y^{\frac{1}{2}}$ for $x, y \geq 0$ to the above inequality yields

$$F(\nabla \bar{u}) = \sqrt{G(x)} \leq \sqrt{NK} + \sqrt{\lambda_1(N+2)}\sqrt{\frac{a}{a-1}}.$$

Choose $x_1, x_2 \in M$ such that $u(x_1) = -1$ and $u(x_2) = \max u$. Let $\gamma : [0,1] \to M$ be a minimal geodesic from x_1 to x_2. Then, by the

Cauchy–Shwarz inequality,

$$\log\left(\frac{a}{a-1}\right) \le \log\left(\frac{a+u(x_2)}{a-1}\right)$$

$$= \int_0^1 d\Big(\log\big(a+u(\gamma(t))\big)\Big)dt$$

$$= \int_0^1 d\bar{u}(\dot{\gamma}(t))dt \le \int_0^1 F(\nabla\bar{u}(\gamma(t)))F(\dot{\gamma}(t))dt$$

$$\le d\left\{\sqrt{NK} + \sqrt{\lambda_1(N+2)}\sqrt{\frac{a}{a-1}}\right\}. \tag{8.3.7}$$

Choose $a > 1$ such that

$$\log\frac{a}{a-1} = 2 + d\sqrt{KN}.$$

From this and (8.3.7), one obtains (8.3.2).

Case 2. $\partial M \neq \emptyset$ and $x_0 \in \partial M$.

Similar to Case 1, $\nabla u(x_0) \neq 0$. Thus, ∇u and G are smooth near the point x_0. Let ν be a normal vector field of ∂M that points outward. By Proposition 2.4.1, there is another outward normal vector field $\nu_{\nabla u}$ on ∂M satisfying (2.4.4). Consequently, $dG(\nu_{\nabla u}) \ge 0$. On the other hand, the Neumann boundary condition $\nabla u \in T(\partial M)$ implies that $g_{\nabla u}(\nu_{\nabla u}, \nabla u)(x) = 0$, equivalently, $du(\nu_{\nabla u})(x) = 0$ for any $x \in \partial M$. Thus, by (6.1.8), we have at x_0

$$d\left(F^2(\nabla\bar{u})\right)(\nu_{\nabla\bar{u}}) = \frac{1}{(a+u)^2}d\left(g_{\nabla u}(\nabla u, \nabla u)\right)(\nu_{\nabla u})$$

$$= \frac{2}{(a+u)^2}g_{\nabla u}\left(D_{\nu_{\nabla u}}^{\nabla u}\nabla u, \nabla u\right)$$

$$= \frac{2}{(a+u)^2}g_{\nabla u}\left(D_{\nabla u}^{\nabla u}\nabla u, \nu_{\nabla u}\right).$$

By the convexity of ∂M and Proposition 2.4.1 again, one obtains that $g_{\nabla u}\left(\nu_{\nabla u}, D_{\nabla u}^{\nabla u}\nabla u\right) \le 0$. Consequently, $d\left(F^2(\nabla\bar{u})\right)(\nu_{\nabla\bar{u}})(x_0) \le 0$, which implies that $dG(\nu_{\nabla u})(x_0) = 0$. Obviously, the tangent derivative of G at x_0 vanishes due to its maximality. Hence, $dG(x_0) = 0$. The rest proof is the same as that of Case 1. This completes the proof. $\square$

8.3.2 Sharp Lower Bound Estimates

For any $K \in \mathbb{R}$ and $N \in (1, \infty]$, let

$$
T(t) = \begin{cases}
\sqrt{(N-1)K} \tan\left(\sqrt{\frac{K}{N-1}}t\right), & \text{for } K > 0,\ 1 < N < \infty, \\
-\sqrt{-(N-1)K} \tanh\left(\sqrt{\frac{K}{N-1}}t\right), & \text{for } K < 0,\ 1 < N < \infty, \\
0, & \text{for } K = 0,\ 1 < N < \infty, \\
Kt, & \text{for } N = \infty.
\end{cases}
$$

$$(8.3.8)$$

Then T satisfies $\dot{T} = K + \frac{T^2}{N-1}$ for $N \in (1, \infty]$, where $\dot{T} = \frac{dT}{dt}$.

Fix $\lambda > 0$, we consider the following initial valued problem (IVP) on a domain of T:

$$
\begin{cases}
\ddot{v} - T\dot{v} = -\lambda v, \\
v(a) = -1, \quad \dot{v}(a) = 0.
\end{cases}
\tag{8.3.9}
$$

It is easy to see the local existence, uniqueness and continuous dependence on the parameters N, K, λ and a for solutions of (8.3.9) defined on the domain of T. The boundary case deserves more attention. For example, although $T(t)$ is singular at the boundary point $-\frac{\pi}{2}$ if $K > 0$, the existence and uniqueness of a solution of the first equation of (8.3.9) on $[-\frac{\pi}{2}, \frac{\pi}{2})$ with boundary value $v(-\pi/2) = -1, \dot{v}(-\pi/2) = 0$ are easily obtained by standard fixed point techniques via using the fact that the first equation of (8.3.9) is equivalent to

$$
\frac{d}{dt}(\mu\dot{v}) = -\lambda\mu v,
\tag{8.3.10}
$$

where μ satisfies $\dot{\mu} = -\mu T$ and μ is the invariant measure associated with the one-dimensional model (8.3.9), i.e., $d\mu = \mu(t)dt$. Let v_a be the solution of (8.3.9). It is clear that $\ddot{v}_a(a) > 0$, and therefore that $\dot{v}(x) > 0$ in a right neighborhood of a. If $b(a)$ is the first zero point of $\dot{v}_a$ after a, then $b(a)$ is finite and is a function continuously depending on the parameters a, λ, K, N (cf. [BQ]). We define $\mathfrak{m}(a) = v_a(b(a))$ to be the maximum value of v_a over the interval $[a, b(a)]$. From (8.3.10), one obtains

$$
\int_a^b \mu v_a dt = 0,
$$

which implies that $\mathfrak{m}(a) > 0$. We always may assume that $\mathfrak{m}(a) < 1$. Otherwise, we use the solution $v_{-b(a)}$ of the following IVP given by

$$v_{-b(a)}(t) = -\frac{v_a(-t)}{\mathfrak{m}(a)}, \quad v_{-b(a)}(-b(a)) = -1$$

instead of the solution v_a of (8.3.9). The following continuity of the maximum of v_a was due to Bakry–Qian in [BQ].

Lemma 8.3.1. *For any $N \in [1, \infty)$ (resp. $N = \infty$) and fix $\lambda > 0$, assume that v_a is a solution of (8.3.9). Then for any $k \in [\mathfrak{m}(a), \frac{1}{\mathfrak{m}(a)}]$ (resp. $(0, \infty)$), there exists an interval which has the first Neumann eigenvalue λ and a corresponding eigenfunction v such that $\min(v) = -1$ and $\max(v) = k$.*

Also, by Theorem 13 in [BQ], we have

Lemma 8.3.2. *For $K \in \mathbb{R}$ and $N \in [1, \infty)$, let $\lambda_1(K, N, a, b)$ denote the first Neumann eigenvalue of $L_{K,N} := \ddot{v} - T\dot{v}$ on the interval (a, b), i.e.,*

$$\begin{cases} \ddot{v} - T\dot{v} = -\lambda v, \\ \dot{v}(a) = \dot{v}(b) = 0. \end{cases} \tag{8.3.11}$$

Then $\lambda_1(K, N, a, b) \geq \lambda_1(K, N, -\frac{b-a}{2}, \frac{b-a}{2}) = \lambda_1(K, N, b - a)$. In other words, the centrally symmetric interval has the lowest Neumann eigenvalue.

The following gradient comparison principle plays a crucial role in subsequent arguments.

Proposition 8.3.1. *Let (M, F, m) be an n-dimensional compact Finsler measure space without boundary or with a convex boundary. Assume that $Ric_N \geq K$ for some $N \in [n, \infty)$ and $K \in \mathbb{R}$. Let λ_1 be the first (nonzero) eigenvalue of Finsler Laplacian, i.e., $\Delta u = -\lambda_1 u$ (in the weak sense) with a Neumann boundary condition $\nabla u \in T_x(\partial M)$ if $\partial M \neq \emptyset$. Assume that $v = v(t)$ is a solution on some interval (a, b) of the one-dimensional model problem*

$$\ddot{v} - T\dot{v} = -\lambda_1 v, \quad \dot{v}(a) = \dot{v}(b) = 0, \quad \dot{v} > 0. \tag{8.3.12}$$

If $[\min(u), \max(u)] \subset [\min(v), \max(v)]$, then

$$F(x, \nabla u(x)) \leq \dot{v}(v^{-1}(u(x))) \tag{8.3.13}$$

for all $x \in M$.

Proof. Since $\int_M u\,dm = 0$, $\min(u) < 0$ while $\max(u) > 0$. To avoid problems at the boundary of $[a, b]$, we may assume that $[\min(u), \max(u)] \subset (\min(v), \max(v))$ by multiplying u by a constant $k \in (0, 1)$. If we prove the result for this u, then letting $k \to 1$ implies the original statement.

Under the assumption that $[\min(u), \max(u)] \subset (\min(v), \max(v))$, v^{-1} is smooth on a neighborhood $\mathcal{U}$ of $[\min(u), \max(u)]$. Consider the function Φ defined on M given by $\Phi(x) = \psi(u)\left(F^2(\nabla u) - \varphi(u)\right)$, where $\psi \in C^\infty(\mathcal{U})$ is a positive smooth function to be determined later and $\varphi(u) = \dot{v}^2(v^{-1}(u(x)))$.

Case 1. Φ attains its maximum at some point $x_0 \in M$.

Without loss of generality, we always assume that $x_0 \in M_u$. Otherwise, there is nothing to prove if $F(x, \nabla u)\big|_{x_0} = 0$. Thus, u is of C^∞ around x_0 by Theorem 8.1.1 and hence Φ is of C^∞ around x_0. We have

$$\nabla^{\nabla u}\Phi(x_0) = 0, \quad L_u\Phi(x_0) \leq 0. \tag{8.3.14}$$

From the first equation of (8.3.14), one obtains, at x_0,

$$\nabla^{\nabla u}\left(F^2(\nabla u) - \varphi(u)\right) = -\frac{\Phi\dot{\psi}}{\psi^2}\nabla u, \tag{8.3.15}$$

equivalently, $d\left(F^2(\nabla u) - \varphi(u)\right)(\nabla u) = -\frac{\Phi\dot{\psi}}{\psi^2}F^2(\nabla u)$. Note that $\nabla^{\nabla u}(F^2(\nabla u)) = 2D^{\nabla u}_{\nabla u}\nabla u$ by (7.1.14). From this and (8.3.15), one obtains

$$H_u = \frac{\nabla^2 u(\nabla u, \nabla u)}{F^2(\nabla u)} = -\frac{1}{2}\left(\frac{\dot{\psi}}{\psi^2}\Phi - \dot{\varphi}\right). \tag{8.3.16}$$

Note that $L_u(u) = \Delta u$ and $F^2(\nabla u) = \Phi/\psi + \varphi$. It follows from (6.1.2) and (8.3.15) that

$$L_u(\Phi) = \Delta^{\nabla u}\Phi = \operatorname{div}_m\left[(F^p(\nabla u) - \varphi)\dot{\psi}\nabla u + \psi\nabla^{\nabla u}(F^2(\nabla u) - \varphi)\right]$$

$$= \frac{\dot{\psi}\Phi}{\psi}\Delta u + \psi L_u(F^2(\nabla u)) - \psi L_u(\varphi)$$

$$+ \frac{\Phi}{\psi^2}\left(\psi\ddot{\psi} - 2\dot{\psi}^2\right)\left(\frac{\Phi}{\psi} + \varphi\right). \tag{8.3.17}$$

Since $\Delta u = -\lambda_1 u$, it is easy to see that

$$L_u(\varphi(u)) = \dot{\varphi}(u)\Delta u + \ddot{\varphi}(u)F(\nabla u)^2 = -\lambda_1 u\dot{\varphi}(u) + \ddot{\varphi}(u)F(\nabla u)^2, \tag{8.3.18}$$

and

$$\frac{1}{2} L_u \left(F^2(\nabla u) \right) \geq d(\mathbf{\Delta} u)(\nabla u) + Ric_N(\nabla u) + \frac{(\mathbf{\Delta} u)^2}{N}$$

$$+ \frac{N}{N-1} \left(\frac{\mathbf{\Delta} u}{N} - H_u \right)^2$$

$$= -\lambda_1 F^2(\nabla u) + Ric_N(\nabla u) + \frac{1}{N} \lambda_1^2 u^2$$

$$+ \frac{N}{N-1} \left(\frac{\lambda_1 u}{N} + H_u \right)^2 \qquad (8.3.19)$$

from (7.1.16). Plugging (8.3.18)–(8.3.19) and (8.3.16) into (8.3.17) yields

$$L_u(\Phi) \geq a_1' \Phi^2 + a_2' \Phi + 2\psi a_3' + 2\psi Ric_N(\nabla u), \qquad (8.3.20)$$

where

$$a_1' := \frac{1}{\psi} \left\{ \frac{\ddot{\psi}}{\psi} + \frac{\dot{\psi}^2}{\psi^2} \left(\frac{N}{2(N-1)} - 2 \right) \right\},$$

$$a_2' := \frac{\varphi}{\psi^2}(\psi\ddot{\psi} - 2\dot{\psi}^2) - \frac{(N+1)\lambda_1 \dot{\psi}}{(N-1)\psi} u - \frac{N\dot{\psi}}{(N-1)\psi} \dot{\varphi} - 2\lambda_1 - \ddot{\varphi};$$

$$a_3' := \frac{\lambda_1^2}{N-1} u^2 + \frac{(N+1)\lambda_1}{2(N-1)} u\dot{\varphi} + \frac{N}{4(N-1)} \dot{\varphi}^2 - \lambda_1 \varphi - \frac{1}{2}\varphi\ddot{\varphi}.$$

In the following, we will prove this theorem by contradiction in this case.
   Assume $F(x, \nabla u) > \dot{v}(v^{-1}(u(x)))$ at x_0. Then, at x_0, we have $\Phi > 0$ and

$$\psi Ric_N(\nabla u) \geq K\psi F^2(\nabla u) = K\psi \left(\frac{\Phi}{\psi} + \varphi \right) = K\Phi + K\psi\dot{v}^2. \qquad (8.3.21)$$

Putting (8.3.21) into (8.3.20) yields

$$L_u(\Phi) \geq a_1 \Phi^2 + a_2 \Phi + a_3,$$

where $a_1 := a_1'$, $a_2 := a_2' + 2K$ and $a_3 := 2\psi(a_3' + K\dot{v}^2)$ are functions of $u(x)$ respectively. Let $t := v^{-1}(u(x))$ and $s := u(x) = v(t)$. Then $\varphi(s) = \dot{v}(t)^2$ and

$$\dot{\varphi} = 2\ddot{v}, \quad \ddot{\varphi} = 2\dot{v}(t)\frac{d\ddot{v}}{dt}.$$

From this, we have

$$\frac{a_3}{2\psi} = \frac{\lambda_1^2 v^2}{N-1} + \frac{N+1}{N-1}\lambda_1 v\ddot{v} + \frac{N}{N-1}\ddot{v}^2 - \lambda_1 \dot{v}^2 - \frac{d\ddot{v}}{dt}\dot{v} + K\dot{v}^2$$

$$= \frac{1}{N-1}\left(\ddot{v} - T\dot{v} + \lambda_1 v\right)\left(N\ddot{v} + T\dot{v} + \lambda_1 v\right) - \dot{v}\frac{d}{dt}\left(\ddot{v} - T\dot{v} + \lambda_1 v\right)$$

$$= 0,$$

where we used equation (8.3.12) with T satisfying $\dot{T} = K + \frac{T^2}{N-1}$. For a_1, a_2, we introduce the following functions:

$$X(t) := \lambda_1 \frac{v(t)}{\dot{v}(t)}, \quad \psi := e^{\int h(v(t))}, \quad f(t) := -h(v(t))\dot{v}(t). \quad (8.3.22)$$

With these notations, we have $\dot{\psi} = \psi h$, $\ddot{\psi} = \psi(h^2 + \dot{h})$ and

$$\dot{f} = -\dot{h}(v)\dot{v}^2 + f(T - X), \quad (8.3.23)$$

where we used equation (8.3.12). From (8.3.22)–(8.3.23), we have

$$a_1 \psi \dot{v}^2 = -\frac{N-2}{2(N-1)}f^2 + f(T - X) - \dot{f} := \eta(f(t), t) - \dot{f},$$

and

$$a_2 = -f^2 + \left(\frac{3N-1}{N-1}T - 2X\right)f - 2T\left(\frac{N}{N-1}T - X\right) - \dot{f}$$

$$:= \beta(f(t), t) - \dot{f}.$$

By Corollary 3 in [BQ], there exists a bounded function f on $[\min u, \max u] \subset (\min v, \max v)$ such that

$$\dot{f} < \min\{\eta(f(t), t), \beta(f(t), t)\},$$

which implies that $a_1 > 0$ and $a_2 > 0$. Therefore, $L_u(\Phi) = a_1\Phi^2 + a_2\Phi + a_3 > 0$ at x_0, which contradicts with the second inequality in (8.3.14). Thus, (8.3.13) follows.

Case 2. $\partial M \neq \emptyset$ and $x_0 \in \partial M$.

If $\nabla u(x_0) = 0$, nothing needs to be proved. Thus, we can assume that $x_0 \in M_u$. Similar to the proof of Theorem 8.3.1, there exists a normal vector field $\nu_{\nabla u}$ on ∂M pointing outward such that $d\Phi(\nu_{\nabla u})(x_0) \geq 0$. On the other hand, at the point x_0,

$$d\Phi(\nu_{\nabla u}) = 2\psi(u)g_{\nabla u}(D_{\nu_{\nabla u}}^{\nabla u}\nabla u, \nabla u)$$

$$= 2\psi(u)g_{\nabla u}(D_{\nabla u}^{\nabla u}\nabla u, \nu_{\nabla u}).$$

By the convexity of ∂M and Proposition 2.4.1, we have $g_{\nabla u}(\nu_{\nabla u}, D_{\nabla u}^{\nabla u}\nabla u) \leq 0$. Thus, $d\Phi(\nu_{\nabla u})(x_0) = 0$ and hence $\nabla^{\nabla u}\Phi(x_0) = 0$. The rest of proof proceeds as Case 1.   □

Remark 8.3.1. If u is the first eigenfunction of Δ corresponding to λ, then λ is the first eigenvalue of $\overleftarrow{\Delta}$ of reverse Finsler structure $\overleftarrow{F}$ corresponding to the eigenfunction $-u$. Moreover, Ric_N and $\overleftarrow{\mathrm{Ric}}_N$ have the same lower bound. Thus, Proposition 8.3.1 still holds on $(M, \overleftarrow{F})$. Hence, we have

$$\overleftarrow{F}\left(x, \overleftarrow{\nabla}(-u(x))\right) \leq \dot{v}\left(v^{-1}(-u(x))\right). \tag{8.3.24}$$

Next, we will compare the maximum of the eigenfunction $u(x)$ with that of the solution $v(t)$ of the one-dimensional model, which is essential for the proof of the main theorem in this section.

Given eigenfunctions u and v as before, let $t_0 \in (a, b)$ be the unique zero of v and let $h = v^{-1} \circ u$. We define the measure ς on $[a, b]$ by $\varsigma(A) := m(h^{-1}(A))$, where m is the measure on (M, F). Equivalently, for any bounded measurable $f : [a, b] \to \mathbb{R}$, we have

$$\int_a^b f(s)d\varsigma(s) = \int_M f(h(x))dm. \tag{8.3.25}$$

By Proposition 5 in [BQ],

$$R(s) = -\exp\left(\lambda \int_{t_0}^s \frac{v}{\dot{v}}dt\right) \int_a^s v(t)d\varsigma(t) \tag{8.3.26}$$

is increasing on $(a, t_0]$ and decreasing on $[t_0, b)$.

Denote by $\mu_{K,N}$ the invariant measure associated to the operator $L_{k,N}(v)(t) = \ddot{v} - T(t)\dot{v}$, that is, a measure satisfying $\int_a^b L_{K,N}(v)d\mu_{K,N} = 0$ for $\dot{v}(a) = \dot{v}(b) = 0$. Equivalently, $\mu_{K,N}$ satisfies $\dot{\mu}_{K,N} + T\mu_{K,N} = 0$. For instance, in the case of $K > 0$, $d\mu_{k,N} = \cos^{N-1}(\sqrt{K/(N-1)}t)dt$. Note that the function v satisfies (8.3.12), which implies that

$$\frac{d}{dt}\log(\mu_{K,N}\dot{v}) = -\lambda\frac{v}{\dot{v}}. \tag{8.3.27}$$

Integrating this on $[a, s]$ and using $\dot{v}(a) = 0$ yield

$$\mu_{K,N}(s)\dot{v}(s) = -\lambda \int_a^s v\mu_{K,N}dt. \qquad (8.3.28)$$

Similarly, from (8.3.27), we have

$$\frac{\mu_{K,N}(t_0)\dot{v}(t_0)}{\mu_{K,N}(s)\dot{v}(s)} = \exp\left(\lambda \int_{t_0}^s \frac{v}{\dot{v}}dt\right). \qquad (8.3.29)$$

Plugging (8.3.28) and (8.3.29) into (8.3.26) yields

$$R(s) = C_0 \frac{\int_a^s v(t)d\varsigma(t)}{\int_a^s v(t)\mu_{K,N}(t)dt} = C_0 \frac{\int_{u \leq v(s)} u(x)dm}{\int_a^s v(t)d\mu_{K,N}}, \qquad (8.3.30)$$

which is increasing on $[a, t_0]$ and decreasing on $[t_0, b]$, where $\lambda C_0 = \mu_{K,N}(t_0)\dot{v}(t_0)$ and $d\mu_{K,N} := \mu_{K,N}(t)dt$ is the measure on $[a, b]$.

Lemma 8.3.3. *Let $x_0 \in M$ with $u(x_0) = -1$. Then $m(B_{r_\epsilon}^{\pm}(x_0)) \leq m(\{u(x) \leq -1 + \epsilon\})$ and $\mu_{K,N}(\{v(t) \leq -1 + \epsilon\}) \leq \mu_{K,N}([a, a + r_\epsilon])$ for any sufficiently small $\epsilon > 0$, where $r_\epsilon := v^{-1}(-1 + \epsilon) - a$ and $B_r^{\pm}(x_0) = B_r^+(x_0) \cup B_r^-(x_0)$.*

Proof. We only prove this lemma for $B_r^+(x_0)$ (a similar argument for $B_r^-(x_0)$). Let $\bar{x} \in B_{r_\epsilon}^+(x_0)$ and $\sigma : [0, \ell] \to M$ be a unit speed minimizing geodesic from x_0 to $\bar{x}$. Define $f(t) := u(\sigma(t))$. Then, by Proposition 8.3.1,

$$|df/dt| = |du(\dot{\sigma})(t)| \leq F(\nabla u)|_{\sigma(t)} \leq \dot{v}|_{v^{-1}(f(t))}. \qquad (8.3.31)$$

Let $h(t) := v^{-1}(f(t))$. Then $|\dot{h}(t)| \leq 1$ from (8.3.31). Consequently, we have $\frac{d}{dt}(v^{-1}(f(t))) \leq 1$, which implies that $a \leq v^{-1}(f(t)) \leq a+t$. Since $\dot{v}$ is increasing in a neighborhood of a, we have $\dot{v}(v^{-1}(f(t))) \leq \dot{v}(a + t)$. Thus,

$$f(t) + 1 \leq \int_0^t \dot{v}(v^{-1}(f(s)))ds \leq \int_0^t \dot{v}(a + s)ds = v(a+t) + 1. \qquad (8.3.32)$$

Since $\bar{x} \in B_{r_\epsilon}^+(x_0)$, $\ell = d(x_0, \bar{x}) \leq r_\epsilon = v^{-1}(-1 + \epsilon) - a$. From this and (8.3.32), one obtains

$$u(\bar{x}) + 1 = f(\ell) + 1 \leq v(a + \ell) + 1 \leq v(v^{-1}(-1 + \epsilon)) + 1 = \epsilon,$$

which means that $\bar{x} \in \{u(x) \leq -1 + \epsilon\}$. Thus, $m(B_{r_\epsilon}^+(x_0)) \leq m(\{u(x) \leq -1 + \epsilon\})$.

Moreover, for any $t \in \{v(t) \leq -1 + \epsilon\}$, we have $t \leq v^{-1}(-1 + \epsilon) = a + r_\epsilon$, which means that $\{v(t) \leq -1 + \epsilon\} \subset [a, a + r_\epsilon]$. Thus, $\mu_{K,N}(\{v(t) \leq 1 + \epsilon\}) \leq \mu_{K,N}([a, a + r_\epsilon])$. $\square$

Proposition 8.3.2. *Let (M, F, m), λ_1, u be as in Proposition 8.3.1 and $1 < N < \infty$. Let $v_{K,N}$ be the solution of (8.3.12) on some interval (a, b) with initial data $v(a) = -1$ and $\dot{v}(a) = 0$, where*

$$a = \begin{cases} -\dfrac{\pi}{2\sqrt{K/(N-1)}}, & \text{if } K > 0, \\ 0, & \text{if } K \le 0, \end{cases}$$

and $b = b(a)$ is the first number after a with $\dot{v}_{K,N}(b) = 0$. Denote $\max(v_{K,N}) = v_{K,N}(b) = \max(v)$. Assume that $\min u = -1$ and $\lambda_1 > \max\{\frac{KN}{N-1}, 0\}$. Then $\max(u(x)) \ge \max(v_{K,N}(t))$.

Proof. We prove this by contradiction. Assume that $\max(u(x)) < \max(v_{K,N}(t))$. Choose $x_0 \in M$ with $u(x_0) = -1$ and choose a constant k such that $k < -\frac{1}{2}$. If $-1 \le u(x) \le k$, then $2|u(x)| > 1$. From (8.3.30), we have

$$m(\{u(x) \le k\}) < 2 \int_{u(x) \le k} |u| \, dm = -2 \int_{u(x) \le k} u \, dm$$

$$\le -2C_0^{-1} R(t_0) \int_{v(s) \le k} v \, d\mu_{K,N}$$

$$= 2C_0^{-1} R(t_0) \int_{v(s) \le k} |v| \, d\mu_{K,N}.$$

Note that $-1 \le \max(u) < v(b) = \max(v(t)) \le k$ on the subset $\{v(s) \le k\}$. We have $\frac{1}{2} < |v(t)| \le 1$ on this subset. Thus, letting $k = -1 + \epsilon$ for some ϵ small enough such that $k < -\frac{1}{2}$ and using Lemma 8.3.3, there is a positive constant C' such that

$$m(\{u(x) \le -1 + \epsilon\}) < 2C_0^{-1} R(t_0) \mu_{K,N}(\{v(t) \le -1 + \epsilon\})$$

$$\le 2C_0^{-1} R(t_0) \mu_{K,N}([a_0, a_0 + r_\epsilon]) \le C' r_\epsilon^N.$$

By Lemma 8.3.1 again, we have

$$m(B_{r_\epsilon}^{\pm}(x_0)) \le m\{u(x) \le -1 + \epsilon\} < C' r_\epsilon^N,$$

which will lead to a contradiction. Since $\max(u) < \max(v_{K,N})$ and $\max(v_{K,N})$ is continuous with respect to (K, N), we also have $\max(u) < \max(v_{K,N'})$ for $N' > N$ close to N. Arguing as before, one obtains that

$$m(B_r^{\pm}(x_0)) < C r^{N'}$$

for some constant $C > 0$ and a sufficiently small number $r > 0$. However, $m(B_r^{\pm}(x_0)) \ge C r^n$ by the manifold structure of M. This is a contradiction.

The previous arguments also work in the case when $x_0 \in \partial M$. The proof is completed. □

Now, we begin to state and prove the main result in this subsection.

Theorem 8.3.2 ([WX]). *Let (M, F, m) be an n-dimensional compact Finsler measure space without boundary or with a convex boundary and d be the diameter of M. Assume that $Ric_N \geq K$ for some $N \in [n, \infty]$ and $K \in \mathbb{R}$. If λ_1 is the first (nonzero) eigenvalue of Finsler Laplacian i.e., $\Delta u = -\lambda_1 u$ in M (in the weak sense) with a Neumann boundary condition $\nabla u \in T_x(\partial M)$ when $\partial M \neq \emptyset$, then*

$$\lambda_1 \geq \lambda_1(K, N, d), \tag{8.3.33}$$

where $\lambda_1(K, N, d)$ represents the first (nonzero) eigenvalue of the one-dimensional model problem on $[-d/2, d/2]$

$$\begin{cases} \ddot{v} - T(t)\dot{v} + \lambda_1(K, N, d)v = 0, \\ \dot{v}(-d/2) = \dot{v}(d/2) = 0, \end{cases} \tag{8.3.34}$$

here $T(t)$ is given by (8.3.8).

Proof. Without loss of generality, we can assume that $\min u = -1$ and $0 < \max u = k \leq 1$ by rescaling. By Theorem 8.3.1, we have $\lambda_1 \geq \frac{NK}{N-1}$ in the case of $K > 0$. Choose $\tilde{K} < K$ close to K, we have $\lambda_1 > \max\{\frac{N\tilde{K}}{N-1}, 0\}$. Therefore, Lemma 8.3.1 and Proposition 8.3.2 imply that there exists an interval $[a, b]$ which has the first Neumann eigenvalue λ_1 and a corresponding eigenfunction v such that $\min v = -1 = \min u$ and $\max v = \max u = k \leq 1$. Choose $x_1, x_2 \in M$ with $u(x_1) = \min u$, $u(x_2) = \max u$ and $\gamma : [0, 1] \to M$ the normal minimal geodesic from x_1 to x_2. Consider the subset I of $[0, 1]$ such that $\frac{d}{dt}u(\gamma(t)) \geq 0$. By using Proposition 8.3.1, we have

$$d \geq \int_0^1 F(\dot{\gamma}(t))dt \geq \int_I F(\dot{\gamma}(t))dt \geq \int_0^1 \frac{1}{F^*(du)}du(\dot{\gamma}(t))dt$$

$$= \int_{-1}^k \frac{1}{F(\nabla u)}du \geq \int_{-1}^k \frac{1}{\dot{v}(v^{-1}(u))}du = \int_a^b dt = b - a. \tag{8.3.35}$$

A general property says that $\lambda_1(\tilde{K}, N, d)$ is monotone decreasing with respect to d. Hence, $\lambda_1(\tilde{K}, N, b - a) \geq \lambda_1(\tilde{K}, N, d)$. Finally, it follows from

Lemma 8.3.2 that

$$\lambda_1 \geq \lambda_1(\tilde{K}, N, b - a) \geq \lambda_1(\tilde{K}, N, d).$$

Letting $\tilde{K} \to K$ yields the conclusion.   □

Theorems 8.3.2 shows that the lower bound for first eigenvalue is sharp in the sense that it can be attained at the first eigenvalue for one-dimensional model (8.3.34). It is difficult to get the exact solutions of (8.3.34). However, from Theorem 8.3.2, one can obtain the following unified lower bound ([Xia4]).

Corollary 8.3.1. *Let* (M, F, m), Ric_N, v *and* d *be the same as in Theorem 8.3.2. If* λ_1 *be the first (nonzero) eigenvalue of Finsler Laplacian, i.e.,* $\boldsymbol{\Delta} u = -\lambda_1 u$ *(in the weak sense) with* $\nabla u \in T_x(\partial M)$ *when* $\partial M \neq \emptyset$*, then*

$$\lambda_1 \geq \sup_{s \in (0,1)} \left\{ 4s(1 - s)\frac{\pi^2}{d^2} + sK \right\}. \tag{8.3.36}$$

Proof. Let $f = \dot{v}$. From (8.3.34), we get the following ODE on $(-d/2, d/2)$:

$$\ddot{f} - T\dot{f} = -\left(\lambda_1(K, N, d) - \dot{T}\right) f, \quad f(-d/2) = f(d/2) = 0. \tag{8.3.37}$$

By the maximum principle, we can prove that f must have fixed sign on $(-d/2, d/2)$. So, we can assume that $f(t) > 0$ for all $t \in (-d/2, d/2)$. For any $a > 1$, by multiplying f^{a-1} on the both sides of (8.3.37) and integrating over $(-d/2, d/2)$, we get

$$\int_{-d/2}^{d/2} \ddot{f} f^{a-1} dt - \int_{-d/2}^{d/2} T f^{a-1} \dot{f} dt = -\int_{-d/2}^{d/2} \left(\lambda_1(N, K, d) - \dot{T}\right) f^a dt. \tag{8.3.38}$$

Since $f(\pm d/2) = 0$, via integrating by parts, we have

$$\int_{-d/2}^{d/2} \ddot{f} f^{a-1} dt = -(a - 1) \int_{-d/2}^{d/2} f^{a-2} \dot{f}^2 dt = -\frac{4(a - 1)}{a^2} \int_{-d/2}^{d/2} \left(\frac{df^{\frac{a}{2}}}{dt}\right)^2 dt, \tag{8.3.39}$$

and

$$\int_{-d/2}^{d/2} T f^{a-1} \dot{f} dt = -\frac{1}{a} \int_{-d/2}^{d/2} \dot{T} f^a dt. \tag{8.3.40}$$

Putting (8.3.39) and (8.3.40) into (8.3.38) yields

$$\frac{4(a-1)}{a^2} \int_{-d/2}^{d/2} \left(\frac{df^{\frac{a}{2}}}{dt} \right)^2 dt = \lambda_1(N, K, d) \int_{-d/2}^{d/2} f^a dt - \frac{a-1}{a} \int_{-d/2}^{d/2} \dot{T} f^a dt$$

$$\leq \left(\lambda_1(N, K, d) - \frac{a-1}{a} K \right) \int_{-d/2}^{d/2} (f^{\frac{a}{2}})^2 dt,$$

(8.3.41)

where we used $\dot{T} = K + \frac{T^2}{N-1} \geq K$ in the inequality. Let $s = 1 - \frac{1}{a} \in (0, 1)$. Then (8.3.41) becomes

$$4s(1-s) \int_{-d/2}^{d/2} \left(\frac{df^{\frac{a}{2}}}{dt} \right)^2 dt \leq \int_{-d/2}^{d/2} (\lambda_1(N, K, d) - sK) (f^{\frac{a}{2}})^2 dt.$$

(8.3.42)

By Wirtinger's inequality, it follows from (8.3.42) that

$$4s(1-s) \left(\frac{\pi}{d} \right)^2 \leq \lambda_1(N, K, d) - sK,$$

that is, $\lambda_1(N, K, d) \geq 4s(1-s) \left(\frac{\pi}{d} \right)^2 + sK$. Therefore, by Theorem 8.3.2, we get the desired estimate. $\qquad \square$

Note that $d \leq \pi\sqrt{(N-1)/K}$ if $K > 0$ by Theorem 6.3.4. A direct calculation gives

$$\sup_{s \in (0,1)} \left\{ 4s(1-s)\frac{\pi^2}{d^2} + sK \right\} = \begin{cases} 0, & \text{if } Kd^2 < -4\pi^2, \\ (\frac{\pi}{d} + \frac{Kd}{4\pi})^2, & \text{if } Kd^2 \in [-4\pi^2, 4\pi^2], \\ K, & \text{if } Kd^2 \in (4\pi^2, (N-1)\pi^2]. \end{cases}$$

In particular, if $K = 0$, then the RHS in (8.3.36) arrives the maximum $\frac{\pi^2}{d^2}$. In this case, $\lambda_1 \geq \frac{\pi^2}{d^2}$. We shall see that this is sharp in the next section. When $K > 0$ or $K < 0$, the RHS in (8.3.36) is not optimal. In fact, when $K > 0$, it is easy to see that

$$v(t) = \sin(t\sqrt{K/(N-1)}), \quad t \in \left[-\frac{\pi}{2\sqrt{K/(N-1)}}, \frac{\pi}{2\sqrt{K/(N-1)}} \right]$$

is a special solution of (8.3.34) associated to the first eigenvalue $\lambda_1(K, N, d) = \frac{NK}{N-1}$. We will see that this is optimal for λ_1 on a compact Finsler manifold without boundary under $\mathrm{Ric}_N \geq K > 0$ for some

$N \in [n, \infty)$ in the next section. However, the lower bounds in Theorems 8.3.1–8.3.2 in the case when $K < 0$ are never attained even in Riemannian case.

Remark 8.3.2. For the generalizations of Theorems 8.3.1–8.3.2 to the case of Finsler p-Laplacian, the readers refer to [Xia5]–[Xia6] for more details.

8.4   Characterizations of Extreme Manifolds

As pointed out at the end of Section 8.3.2, λ_1 may attains its extremum in the case when $\text{Ric}_N \geq K$ for $K \geq 0$. We shall characterize Finsler manifolds on which the first eigenvalue λ_1 attains its extremum. We argue this according to two cases when $K = 0$ and $K > 0$. The related arguments and results are from [Xia4] and [Xia6] respectively.

8.4.1   *Case 1.* $K = 0$

Recall that the solution T of the equation $\dot{T} = K + \frac{T^2}{N-1}$ for $N \in (1, \infty]$ is given by (8.3.8). When $K = 0$, it is obvious that $T = 0$ and $T = -\frac{N-1}{t}$.

If $T = 0$, it is easy to see that the first eigenfunction of (8.3.12) is $v(t) = \sin(\sqrt{\lambda}t)$ up to translations and dilatations. In this case, $\delta(a) := b(a) - a = \frac{\pi}{\sqrt{\lambda_1}}$ for any $a \in \mathbb{R}$. If $T = -\frac{N-1}{t}$, we also denote $\delta(a) = b(a) - a$ as a function of $a \in [0, +\infty)$ with respect to the one-dimensional model problem (8.3.12) with $T = -\frac{N-1}{t}$. In this case, we have the following lemma.

Lemma 8.4.1. *The function* $\delta : [0, +\infty] \to \mathbb{R}^+$ *is a continuous function such that*

$$\delta(a) > \frac{\pi}{\sqrt{\lambda_1}}, \quad \text{for } a \in [0, +\infty), \tag{8.4.1}$$

$$\delta(a) = \frac{\pi}{\sqrt{\lambda_1}}, \quad \text{for } a = +\infty, \tag{8.4.2}$$

$\mathfrak{m}(a) := v_a(b(a)) < 1, \lim_{a \to +\infty} \mathfrak{m}(a) = 1$ *and* $\mathfrak{m}(a) = 1$ *if and only if* $a = +\infty$.

The proof can be found in [Va]. We omit it here. It follows from Lemma 8.4.1 and the proof of Theorem 8.3.2 (see (8.3.35)) that $d \geq \delta(a) \geq \frac{\pi}{\sqrt{\lambda_1}}$, i.e., $\lambda_1 \geq \frac{\pi^2}{d^2}$ and the equality holds only if $a = +\infty$, that is, $\max(u) = -\min(u) = \max(v) = -\min(v)$. Consequently, one obtains the following.

Lemma 8.4.2. *Let (M, F, m) be an n-dimensional compact Finsler measure space without boundary or with a convex boundary and d be the diameter of M. Assume that $Ric_N \geq 0$ for some $N \in [n, \infty]$ and $\Delta u = -\lambda_1 u$ (in the weak sense) with a Neumann boundary condition $\nabla u \in T_x(\partial M)$ when $\partial M \neq \emptyset$. Then $\lambda_1 \geq \frac{\pi^2}{d^2}$. Moreover, a necessary (but not necessarily sufficient) condition for equality to hold is that $\max u = -\min(u)$ for any eigenfunction $u(x)$.*

Without loss of generality, we always assume that $\min(u) = -1$ by rescaling.

Proposition 8.4.1. *Under the same assumptions as in Lemma 8.4.2, if $\lambda_1 = \frac{\pi^2}{d^2}$, then*

(1) *the function $P(x) := F^2(x, \nabla u) + \lambda_1 u(x)^2$ is a constant on M. In particular, $P(x) = \lambda_1$ on M. Moreover, $M_u = \{x \in M \mid u(x) \neq \pm 1\}$.*
(2) *the vector field $X := \frac{\nabla u}{F(\nabla u)}$ is a geodesic field of F on M_u.*

Proof. (1) Since $\lambda_1 = \frac{\pi^2}{d^2}$, the one-dimensional model function $v(t)$ satisfies $-\min(v) = \max v = \max u = -\min u = 1$ and $a = +\infty$ by Lemma 8.4.2. In this case, $v(t) = \sin(\sqrt{\lambda_1} t)$ up to translations and dilations. Thus, $\dot{v}^2 = \lambda_1(1 - v^2)$. From Proposition 8.3.1, one obtains

$$F^2(\nabla u) \leq [\dot{v}(v^{-1}(u(x)))]^2 = \lambda_1(1 - u^2), \qquad (8.4.3)$$

which implies that $P(x) \leq \lambda_1$ everywhere on M.

Choose $x_1, x_2 \in M$ such that $u(x_1) = -1$ and $u(x_2) = 1$. Let $\gamma : [0, 1] \to M$ be a normal minimal geodesic from x_1 to x_2 as in the proof of Theorem 8.3.2. From (8.3.35) and $\lambda_1 = \frac{\pi^2}{d^2}$, we get $F(\nabla u) = \dot{v}(v^{-1}(u(x)))$ along γ, which means that $P(x) = \lambda_1$ along γ from (8.4.3). To prove that $P(x) = \lambda_1$ on M, let us consider the following linear operator

$$\mathfrak{L}(\phi) := L_u(\phi) - \frac{1}{2F^2(\nabla u)} g_{\nabla u}\left(\nabla^{\nabla u}(F^2(\nabla u) - \lambda_1 u^2), \nabla^{\nabla u}\phi\right), \qquad (8.4.4)$$

for any smooth function ϕ on M. Obviously, $\mathfrak{L}$ is uniformly elliptic on M_u. Note that the first order part of $\mathfrak{L}$ plays no role in the maximum principle. By a direct calculation and (7.1.14), we have on M_u

$$\nabla^{\nabla u} F^2(\nabla u) = 2\nabla^2 u(\nabla u), \qquad \nabla^{\nabla u} P = 2\nabla^2 u(\nabla u) + 2\lambda_1 u \nabla u, \qquad (8.4.5)$$

and
$$L_u(P) = \frac{1}{2}L_u(F^2(\nabla u)) + \frac{\lambda_1}{2}L_u(u^2) = \mathrm{Ric}_\infty(\nabla u) + \|\nabla^2 u\|^2_{HS(\nabla u)} - \lambda_1^2 u^2$$

$$\geq \mathrm{Ric}_N(\nabla u) + \|\nabla^2 u\|^2_{HS(\nabla u)} - \lambda_1^2 u^2 \tag{8.4.6}$$

where we used Theorem 7.1.1 and $\boldsymbol{\Delta} u = -\lambda_1 u$ on M_u. Observe that

$$\|\nabla^2 u(\nabla u)\|^2_{HS(\nabla u)} = g_{\nabla u}\left(\nabla^2 u(\nabla u), \nabla^2 u(\nabla u)\right) \leq F^2(\nabla u)\|\nabla^2 u\|^2_{HS(\nabla u)}.$$

$$\tag{8.4.7}$$

Thus, from (8.4.4)–(8.4.7), we get

$$\frac{1}{2}\mathcal{L}(P) \geq \mathrm{Ric}_N(\nabla u) + \|\nabla^2 u\|^2_{HS(\nabla u)} - \frac{1}{F^2(\nabla u)}\|\nabla^2 u(\nabla u)\|^2_{HS(\nabla u)} \geq 0.$$

$$\tag{8.4.8}$$

By the maximum principle, the set $\tilde{M} := \{x \in M \mid P(x) = \lambda_1\}$ is open in $\Omega := \{x \in M \mid u(x) \neq \pm 1\}$. It is obvious that $\tilde{M}$ is closed in Ω. So, $P = \lambda_1$ on the connected component Ω_1 of Ω containing γ.

Let Ω_2 be another connected component of Ω. We can choose $x_i \in \Omega_i$ with $u(x_i) = 0$ $(i = 1, 2)$. Let $c : [0, \ell] \to M$ be a unit speed minimal geodesic from x_1 to x_2. Then there exists a point t_0 such that $c(t_0) \in u^{-1}\{-1, 1\}$. Otherwise, $\Omega_1 = \Omega_2$. Without loss of generality, let $u(c(t_0)) = 1$. Choose the subset $I_1 \subset [0, t_0]$ such that $\frac{d}{dt}u(c(t)) > 0$ and the subset $I_2 \subset [t_0, \ell]$ such that $\frac{d}{dt}u(c(t)) < 0$. Note that $F(\nabla u)|_{c(t)} = F^*(du(\dot{c}(t)))$ and

$$\overleftarrow{F}(\overleftarrow{\nabla} u)|_{c(t)} = F(-\overleftarrow{\nabla} u)|_{c(t)} = F(\nabla(-u))|_{c(t)} = F^*(-du(\dot{c}(t))).$$

By the Cauchy–Schwarz inequality, Lemmas 8.4.1–8.4.2, and Proposition 8.3.1 (also (8.3.24)), we have

$$d \geq \int_0^{t_0} F(\dot{c}(t))dt + \int_{t_0}^\ell F(\dot{c}(t))dt \geq \int_{I_1} F(\dot{c}(t))dt + \int_{I_2} F(\dot{c}(t))dt$$

$$\geq \int_0^{t_0} \frac{1}{F^*(du)}du(\dot{c}(t))dt - \int_{t_0}^\ell \frac{1}{F^*(-du)}du(\dot{c}(t))dt$$

$$= \int_0^1 \frac{du}{F(\nabla u)} + \int_1^0 \frac{-du}{\overleftarrow{F}(\overleftarrow{\nabla} u)}$$

$$= \int_0^1 \frac{du}{F(\nabla u)} + \int_{-1}^0 \frac{du}{\overleftarrow{F}(\overleftarrow{\nabla}(-u))} \geq 2\int_0^1 \frac{du}{v(v^{-1}(u))} = 2\int_0^b dt = \frac{\pi}{\sqrt{\lambda_1}},$$

$$\tag{8.4.9}$$

where $b = \frac{\pi}{2\sqrt{\lambda_1}}$. Since $d = \frac{\pi}{\sqrt{\lambda_1}}$, all inequalities become equalities in (8.4.9). Thus, we have $\frac{d}{dt}u(c(t)) > 0$ a.e. on $[0, t_0]$ and $\frac{d}{dt}u(c(t)) < 0$ a.e. on $[t_0, \ell]$. Moreover, $\frac{d}{dt}[u(c(t))] = F(\nabla u) = \dot{v}(v^{-1}(u))|_{c(t)}$ a.e. on $[0, t_0]$ and $-\frac{d}{dt}[u(c(t))] = \overleftarrow{F}(\overleftarrow{\nabla} u) = \dot{v}(v^{-1}(-u))|_{c(t)}$ a.e. on $[t_0, \ell]$, which implies that, up to a translation in the domain of definition, $\pm u(c(t)) = v(t) = \sin\sqrt{\lambda_1}t$. Thus, for the connected component Ω_2, there exists an interior point x_0 in Ω_2 such that $P(x_0) = \lambda_1$ by (8.4.3). By the maximum principle, $P(x) = \lambda_1$ on Ω_2 and hence on Ω, which also shows that $u(x) \neq \pm 1$ if and only if $\nabla u(x) \neq 0$ on Ω, that is, $M_u = \Omega$. Obviously, $P(x) = \lambda_1$ on $M \backslash M_u$. Thus, $P(x) = \lambda_1$ on the whole manifold M.

(2) By (1), $P(x) = \lambda_1$ on M. Thus, $\mathcal{L}(P) = L_u(P) = 0$ and all inequalities in (8.4.6)–(8.4.8) become equalities, which mean that

$$\text{Ric}_\infty(\nabla u) = \text{Ric}_N(\nabla u) = 0, \tag{8.4.10}$$

$$\|\nabla^2 u(\nabla u)\|^2_{HS(\nabla u)} = F^2(\nabla u)\|\nabla^2 u\|^2_{HS(\nabla u)} \quad \text{on } M_u. \tag{8.4.11}$$

On the other hand, from (8.4.5), on M_u, one obtains

$$\nabla^2 u(\nabla u) = -\lambda_1 u \nabla u, \tag{8.4.12}$$

which implies that

$$\|\nabla^2 u\|^2_{HS(\nabla u)} = \lambda_1^2 u^2 \tag{8.4.13}$$

because of (8.4.11). Hence, by (8.4.12), we have

$$D_X^X X = \frac{1}{F(\nabla u)} D_{\nabla u}^{\nabla u}\left(\frac{\nabla u}{F(\nabla u)}\right)$$

$$= -\frac{1}{F^4(\nabla u)}\left[g_{\nabla u}(\nabla^2 u(\nabla u), \nabla u)\nabla u - F^2(\nabla u)\nabla^2 u(\nabla u)\right] = 0, \tag{8.4.14}$$

that is, X is a geodesic field of F on M_u. $\qquad\square$

Note that $\overleftarrow{\Delta}(-u) = -\lambda_1(-u)$, $\overleftarrow{\nabla} u = -\nabla(-u)$ and $\overleftarrow{\text{Ric}}_N(x, V) = \text{Ric}_N(x, -V)$ for any vector $V \in T_x M$. From Proposition 8.4.1, one obtains the following result.

Proposition 8.4.1'. *Let (M, F, m), Ric_N, λ_1 and d be the same as in Lemma 8.4.3. Assume $\overleftarrow{F}$ is the reverse Finsler structure of F. Then*

(1) *the function $\overleftarrow{P}(x) := \overleftarrow{F}^2(x, \overleftarrow{\nabla}(-u)) + \lambda_1 u(x)^2 = P(x)$ is a constant on M. In particular, $\overleftarrow{P}(x) = \lambda_1$ on M.*

(2) *the vector field $\overleftarrow{X} := \frac{\overleftarrow{\nabla}(-u)}{\overleftarrow{F}(\overleftarrow{\nabla}(-u))} = -X$ is a geodesic field of $\overleftarrow{F}$ on M_u.*

Proposition 8.4.2. *Under the same assumptions as in Proposition 8.4.1, the weighted Riemannian manifold (M_u, g_X, m) splits isometrically as $((-d/2, d/2) \times M_0, dt^2 \otimes h_X, m)$ with $M_0 = u^{-1}(0)$, where h_X is the induced Riemannian metric on M_0 from g_X defined by (3.1.16), and Ψ is constant along each integral curve $\varphi_s(x) = \varphi(s, x)$ of X for each $x \in M_0$.*

Proof. Let $\varphi_s = \exp(sX)$ be the one-parameter transformation group generated by X on M_u. Since X is a geodesic vector field, X is also a geodesic field of the Riemannian metric g_X on M_u by Lemma 4.1.2 and $\varphi_s(x) = \exp_x(sX)$ is a geodesic of F (also g_X) emanating from $x \in M_u$ with the initial tangent vector X_x. Note that $F(\nabla u) \neq 0$ on the level set $M_0 = u^{-1}(0)$. Then M_0 is a hypersurface (which may have more than one components) of M and $g_X(X, Y) = \frac{1}{F(\nabla u)} g_{\nabla u}(\nabla u, Y) = \frac{1}{F(\nabla u)} du(Y) = 0$ for any $x \in M_0$ and $Y \in T_x M_0$. Define a map $\varphi : I \times M_0 \to M_u$ by $\varphi(s, x) = \varphi_s(x)$. We claim that $I = (-d/2, d/2)$ and φ is a diffeomorphism.

In fact, for any $z \in M_u$, we can assume $0 \le u(z) < 1$. Otherwise, if $-1 < u(z) \le 0$, then we use $\overleftarrow{F}$, $\overleftarrow{g}$ and $\overleftarrow{X}$ instead of F, g and X in the following arguments. Let $x_0 \in M_0$ be the nearest point from M_0 to z with respect to F and $c : [0, \ell] \to M_u$ a normal minimizing geodesic of F from x_0 to z. By the first variation formula of c, one obtains that $g_{\dot c(0)}(\dot c(0), Y) = 0$ for any $Y \in T_{x_0} M_0$. Consequently, $g_X(X, Y) = g_{\dot c(0)}(\dot c(0), Y) = 0$ at x_0 for all $Y \in T_{x_0} M_0$, which implies that $X(x_0) = \dot c(0)$ by Corollary 1.1.3. Similarly, we have $\dot \tau(0) = \overleftarrow{X}(x_0) = -X(x_0)$ if $-1 < u(z) \le 0$. Thus, $c(s) = \varphi_{\pm s}(x_0)$ ($s \in [0, \ell]$) by the uniqueness of geodesics. In particular, $z = c(\ell) = \varphi_{\pm \ell}(x_0)$, which means that φ is surjective. Observe that $\dot c(s) = \dot \varphi_{\pm s} = \pm X$. We have $du(\dot c(s)) = g_{\nabla u}(\nabla u, X) = F(\nabla u)$ if $\dot c(s) = X$ and $-du(\dot c(s)) = g_{\nabla u}(\nabla u, -X) = \overleftarrow{g}_{\overleftarrow{\nabla}(-u)}(\overleftarrow{\nabla}(-u)), X) = \overleftarrow{F}(\overleftarrow{\nabla}(-u))$ if $\dot c(s) = \overleftarrow{X}$. Let $f(s) := \pm u(c(s))$. It follows from (1) of Proposition 8.4.3 and Proposition 8.4.3', respectively, that

$$\begin{cases} (f')^2 + \lambda_1 f^2 = \lambda_1, \\ f(0) = 0, \quad f'(0) = \sqrt{\lambda_1}. \end{cases} \tag{8.4.15}$$

Differentiating the first equation of (8.5.16) and using the initial conditions, we get the solution of (8.5.16), that is, $f(s) = \pm u(c(s)) = \sin \sqrt{\lambda_1} s$. Since $|f(s)| < 1$, $c(s)$ is minimal and $\lambda_1 = \frac{\pi^2}{d^2}$, we have $0 \le s \le \ell < \frac{d}{2}$. Note that $\varphi_{-s} = \exp(-sX) = \exp(s\overleftarrow{X})$ is a geodesic of $\overleftarrow{F}$, which is also a geodesic of the Riemannian metric $\overleftarrow{g}_{\overleftarrow{X}} = g_X$. Thus, $\varphi_s = \exp(sX)$ can be defined on the interval $I = (-d/2, d/2)$ as a geodesic of g_X and $\pm u(\varphi_s) = v(s) = \sin\sqrt{\lambda_1}s$ for $s \in (-d/2, d/2)$.

If $\varphi_s(x_1) = \varphi_t(x_2)$ for $s, t \in I$ and $x_1, x_2 \in M_0$. Then $v(s) = \pm u(\varphi_s(x_1)) = \pm u(\varphi_t(x_2)) = v(t)$ implies $s = t$ since $v(s)$ is a strictly increasing function on $[-d/2, d/2]$. Moreover, since the one-parameter transformation φ_s is injective, we have $x_1 = x_2$. So, φ is also injective. Since the exponential map is differentiable, φ is a diffeomorphism. Thus, the claim follows.

Now, we prove that φ is an isometry with respect to g_X. Choose a local orthonormal basis $\{e_i\}_{i=1}^n$ with respect to g_X on M_u such that $e_n = X$. From (8.4.12)–(8.4.13), we have

$$\nabla^2 u(e_i) = 0 \ (1 \le i \le n-1), \quad \nabla^2 u(e_n) = -\lambda_1 u e_n. \qquad (8.4.16)$$

From this and (8.4.14), one obtains that $D_{e_i}^X X = 0$ for all $1 \le i \le n$, which implies that X is a parallel (and hence Killing) unit vector field on M_u with respect to g_X. Thus, φ maps M_0 isometrically (with respect to g_X) onto $\varphi(M_0)$ via φ_s and $g_X(d\varphi(\partial/\partial s), d\varphi(\partial/\partial s)) = g_X(X, X) = 1$. So, to prove that φ is an isometry with respect to g_X, it suffices to check that

$$g_X(d\varphi(s, \cdot)(Y), d\varphi(\partial/\partial s)) = g_X(d\varphi_s(Y), X) = \frac{1}{F(\nabla u)} g_{\nabla u}(\nabla u, d\varphi_s(Y))$$

$$= \frac{1}{F(\nabla u)} Y(u \circ \varphi_s) = \frac{1}{F(\nabla u)} Y(\pm v(s))$$

$$= 0, \qquad (8.4.17)$$

where $Y \in T_x M_0$. Consequently, (M_u, g_X) is isometric to $((-d/2, d/2) \times M_0, dt^2 \otimes h_X)$, where h_X is the Riemannian metric on M_0 induced from g_X.

Since $\mathrm{Ric}_\infty(X) = \mathrm{Ric}_N(X) = 0$ from (8.4.10), by definition, we get $(\Psi \circ \varphi(t))'(0) = 0$, where $\varphi(t) = \varphi_t$ is a geodesic with $\dot\varphi(0) = X$. Thus,

$$\frac{d}{ds}(\Psi \circ \varphi_t)|_{t=s} = \frac{d}{ds}(\varphi_t^* \Psi)|_{t=s} = \lim_{t \to 0} \frac{\varphi_{t+s}^* \Psi - \varphi_s^* \Psi}{t}$$

$$= \lim_{t \to 0} \frac{\varphi_t^* \circ \varphi_s^* \Psi - \varphi_s^* \Psi}{t} = \varphi_s^*\left(\frac{d}{dt}(\varphi_t^* \Psi)|_{t=0}\right) = 0, \qquad (8.4.18)$$

that is, Ψ is constant along each curve φ_s. $\qquad\square$

Since the volume form dm on M can be decomposed as $dm = e^{-\Psi(\varphi_s)}\text{Vol}_{\dot{\varphi}_s}$ along the geodesic φ_s, from Proposition 8.4.2, one obtains the following

Corollary 8.4.1. *Under the same assumptions as in Proposition 8.4.1, (M_u, m) admits a diffeomorphic, measure-preserving splitting $(M_u, m) = ((-d/2, d/2) \times M_0, m_1 \times m_0)$, where m_1 is the one-dimensional Lebesgue measure and $m_0 := m|_{M_0}$.*

Based on Propositions 8.4.1–8.4.2 and 8.4.1′, we prove the following sharp spectral gap, which generalizes the well known Zhong-Yang's result in Riemannian geometry ([ZY]).

Theorem 8.4.1. *Let (M, F, m) be an n-dimensional compact Finsler measure space without boundary or with a convex boundary and d be the diameter of M. Assume that $Ric_N \geq 0$ for some $N \in [n, \infty]$ and $\Delta u = -\lambda_1 u$ in M (in the weak sense) with a Neumann boundary condition $\nabla u \in T_x(\partial M)$ when $\partial M \neq \emptyset$. Then $\lambda_1 \geq \frac{\pi^2}{d^2}$ and the equality holds if and only if M is a one-dimensional segment or circle.*

Proof. The first claim follows from Lemma 8.4.2. As for the equality, it suffices to prove that $\dim M = 1$. Once $\dim M = 1$, it is well known that the only one-dimensional connected compact manifolds are circles (without boundary) or segments (with boundary). Next we prove that $\dim M = 1$ in two cases, according to the number of connected components of the hypersurface M_0.

Case 1. M_0 has more than one connected component, say, two connected components, $M_0^{(1)}$ and $M_0^{(2)}$. In this case, let x_1 and x_2 be in two different components $M_0^{(1)}$ and $M_0^{(2)}$ respectively and $\gamma(t)$ a unit speed minimizing geodesic of F or $\overleftarrow{F}$ from x_1 to x_2 depending on $du(\dot{\gamma}) > 0$ or < 0 around $t = 0$. Observe that $\varphi_t(M_0^{(1)})$ and $\varphi_t(M_0^{(2)})$ are two different connected components of M_u since φ_t is a diffeomorphism, where $t \in (-d/2, d/2)$. Then γ must pass through a maximum or a minimum. Otherwise, $\varphi_t(M_0^{(1)}) \cup \varphi_t(M_0^{(2)})$ becomes a path connected subset of M_u via γ and hence connected, which is a contradiction. Since the length of γ is less than or equal to the diameter d and $u(x_1) = u(x_2) = 0$, we obtain that $\pm u(\gamma(t)) = \sin\sqrt{\lambda}t$ on $[0, d]$ by the same argument as (8.4.9). Thus, we have $g_X(\nabla u, \dot{\gamma}) = \sqrt{\lambda} = F(\nabla u)$ or $\overleftarrow{g}_X(\overleftarrow{\nabla}u, \dot{\gamma}) = \sqrt{\lambda} = \overleftarrow{F}(\nabla u)$ at $t = 0$, which means that the equality holds in the Cauchy–Schwartz inequality. Consequently, $\dot{\gamma}(0) = \pm X_{x_1}$ and $\gamma(t) = \exp(\pm tX)$, which implies that there are at most two points $y = \exp_{x_1}(\pm dX)$. Therefore, the connected components of M_0 are discrete and the manifold M is one-dimensional.

Case 2. M_0 has only one connected component. In this case, define a map $\Phi : [-\frac{d}{2}, \frac{d}{2}] \times M_0 \to M$ by

$$\Phi(s, x) = \exp_x(sX), \quad \text{for} \quad x \in M_0, \quad X \in T_x M_0. \qquad (8.4.19)$$

Obviously, φ is exactly the restriction of the map Φ to $(-d/2, d/2) \times M_0$. Next we will prove that Φ is a unique differential extension of φ and Φ is a diffeomorphism.

For any $y \in M$ with $u(y) \neq \pm 1$, then $y \in M_u$ and there is a point $x \in M_0$ and $s \in (-d/2, d/2)$ such that $y = \varphi(s, x)$ by Corollary 8.4.1. If $y \in M \backslash M_u$, then $u(y) = 1$ or -1. Without loss of generality, assume that $u(y) = 1$ (it is similar for the case $u(y) = -1$). Note that the set of $u^{-1}(1)$ has empty interior point of M. Each geodesic ball $B_\epsilon^{\pm}(y)$ contains a point y_ϵ such that $v(d/2 - \epsilon) < u(y_\epsilon) < 1$ (only if $s = v^{-1}(u(y_\epsilon))$ is close to $\frac{d}{2}$). From the proof of Proposition 8.4.2, there is a unique point $x_\epsilon \in M_0$ such that $y_\epsilon = \varphi_{s_0}(x_\epsilon)$ for some $s_0 \in [0, d/2)$. Moreover, $\gamma_\epsilon(s) := \varphi_s(x_\epsilon)$ is a geodesic of g_X (also the Finsler metric F) and $u(\gamma_\epsilon(s)) = v(s) = \sin\sqrt{\lambda}s$ for $s \in [0, d/2)$. Let z_ϵ be a limit of $\gamma_\epsilon(s)$ as $s \to \frac{d}{2}$. Then $u(z_\epsilon) = 1$ and

$$d(y_\epsilon, z_\epsilon) \leq L(\gamma_\epsilon|_{[s_0, d/2]}) \leq v^{-1}(1) - v^{-1}(v(d/2 - \epsilon)) = \epsilon.$$

Let ϵ go to zero and take a convergent subsequence of x_ϵ with limit x since M_0 is a closed subset. Then, by continuity of the exponential map, $\Phi(\frac{d}{2}, x) = y$, which means that Φ is surjective. This also shows that Φ is a unique differentiable extension of φ since the exponential map is differentiable. Since the differential of Φ has determinant 1 in $(-d/2, d/2) \times M_0$, it is 1 everywhere and Φ is a local diffeomorphism.

By a similar density argument, Φ is also a local Riemannian isometry with respect to g_X. Consider two points $x_1, x_2 \in M_0$ such that $\Phi(d/2, x_1) = \Phi(d/2, x_2) = z \in M \backslash M_u$. We choose points z_1^ϵ on the geodesic $\varphi_1(t, x_1) = \exp_{x_1} tX$ and z_2^ϵ on the geodesic $\varphi_2(t, x_2) = \exp_{x_2} tX$ such that z_1^ϵ and z_2^ϵ in a geodesic ball $B_{\epsilon/2}(z)$ (with respect to g_X) with $u(z_1^\epsilon) < 1$ and $u(z_2^\epsilon) < 1$. Since $d_X(z_1^\epsilon, z_2^\epsilon) < \epsilon$, where d_X is the distance induced by g_X, we have $d_X(x_1, x_2) = d_X(z_1^\epsilon, z_2^\epsilon) < \epsilon$ by Proposition 8.4.2. Letting $\epsilon \to 0$ yields $x_1 = x_2$. Thus, Φ is injective.

Next, we prove that the maximal point and the minimal point are unique respectively. In fact, assume that $p_i (i = 1, 2)$ and q are the minimal points and the maximal point of u, respectively, i.e., $u(p_i) = -1 (i = 1, 2)$ and $u(q) = 1$. Let $\gamma_i : [0, \ell_i] \to M$ be unit speed minimal geodesics from p_i to

q respectively. Consider the subsets I_i of $[0, \ell_i]$ such that $\frac{d}{dt} u(\dot\gamma(t)) \geq 0$. By Propositions 8.4.1 and 8.4.1′, we have

$$\sqrt{\lambda} d \geq \sqrt{\lambda} L(\gamma_i) = \sqrt{\lambda} \int_0^{\ell_i} F(\dot\gamma_i) dt \geq \sqrt{\lambda} \int_{I_i} F(\dot\gamma_i) dt$$

$$= \int_{I_i} \frac{F(\dot\gamma_i) F(\nabla u)}{\sqrt{1 - [u(\gamma_i(t))]^2}} dt$$

$$\geq \int_0^{\ell_i} \frac{du(\dot\gamma_i(t))}{\sqrt{1 - [u(\gamma_i(t))]^2}} dt = \int_{-1}^1 \frac{du}{\sqrt{1 - u^2}} = \pi. \qquad (8.4.20)$$

The assumption $d = \frac{\pi}{\sqrt{\lambda}}$ forces that lengths of γ_i satisfy $L(\gamma_i) = d(p_i, q) = d$, $\frac{d}{dt} u(\dot\gamma_i(t)) \geq 0$ a.e. along $\gamma_i(t)$ and the equality holds in the Cauchy–Schwartz inequality, which means that $\dot\gamma_i(t) = X$. From this and a similar argument to (8.4.20), we get that $L(\gamma_i|_{\widehat{x_i q}}) = d/2$, where $x_i = \gamma_i \cap M_0$. Thus, $q = \exp_{x_1}\left(\frac{d}{2} X\right) = \exp_{x_2}\left(\frac{d}{2} X\right)$, which implies that $x_1 = x_2$ from the injectivity of Φ. Since $\gamma_i(t) (i = 1, 2)$ have the same tangent vector $\dot\gamma_i(t) = X$ at $x_1 = x_2$, we have $\gamma_1 = \gamma_2$ by the uniqueness of geodesics. So, $p_1 = p_2$ and hence the minimal point is unique. Similarly, the maximal point is also unique. Since Φ is bijective, there is only one point in M_0. Hence, $\dim M = 1$. This completes the proof. $\qquad \square$

8.4.2 Case 2. $K > 0$

We shall give the sharp lower bound for the first eigenvalue on compact Finsler manifolds without boundary with $\mathrm{Ric}_N \geq K > 0$. This generalizes the well known Lichnerowicz–Obata's theorem in Riemannian geometry ([Lic], [Ob]). The following $(p, 2)$-Sobolev inequality is crucial to study the sharp lower bound for λ_1.

Proposition 8.4.3. *Let (M, F, m) be an $n(\geq 2)$-dimensional compact Finsler measure space without boundary. Assume that $\mathrm{Ric}_N \geq K > 0$ for some $N \in [n, \infty)$ and $m_0 := m(M) > 0$. Then, for any $u \in W^{1,2}(M)$ and $\nu \geq N$ ($\nu > 2$ if $N = n = 2$),*

$$\left(\int_M |u|^p dm \right)^{\frac{2}{p}} \leq m_0^{-\frac{2}{\nu}} \int_M |u|^2 dm + (p-2) m_0^{-\frac{2}{\nu}} \frac{N-1}{KN} \int_M [F^*(du)]^2 dm,$$

$$(8.4.21)$$

where $p = \frac{2\nu}{\nu - 2}$.

Proof. The proof follows the line of Theorem 5.6 in [Oh3] but is more delicate than it. Since M is compact, the reversibility Λ and $m(M)$ are finite. Observe that (8.4.21) is invariant by replacing u (resp., m) with cu (resp., cm) for some constant $c > 0$. We can normalize u (resp., m) such that $\|u\|_{L^p} = 1$ (resp., $m_0 = 1$).

Let us first consider nonnegative u. By truncation, we can assume that $u \in L^\infty(M)$ and $\inf_M u > 0$. It suffices to prove that (8.4.21) is true for nonnegative u with respect to F. In fact, we can prove (8.4.21) for nonpositive u with respect to $\overleftarrow{F}$ in the same way. In general, we divide u into $u_+ := \max\{u, 0\}$ and $u_- := \max\{-u, 0\}$. Then

$$\|u\|_{L^p}^2 = (\|u_+\|_{L^p}^p + \|u_-\|_{L^p}^p)^{\frac{2}{p}} \le \|u_+\|_{L^p}^2 + \|u_-\|_{L^p}^2$$

yields the conclusion.

First, by Theorem 1.4 in [Xia6], we can take the smallest positive constant C satisfying

$$\|u\|_{L^p}^2 - \|u\|_{L^2}^2 \le C(p-2)\|F^*(du)\|_{L^2}^2, \quad p = \frac{2\nu}{\nu - 2} > 2, \quad (8.4.22)$$

for any nonnegative function $u \in W^{1,2}(M)$ and $\nu \ge N$ ($\nu > 2$ if $N = n = 2$). By the variational principle, for any $\phi \in C^\infty(M)$ and $\varepsilon > 0$, we have

$$\frac{\|u + \varepsilon\phi\|_{L^p}^2 - \|u\|_{L^p}^2 - \|u + \varepsilon\phi\|_{L^2}^2 + \|u\|_{L^2}^2}{p - 2}$$

$$\le C\Big\{\|F^*(du + \varepsilon\phi)\|_{L^2}^2 - \|F^*(du)\|_{L^2}^2\Big\}.$$

Dividing both sides by ε and letting $\varepsilon \to 0$, one obtains

$$\frac{2}{p-2}\int_M \phi\left(u^{p-1} - u\right) dm \le 2C\int_M d\phi(\nabla u)dm = -2C\int_M \phi\Delta u dm.$$

Since ϕ is arbitrary, it follows that Δu is well defined and

$$u^{p-1} - u = -C(p-2)\Delta u. \quad (8.4.23)$$

Substituting u with e^u in (8.4.23) gives

$$e^{(p-2)u} = 1 - C(p-2)\left(\Delta u + F^2(\nabla u)\right). \quad (8.4.24)$$

Next, fix $h \in C^\infty(M)$ and consider the function e^{ah} for $a \in \mathbb{R}$. Observe that

$$\nabla e^{ah} = ae^{ah}\nabla h, \quad \Delta e^{ah} = ae^{ah}\left(\Delta h + aF^2(\nabla h)\right), \quad \text{if } a \ge 0, \quad (8.4.25)$$

$$\nabla e^{ah} = ae^{ah}\overleftarrow{\nabla} h, \quad \Delta e^{ah} = ae^{ah}\left(\overleftarrow{\Delta} h + a\overleftarrow{F}^2(\overleftarrow{\nabla} h)\right), \quad \text{if } a \le 0, \quad (8.4.26)$$

where $\overleftarrow{\cdot}$ means the quantity corresponding to $\overleftarrow{F}$. In particular, if F is reversible, then $\overleftarrow{\nabla}u = \nabla u$ and $\overleftarrow{\Delta}u = \Delta u$. In this case, (8.4.25) and (8.4.26) are equivalent. Let

$$\Gamma_2(u) := \Delta^{\nabla u}\left(\frac{1}{2}F^2(\nabla u)\right) - d(\Delta u)(\nabla u).$$

Correspondingly, we have $\overleftarrow{\Gamma}_2(u)$. When $a \geq 0$, we have

$$\Gamma_2\left(e^{ah}\right) = \frac{1}{2}\Delta^{\nabla h}\left(a^2 e^{2ah}F^2(\nabla h)\right) - a^2 e^{ah}d\left(e^{ah}\left\{\Delta h + aF^2(\nabla h)\right\}\right)(\nabla h)$$

$$= \frac{1}{2}a^2 \operatorname{div}_m\left(e^{2ah}\nabla^{\nabla h}\left(F^2(\nabla h)\right) + 2ae^{2ah}F^2(\nabla h)\nabla h\right)$$

$$\quad - a^2 e^{2ah}\Big\{a\left\{\Delta h + aF^2(\nabla h)\right\}F^2(\nabla h) + d(\Delta h)(\nabla h)$$

$$\quad + ad\left(F^2(\nabla h)\right)(\nabla h)\Big\}$$

$$= \frac{1}{2}a^2 e^{2ah}\Big\{\Delta^{\nabla h}\left(F^2(\nabla h)\right) + ad\left(F^2(\nabla h)\right)(\nabla h) + a^2 F^4(\nabla h)$$

$$\quad - d(\Delta h)(\nabla h)\Big\}$$

$$= a^2 e^{2ah}\Big\{\Gamma_2(h) + ad\left(F^2(\nabla h)\right)(\nabla h) + a^2 F^4(\nabla h)\Big\}. \qquad (8.4.27)$$

On the other hand, it follows from the integration by parts that

$$\int_M \Gamma_2\left(e^{ah}\right)dm$$

$$= \left\|\Delta\left(e^{ah}\right)\right\|_{L^2}^2 = a^2\int_M e^{2ah}\Big\{(\Delta h)^2 + 2aF^2(\nabla h)\Delta h + a^2 F^4(\nabla h)\Big\}dm$$

$$= a^2\int_M e^{2ah}\Big\{(\Delta h)^2 - 2ad\left(F^2(\nabla h)\right)(\nabla h) - 3a^2 F^4(\nabla h)\Big\}dm.$$

Combining this with (8.4.27) yields

$$\int_M e^{2ah}(\Delta h)^2 dm = \int_M e^{2ah}\Big\{\Gamma_2(h) + 3ad(F^2(\nabla h))(\nabla h) + 4a^2 F^4(\nabla h)\Big\}dm. \qquad (8.4.28)$$

Equivalently, for any $b \geq 0$, we have

$$\int_M e^{bh}(\Delta h)^2 dm = \int_M e^{bh}\left\{\Gamma_2(h) + \frac{3b}{2}d(F^2(\nabla h))(\nabla h) + b^2 F^4(\nabla h)\right\}dm. \qquad (8.4.29)$$

Moreover, applying Corollary 7.1.2 to $e^{ah}(a \geq 0)$ and using (8.4.25), we get in the weak sense

$$\Gamma_2(h) + a\left(d(F^2(\nabla h))\right)(\nabla h) + a^2 F^4(\nabla h)$$
$$\geq KF^2(\nabla h) + \frac{1}{N}\left\{(\Delta h)^2 + 2aF^2(\nabla h)\Delta h + a^2 F^4(\nabla h)\right\}. \quad (8.4.30)$$

Note that $\overleftarrow{\mathrm{Ric}}_N \geq K$ if $\mathrm{Ric}_N \geq K$. Therefore, in the same way, we also have the same type inequality as (8.4.30) for $a \geq 0$, in which $\Gamma_2, F, \nabla, \Delta$ are replaced by $\overleftarrow{\Gamma}_2, \overleftarrow{F}, \overleftarrow{\nabla}, \overleftarrow{\Delta}$. Substituting h with $-h$ in this inequality yields

$$\Gamma_2(h) - a\left(d(F^2(\nabla h))\right)(\nabla h) + a^2 F^4(\nabla h)$$
$$\geq KF^2(\nabla h) + \frac{1}{N}\left\{(\Delta h)^2 - 2aF^2(\nabla h)\Delta h + a^2 F^4(\nabla h)\right\} \quad (8.4.31)$$

for $a \geq 0$, where we used $\overleftarrow{\Gamma}_2(-h) = \Gamma_2(h)$. Combining (8.4.30) with (8.4.31) implies that (8.4.30) holds for any $a \in \mathbb{R}$.

On one hand, for any $b \geq 0$, multiplying by $e^{bu}\Delta u$ on both sides of (8.4.24) and integrating by parts, together with (8.4.29), yield

$$C\int_M e^{bu}\Gamma_2(u)dm = \int_M e^{bu}F^2(\nabla u)dm + C(p-2)(b-1)\int_M e^{bu}F^4(\nabla u)dm$$
$$+C\left(p-1-\frac{b}{2}\right)\int_M e^{bu}d(F^2(\nabla u))(\nabla u)dm. \quad (8.4.32)$$

On the other hand, multiplying by e^{bu} on the both sides of (8.4.30) and integrating it give, together with (8.4.29),

$$\left(1-\frac{1}{N}\right)\int_M e^{bu}\Gamma_2(u)dm$$
$$\geq K\int_M e^{bu}F^2(\nabla u)dm + \left(\frac{(a-b)^2}{N} - a^2\right)\int_M e^{bu}F^4(\nabla u)dm$$
$$+ \left(\frac{3b-4a}{2N} - a\right)\int_M e^{bu}d[F^2(\nabla u)](\nabla u)dm. \quad (8.4.33)$$

Comparing the coefficients in (8.4.32) with (8.4.33), we would like to choose a and b enjoying

$$p-1-\frac{b}{2} = \frac{3b-2(N+2)a}{2(N-1)}, \quad (p-2)(b-1) = \frac{(a-b)^2 - Na^2}{N-1}$$

as well as $b \geq 0$. It is easy to see that $a = \frac{b}{2} - (p-1)\frac{N-1}{N+2}$ and b satisfies

$$\frac{b^2}{4} + \left(\frac{p-1}{N+2} - 1\right)b - (p-2) + (p-1)^2\left(\frac{N-1}{N+2}\right)^2 = 0.$$

Note that the discriminant of the above equation is $\frac{N}{(N+2)^2}(p-1)[2N - (N-2)p]$, which is nonnegative for $2 \leq p \leq \frac{2N}{N-2}$, equivalently, $\frac{1}{p} = \frac{1}{2} - \frac{1}{\nu}$ for $\nu \geq N$. Thus, we choose $a = -\frac{2}{N-2} \leq 0$ and $b = \frac{2(N-3)}{N-2} \geq 0$ as required when $N \geq n \geq 3$. Hence, we have $C \leq \frac{N-1}{NK}$.   □

Let us look at an application of (8.4.21). Let

$$\varphi_p(\varepsilon) = \int_M |1 + \varepsilon\varphi|^p dm,$$

where $\varepsilon > 0$ and $\int \varphi dm = 0$. By Taylor's expansion at $\varepsilon = 0$, we have

$$\varphi_p(\varepsilon) = m_0 + p\left(\int_M \varphi dm\right)\varepsilon + \frac{1}{2}p(p-1)\left(\int_M \varphi^2 dm\right)\varepsilon^2 + o(\varepsilon^2).$$

Hence,

$$(\varphi_p(\varepsilon))^{\frac{2}{p}} = m_0^{\frac{2}{p}} + 2m_0^{\frac{2}{p}-1}\left(\int_M \varphi dm\right)\varepsilon + (p-1)m_0^{\frac{2}{p}-1}\left(\int_M \varphi^2 dm\right)\varepsilon^2$$

$$+ (2-p)m_0^{\frac{2}{p}-2}\left(\int_M \varphi dm\right)^2\varepsilon^2 + o(\varepsilon^2).$$

Applying (8.4.21) to $u = 1 + \varepsilon\varphi$, and using the above formulae yield

$$(p-2)m_0^{\frac{2}{p}-1}\left(\int_M \varphi^2 dm\right) \leq (p-2)m_0^{-\frac{2}{\nu}}\frac{N-1}{NK}\int_M F^2(\nabla\varphi)dm,$$

which means that $\lambda_1 \geq \frac{KN}{N-1}$ since $\frac{1}{p} = \frac{1}{2} - \frac{1}{\nu}$. It is surprising that the $(p,2)$-Sobolev inequality implies the lower bound estimate for λ_1 independent of the choice of ν. From this, we can see that the equality in (8.4.21) holds if and only if $\lambda_1 = \frac{KN}{N-1}$. To characterize the equality, we need study the existence of extreme functions such that the equality holds in (8.4.21). First, following the proof of Theorem 4 in [BL] with N instead of n, we have the following sharp inequality.

Lemma 8.4.3. *Let (M, F) be an $n(> 2)$-dimensional compact Finsler measure space without boundary. Assume that, for all $f \in W^{1,2}(M)$,*

$$\|f\|_{L^{2N/(N-2)}}^2 \leq \int_M f^2 dm + \frac{4}{N(N-2)}\int_M [F^*(df)]^2 dm, \quad (8.4.34)$$

and that there exists a real-valued Lipschitz function f on M with $\max\{F(\nabla f), F(\nabla(-f))\} \le 1$ *and* $\max_{x_1,x_2 \in M}|f(x_2) - f(x_1)| = \pi$. *Then there exist nonconstant extremum functions f_t for (8.4.34). More precisely, if we translate f such that $\int_M \sin f \, dm = 0$, then, for each $t \in (-1, 1)$ and $N \ge n$,*

$$\|f_t\|^2_{L^{2N/(N-2)}} = \int_M f_t^2 \, dm + \frac{4}{N(N-2)} \int_M [F^*(df_t)]^2 \, dm, \quad (8.4.35)$$

where $f_t = (1 - t \sin f)^{1-N/2}$.

Assume that

$$\|f\|^2_{L^p} \le m_0^{-\frac{2}{\nu}} \int_M f^2 \, dm + (p-2)m_0^{-\frac{2}{\nu}}\frac{N-1}{NK} \int_M [F^*(df)]^2 \, dm$$

$$(8.4.36)$$

for all $f \in W^{1,2}(M)$ and $\nu \ge N$ ($\nu > 2$ if $N = n = 2$), where $p = \frac{2\nu}{\nu-2}$. By rescaling as in the proof of Proposition 8.4.3, one obtains

$$\|f\|^2_{L^{2\nu/(\nu-2)}} \le \int_M f^2 \, dm + \frac{4}{\nu(\nu-2)} \int_M \left[\tilde{F}^*(df)\right]^2 \, dm,$$

where $\tilde{F}^*$ is the dual of $\tilde{F} = \sqrt{\nu(N-1)/KN}F$. Denote by $\tilde{\nabla}$ the gradient with respect to $\tilde{F}$. If there exists a real-valued Lipschitz function f on M with

$$\max_{x_1,x_2 \in M}|f(x_2) - f(x_1)| = \pi, \quad (8.4.37)$$

$$\max\{F(\nabla f), \overleftarrow{F}(\overleftarrow{\nabla} f)\} = \max\{F^*(df), \overleftarrow{F}^*(df)\} \le \sqrt{\nu(N-1)/NK},$$

$$(8.4.38)$$

equivalently, $\max\{\tilde{F}(\tilde{\nabla} f), \tilde{F}(\tilde{\nabla}(-f))\} = \max\{\tilde{F}^*(df), \overleftarrow{\tilde{F}}^*(df)\} \le 1$, then, by Lemma 8.4.3 and rescaling back again, there exist nonconstant extremum functions $f_t = (1 - t \sin f)^{1-\nu/2}$ satisfying

$$\|f_t\|^2_{L^p} = m_0^{-\frac{2}{\nu}} \int_M f_t^2 \, dm + (p-2)m_0^{-\frac{2}{\nu}}\frac{N-1}{KN} \int_M [F^*(df_t)]^2 \, dm.$$

It is worthwhile mentioning that the existence of f satisfying (8.4.37)–(8.4.38) is equivalent to that of f satisfying

$$\max\{F(\nabla f), F(\nabla(-f))\} \le 1,$$

$$\max_{x_1,x_2 \in M}|f(x_2) - f(x_1)| = \pi\sqrt{\nu(N-1)/NK}, \quad (8.4.39)$$

which imply that $d \geq \pi\sqrt{(N-1)/K}$. In fact, for any $x_1, x_2 \in M$, choose a minimal geodesic $\gamma : [0, \ell] \to M$ from $x_1 = \gamma(0)$ to $x_2 = \gamma(\ell)$. Assume that there is a function f satisfying (8.4.39). Without loss of generality, we may assume $df(\dot{\gamma}(t)) < 0$ on a subset $[0, t_0]$ and $df(\dot{\gamma}(t)) > 0$ on a subset $[t_0, \ell]$. Then

$$d = \max_{x_1,x_2 \in M} \int_0^\ell F(\dot{\gamma}(t)) dt \geq \max_{x_1,x_2 \in M} \left\{ \int_0^{t_0} \frac{-df(\dot{\gamma}(t))}{F(\nabla(-f))} dt + \int_{t_0}^\ell \frac{df(\dot{\gamma}(t))}{F(\nabla f)} dt \right\}$$

$$\geq \max_{x_1,x_2 \in M} \left\{ -\int_0^{t_0} df(\dot{\gamma}(t)) dt + \int_{t_0}^\ell df(\dot{\gamma}(t)) dt \right\}$$

$$= \max_{x_1,x_2 \in M} \left\{ |f(\gamma(t_0)) - f(\gamma(0))| + |f(\gamma(\ell)) - f(\gamma(t_0))| \right\}$$

$$\geq \max_{x_1,x_2 \in M} |f(x_2) - f(x_1)| = \pi\sqrt{\nu(N-1)/NK}$$

$$\geq \pi\sqrt{(N-1)/K}. \tag{8.4.40}$$

Summing up, we get the following result.

Proposition 8.4.4. *Let (M, F, m) be an $n(\geq 2)$-dimensional compact Finsler measure space without boundary. Assume that (8.4.36) holds for all $f \in W^{1,2}(M)$ and $\nu \geq N$ ($\nu > 2$ if $N = n = 2$) and there is a real-valued Lipschitz function f on M with $\max\{F(\nabla f), F(\nabla(-f))\} \leq 1$ and $\max_{x_1,x_2 \in M} |f(x_2) - f(x_1)| = \pi\sqrt{\nu(N-1)/NK}$. Then there exist nonconstant extremum functions f_t for (8.4.36). More precisely, if we translate f such that*

$$\int_M \sin\left(f\sqrt{NK/\nu(N-1)}\right) dm = 0,$$

then, for each $t \in (-1, 1)$ and $\nu \geq N$ ($\nu > 2$ if $N = n = 2$), there are functions $f_t = \left(1 - t\sin\left(f\sqrt{NK/\nu(N-1)}\right)\right)^{1-\nu/2}$ satisfying

$$\|f_t\|_{L^p}^2 = m_0^{-\frac{2}{\nu}} \int_M f_t^2 dm + (p-2) m_0^{-\frac{2}{\nu}} \frac{N-1}{KN} \int_M [F^*(df_t)]^2 dm,$$

$$\tag{8.4.41}$$

where $p = \frac{2\nu}{\nu-2}$.

Moreover, similar to Proposition 8.4.1, we have the following result.

Proposition 8.4.5. *Let (M, F, m) be an n-dimensional compact Finsler measure space without boundary. Assume that $Ric_N \geq K > 0$ for some $N \in [n, \infty)$ and u is an eigenfunction with $\Delta u = -\lambda_1 u$ (in the weak sense), where $\lambda_1 = \frac{KN}{N-1}$. Then*

(1) *the function $P(x) := F^2(x, \nabla u) + \frac{\lambda_1}{N} u(x)^2$ is a constant on M. Further, we can normalize u such that $\max_M u = -\min_M u = 1$ and hence $P(x) = \frac{\lambda_1}{N}$ on M. Moreover, $M_u = \{x \in M \mid u(x) \neq \pm 1\}$.*
(2) *the vector field $X := \frac{\nabla u}{F(\nabla u)}$ is a geodesic field of F on M_u.*

Proof. (1) Let φ be any nonnegative function in $W^{1,2}(M) \cap L^\infty(M)$. By Corollary 7.1.2, we have

$$-\frac{1}{2} \int_M d\varphi \left(\nabla^{\nabla u} P\right) dm = -\frac{1}{2} \int_M d\varphi \left[\nabla^{\nabla u} \left(F^2(\nabla u)\right)\right] dm$$

$$-\frac{\lambda_1}{N} \int_M d\varphi \left(u \nabla u\right) dm$$

$$\geq \int_M \varphi \left(d(\Delta u)(\nabla u) + Ric_N(\nabla u) + \frac{(\Delta u)^2}{N}\right) dm$$

$$-\frac{\lambda_1}{N} \int_M d(u\varphi)(\nabla u) dm + \frac{\lambda_1}{N} \int_M \varphi F^2(\nabla u) dm$$

$$\geq \int_M \left(-\frac{N-1}{N}\lambda_1 + K\right) \varphi F^2(\nabla u) dm = 0,$$

$$(8.4.42)$$

which means that $\Delta^{\nabla u} P \geq 0$ on M in the weak sense. Since M is compact and $\Delta^{\nabla u}$ is a linear elliptic operator, by the maximum principle, P must be a constant on M. Since $F(\nabla u) = 0$ at the maximum and minimum points of u, $P = \frac{\lambda_1}{N}(\max |u|)^2$. Since u change its sign, we have $\max u = -\min u$. We rescale u such that $\max u = 1$. Thus, we conclude that $\max u = 1 = -\min u$ and

$$P = F^2(\nabla u) + \frac{\lambda_1}{N} u^2 = \frac{\lambda_1}{N} \quad \text{on} \quad M, \qquad (8.4.43)$$

which implies that $u(x) \neq \pm 1$ if and only if $\nabla u \neq 0$, i.e., $M_u = \{x \in M \mid u(x) \neq \pm 1\}$.

(2) Note that P is smooth on M_u. Taking derivative on both sides of (8.4.43) and using (7.1.14) with $p = 2$ yield $\nabla^2 u(\nabla u) = -\frac{\lambda_1}{N} u \nabla u$, which implies that $\nabla^2 u(\nabla u, \nabla u) = -\frac{\lambda_1}{N} u F^2(\nabla u)$ on M_u. From this, we have

$$D_X^X X = \frac{1}{F(\nabla u)} D_{\nabla u}^{\nabla u} \left(\frac{\nabla u}{F(\nabla u)} \right)$$

$$= -\frac{1}{F^4(\nabla u)} \left[g_{\nabla u}(\nabla^2 u(\nabla u), \nabla u)\nabla u - F^2(\nabla u)\nabla^2 u(\nabla u) \right] = 0.$$

Hence, X is a geodesic field on M_u.   □

In the same way as Proposition 8.4.1′, we have

Proposition 8.4.5′. *Let* (M, F, m), Ric_N, λ_1 *and* u *be the same as in Proposition 8.4.5. Let* $\overleftarrow{F}$ *be the reverse Finsler structure of* F. *Then*

(1) *the function* $\overleftarrow{P}(x) := \overleftarrow{F}^2(x, \overleftarrow{\nabla}(-u)) + \frac{\lambda_1}{N} u(x)^2 = P(x)$ *is a constant on* M. *Further, we can normalize* u *such that* $\max u = -\min u = 1$ *and* $\overleftarrow{P}(x) = \frac{\lambda_1}{N}$ *on* M.

(2) *the vector field* $\overleftarrow{X} := \frac{\overleftarrow{\nabla}(-u)}{\overleftarrow{F}(\overleftarrow{\nabla}(-u))} = -X$ *is a geodesic field of* $\overleftarrow{F}$ *on* M_u.

With Propositions 8.4.4–8.4.5, we can prove the following rigidity result.

Theorem 8.4.2. *Let* (M, F, m) *be an* $n(\geq 2)$-*dimensional compact Finsler measure space without boundary. Assume that* $Ric_N \geq K > 0$ *for some* $N \in [n, \infty)$ *and* $m_0 > 0$. *Then* $\lambda_1 = \frac{NK}{N-1}$ *if and only if* $d = \pi\sqrt{(N-1)/K}$, *where* d *is the diameter of* M. *In this case,* $N = n$. *Further, if the measure* m *can be normalized such that* $\lim_{r \to 0} \frac{m(B_r^\pm(x))}{V_{K/(N-1), N}(r)} = 1$ *for any* $x \in M$, *then* F *is of positive constant flag curvature* $\frac{K}{N-1}$ *and zero S-curvature. In particular,* F *is of positive constant flag curvature* $\frac{K}{N-1}$ *and zero S-curvature with respect to the Busemann–Hausdorff measure* m_{BH}.

Proof. Assume that $\lambda_1 = \frac{KN}{N-1}$ and u is an eigenfunction corresponding to λ_1. By Propositions 8.4.5 and 8.4.5′, we may assume that $u(x_1) = \min u = -1$ and $u(x_2) = \max u = 1$ (the case when $u(x_1) = 1$ and $u(x_2) = -1$ can be argued by substituting u with $-u$ and F with $\overleftarrow{F}$), and hence

$$F^2(\nabla u) = \frac{K}{N-1}(1 - u^2).$$

Let $\gamma : [0, \ell] \to M$ be a minimal geodesic $\gamma(t)$ from $x_1 = \gamma(0)$ to $x_2 = \gamma(\ell)$. Consider a subset I of $[0, \ell]$ such that $du(\dot{\gamma}(t)) \geq 0$. Then

$$
d \geq \int_0^\ell F(\dot{\gamma}(t))dt \geq \int_I F(\dot{\gamma}(t))dt
$$

$$
\geq \int_0^\ell \frac{du(\dot{\gamma}(t))}{F(\nabla u)}dt = \sqrt{\frac{N-1}{K}} \int_{-1}^1 \frac{du}{\sqrt{1-u^2}} = \pi\sqrt{\frac{N-1}{K}}. \qquad (8.4.44)
$$

Thus, $d = \pi\sqrt{(N-1)/K}$ by Theorem 6.3.4. Conversely, if $d = \pi\sqrt{(N-1)/K}$, there are points $x_1, x_2 \in M$ such that $d_F(x_1, x_2) = \pi\sqrt{(N-1)/K}$. Then by taking $f = d_F$, $\nu = N$ ($\nu = N+1$ if $N = n = 2$) and using Propositions 8.4.3–8.4.4, there exist nonconstant extremum functions f_t for (8.4.36), which implies that $\lambda_1 = \frac{KN}{N-1}$ by Taylor's expansion for f_t as before. This proves the first claim.

In the case when $\lambda_1 = \frac{KN}{N-1}$, equivalently, $d = \pi\sqrt{(N-1)/K}$, there exist $x_1, x_2 \in M$ such that $d_F(x_1, x_2) = d$. We scale u such that $u(x_1) = \min u = -1$ since $\overleftarrow{\Delta}(-u) = -\lambda_1(-u)$ and choose a minimal geodesic $\gamma(t)$ from x_1 to x_2 as above. Since the equality in (8.4.44) holds, all inequalities become equalities. Thus, we have $u(x_2) = \max u = 1$. By (8.4.44) again, we have $du(\dot{\gamma}(t)) = F(\nabla u)F(\dot{\gamma}(t)) \geq 0$ along $\gamma(t)$. Equivalently, ∇u and $\dot{\gamma}(t)$ are linearly dependent. Note that ∇u points into the direction in which u increases the most. We can reparameterize $\gamma(t)$ with $F(\dot{\gamma}) = 1$ such that $\nabla u = F(\nabla u)\dot{\gamma}(t)$.

Step 1. Prove that $S(\dot{\gamma}(t)) = 0$ and $\mathbf{K}(\Pi, \dot{\gamma}(t)) = \frac{K}{N-1}$ for the section $\Pi = \mathrm{span}\{\dot{\gamma}, y\}$, where $y \in T_\gamma M$ is an arbitrary unit vector perpendicular to $\dot{\gamma}$ with respect to $g_{\dot{\gamma}}$.

Choose a local orthonormal basis $\{e_i\}$ at any $x \in M_u$ with respect to $g_{\nabla u}$ such that $e_1 = X$. Then $\nabla u = u_1 e_1$, $H_u = u_{1;1}$, where $u_1 = F(\nabla u)$. Since P is constant, all inequalities in (8.4.42) become equalities. This is equivalent to that $\Delta u = NH_u$ and the inequality in (7.1.16) becomes equality. From the proof of Lemma 6.3.2 (see (6.3.16)–(6.3.17)), we have $u_{1;1} = H_u = -\frac{\lambda_1}{N}u$, $u_{2;2} = \cdots = u_{n;n}$, $u_{i;j} = 0 (i \neq j)$, and

$$
S(\nabla u) = -\frac{N-n}{n-1}(tr_{\nabla u}(\nabla^2 u) - u_{1;1})
$$

$$
= -\frac{N-n}{n-1}(\Delta u + S(\nabla u) - u_{1;1}) = -\frac{N-n}{n-1}(-Ku + S(\nabla u)),
$$

which means that

$$S(\nabla u) = \frac{N-n}{N-1} K u \tag{8.4.45}$$

on M_u, where we used $\mathbf{\Delta} u = -\frac{NK}{N-1} u$ on M_u. We claim that $S(\nabla u) = 0$ and hence $N = n$. Obviously, it is true on $M \backslash M_u$. Assume that $S(\nabla u) \neq 0$ on M_u. Note that $\nabla u = F(\nabla u)\dot\gamma(t)$ and hence $S(\nabla u) = F(\nabla u)S(\dot\gamma(t)) \neq 0$ along γ. For any two nonconstant eigenfunctions $u \neq \bar{u}$ corresponding to λ_1, it follows from (8.4.43) and (8.4.45) that

$$\frac{F^2(\nabla \bar{u})}{F^2(\nabla u)} = \frac{1 - \bar{u}^2}{1 - u^2} = \frac{\bar{u}^2}{u^2} \quad \text{on } M_u.$$

Thus, $u^2 = \bar{u}^2$ and $M_u = M_{\bar{u}}$. However, $du(\dot\gamma(t)) \geq 0$ and $d\bar{u}(\dot\gamma(t)) \geq 0$ along $\gamma(t)$. This is impossible. Consequently, $S(\nabla u) = S(\dot\gamma(t)) = 0$ and hence $N = n$. In this case, it is easy to see that $u_{1;1} = u_{2;2} = \cdots = u_{n;n} = -\frac{\lambda_1}{N} u$ and $u_{i;j} = 0 (1 \leq i \neq j \leq n)$, that is,

$$\nabla^2 u(\nabla u) = -\frac{\lambda_1 u}{N}\nabla u, \quad \nabla^2 u(e_\alpha) = -\frac{\lambda_1 u}{N} e_\alpha, \quad \nabla^2 u(\nabla u, e_\alpha) = 0 \tag{8.4.46}$$

for $2 \leq \alpha \leq n$. Since ∇u is a geodesic field, the flag curvature $\mathbf{K}(\cdot, \nabla u)$ is exactly the sectional curvature $\mathbf{K}^{\nabla u}(\nabla u, \cdot)$ of $g_{\nabla u}$ for the section containing ∇u by Lemma 4.1.2. Thus, by a direct calculation, one obtains that

$$\mathbf{K}(\Pi_\alpha, \nabla u) = \mathbf{K}^{\nabla u}(\nabla u, e_\alpha) = \frac{1}{F^2(\nabla u)} R^{\nabla u}(\nabla u, e_\alpha, \nabla u, e_\alpha)$$

$$= \frac{1}{F^2(\nabla u)} g_{\nabla u}\left(R^{\nabla u}(e_\alpha, \nabla u)\nabla u, e_\alpha\right) = \frac{\lambda_1}{N} = \frac{K}{N-1}, \tag{8.4.47}$$

where $\Pi_\alpha = \text{span}\{e_\alpha, \nabla u\}$. Thus, Step 1 is completed.

Step 2. Prove that there exists $x' \in M$ such that $d_F(x, x') = d$ for any $x \in M$.

Suppose to the contrary, there exists a small number $\varepsilon_0 > 0$ such that $B^+_{d-\varepsilon_0}(x) = M$ (similar discussions for $B^-_{d-\varepsilon_0}(x) = M$) for any $x \in M$. By Theorem 6.4.1 and the assumption, we have

$$\frac{m_0}{V_{c,N}(d - \varepsilon_0)} = \frac{m(B^+_{d-\varepsilon_0}(x))}{V_{c,N}(d - \varepsilon_0)} \leq \lim_{\delta \to 0} \frac{m(B^+_\delta(x))}{V_{c,N}(\delta)} = 1, \tag{8.4.48}$$

where $c = \frac{K}{N-1}$. On the other hand, since $d = \pi\sqrt{(N-1)/K}$, there are two points $z_1, z_2 \in M$ such that $d_F(z_1, z_2) = d$. Then, for any $\varepsilon > 0$, we have $B^+_{d-\varepsilon}(z_1) \cap B^-_\varepsilon(z_2) = \emptyset$. Otherwise, $d_F(z_1, z_2) < d$ by the triangle inequality. This is impossible. Thus,

$$m_0 \geq m\left(B^+_{d-\varepsilon}(z_1) \cup B^-_\varepsilon(z_2)\right)$$
$$= m\left(B^+_{d-\varepsilon}(z_1)\right) + m\left(B^-_\varepsilon(z_2)\right) > m\left(B^+_{d-\varepsilon}(z_1)\right). \qquad (8.4.49)$$

For any $x \in M$, it is obvious that $d_F(z_1, x) + d_F(x, z_2) \geq d$. Now, we assume that $d_F(z_1, x) + d_F(x, z_2) = d + 2\varepsilon'$ for some $\varepsilon' \geq 0$. Let $d_1 := d_F(z_1, x)$, $d_2 := d_F(x, z_2)$ and $B^\pm_\bullet(x) := B^+_\bullet(x) \cap B^-_\bullet(x)$. From the same logic as above, we have $B^+_{d_1 - \varepsilon'}(z_1) \cap B^\pm_{\varepsilon'}(x) = \emptyset$ and $B^\pm_{d_2 - \varepsilon'}(x) \cap B^-_{\varepsilon'}(z_2) = \emptyset$. Moreover, note that $(d_1 - \varepsilon') + (d_2 - \varepsilon') = d$, which implies that

$$V_{c,N}(d) = V_{c,N}(d_1 - \varepsilon') + V_{c,N}(d_2 - \varepsilon').$$

Hence, by Theorem 6.4.1 again, we have

$$m_0 \geq m\left(B^+_{d_1 - \varepsilon'}(z_1)\right) + m\left(B^\pm_{\varepsilon'}(x)\right) + m\left(B^-_{d_2 - \varepsilon'}(z_2)\right)$$
$$\geq m_0 \frac{V_{c,N}(d_1 - \varepsilon')}{V_{c,N}(d)} + m_0 \frac{V_{c,N}(d_2 - \varepsilon')}{V_{c,N}(d)} + m\left(B^\pm_{\varepsilon'}(x)\right)$$
$$= m_0 + m\left(B^\pm_{\varepsilon'}(x)\right),$$

which means $\varepsilon' = 0$. Consequently, $d_F(z_1, x) + d_F(x, z_2) = d$ for any $x \in M$. From this, we have

$$\boldsymbol{\Delta} d_F(z_1, x) = \boldsymbol{\Delta}\left(-d_F(x, z_2)\right) = -\overleftarrow{\boldsymbol{\Delta}}\left(\overleftarrow{d}_{\overleftarrow{F}}(z_2, x)\right). \qquad (8.4.50)$$

Note that $\overleftarrow{\mathrm{Ric}}_N \geq K$ if $\mathrm{Ric}_N \geq K$. Let $r(x) := d_F(z_1, x)$ and $\overleftarrow{r}(x) := \overleftarrow{d}_{\overleftarrow{F}}(z_2, x)(= d_F(x, z_2))$ be the distance functions from z_1 and z_2 with respect to F and $\overleftarrow{F}$, respectively. Then $r(x) + \overleftarrow{r}(x) = d$. By Theorem 6.3.5, we have

$$\boldsymbol{\Delta} r \leq \sqrt{(N-1)K} \cot\left(r\sqrt{K/(N-1)}\right) = \frac{\partial}{\partial r} \log \mathfrak{s}_c(r)^{N-1}, \qquad (8.4.51)$$

$$\overleftarrow{\boldsymbol{\Delta}} \overleftarrow{r} \leq \sqrt{(N-1)K} \cot\left(\overleftarrow{r}\sqrt{K/(N-1)}\right)$$
$$= -\sqrt{(N-1)K} \cot\left(r\sqrt{K/(N-1)}\right) \qquad (8.4.52)$$

pointwise on $D_{z_1} = M\backslash(\{z_1\} \cup Cut(z_1))$ or on $M\backslash\{z_1\}$ in the weak sense. From (8.4.50)–(8.4.52), one obtains

$$\boldsymbol{\Delta} r = \frac{\partial}{\partial r} \log \mathfrak{s}_c(r)^{N-1}. \qquad (8.4.53)$$

Now, we choose the geodesic polar coordinate (r, ξ) centered at z_1 such that $r(z) = F(v)$ and $\xi^\alpha(z) = \xi^\alpha \left(\frac{v}{F(v)} \right)$, where $v = \exp^{-1}(z) \in T_{z_1} M \backslash \{0\}$. Thus, we can write $dm|_{\exp_{z_1}(r\xi)} = \sigma(r, \xi) dr d\xi$, where $\xi \in I_{z_1} = \{\xi \in T_{z_1} M | F(\xi) = 1\}$ (cf. Section 5). By (6.1.5), we have

$$\mathbf{\Delta} r = \frac{\partial}{\partial r} \log \sigma(r, \xi).$$

Consequently,

$$\frac{\partial}{\partial r} \log \sigma(r, \xi) = \frac{\partial}{\partial r} \log \mathfrak{s}_c^{N-1}(r) \quad \text{on } D_{z_1}.$$

Integrating this on both sides from s to t yields

$$\mathfrak{s}_c^{N-1}(s) \sigma(t, \xi) = \mathfrak{s}_c^{N-1}(t) \sigma(s, \xi).$$

Integrating in t from r_1 to r_2 and then in s from 0 to r_1, one obtains

$$\frac{m\left(B_{r_2}^+(z_1)\right)}{m\left(B_{r_1}^+(z_1)\right)} = \frac{V_{c,N}(r_2)}{V_{c,N}(r_1)},$$

where we used (6.4.2) and (6.4.3). From this and (8.4.49), we have

$$\frac{m_0}{V_{c,N}(d - \varepsilon)} > \frac{m\left(B_{d-\varepsilon}^+(z_1)\right)}{V_{c,N}(d - \varepsilon)} = \lim_{\delta \to 0} \frac{m\left(B_\delta^+(z_1)\right)}{V_{c,N}(\delta)} = 1.$$

In particular, the above inequality holds when $\varepsilon = \varepsilon_0$, which contradicts with (8.4.48). This prove the claim.

Step 3. Prove that $S \equiv 0$ and $\mathbf{K} = \frac{K}{N-1}$.

For any $x \in M$, there exists $x' \in M$ such that $d_F(x, x') = d$. Let $\sigma(t)$ be a unit speed minimal geodesic from x to x' and assume that $u(x) = -1$ and $u(x') = 1$ as in the paragraph before Step 1. From Step 1, we have $S(\dot\sigma) = 0$ and $\mathbf{K}(\dot\sigma, y) = c$ for any unit vector $y \in T_\sigma M$ perpendicular to $\dot\sigma$ with respect to $g_{\dot\sigma}$. Thus, $S \equiv 0$ and $\mathbf{K} = \frac{K}{N-1}$ by the arbitrary of x.

Since $N = n$, for the Busemann–Hausdorff measure $m = m_{BH}$, it holds

$$\lim_{r \to 0} \frac{m_{BH}(B_r^\pm(x))}{V_{K/(N-1),N}(r)} = 1$$

by Lemma 6.4.2. The conclusion follows from the previous claim. $\square$

If F is reversible, and F is of positive constant flag curvature and zero S-curvature with respect to m_{BH}, then F is Riemannian by the main theorem in [KY]. Thus, by Obata's Theorem ([Ob]), we have

Corollary 8.4.2. *Let* (M, F, m^{BH}) *be an* $n(\geq 2)$*-dimensional compact reversible Finsler measure space without boundary. Assume that* $\mathrm{Ric}_N \geq K > 0$ *for some* $N \in [n, \infty)$ *and* $m_0 > 0$. *Then* $\lambda_1 = \frac{NK}{N-1}$ *if and only if* $d = \pi\sqrt{(N-1)/K}$. *In this case,* $N = n$ *and* M *is isometric to the Euclidean sphere* $\mathbb{S}^n(r)$ *of radius* $r = \sqrt{(N-1)/K}$.

Obviously, Corollary 8.4.2 in Riemannian case is exactly Lichnerowicz–Obata's results ([Lic], [Ob]). All above results on compact Finsler manifolds with $\mathrm{Ric}_N \geq K > 0$ are also true for forward (resp., backward) Finsler manifolds because of Bonnet–Myers Theorem (see Theorem 6.3.4). The following example shows that there are many non-Riemannian Finsler metrics on M with positive constant flag curvature and zero S-curvature.

Example 8.4.1 ([BS]). Let $\mathbb{S}^3$ be a standard unit sphere in $\mathbb{R}^4$. Then $\mathbb{S}^3$ is a compact Lie group on which there exist the standard right invariant 1-forms $\theta^1, \theta^2, \theta^3$ satisfying

$$d\theta^1 = 2\theta^2 \wedge \theta^3, \quad d\theta^2 = 2\theta^3 \wedge \theta^1, \quad d\theta^3 = 2\theta^1 \wedge \theta^2.$$

For any $k \geq 1$, define a family of Riemannian metrics α_k and 1-forms β_k on $\mathbb{S}^3$ by

$$\alpha_k^2 = k^2\theta^1 \otimes \theta^1 + k\theta^2 \otimes \theta^2 + k\theta^3 \otimes \theta^3, \quad \beta_k = \sqrt{k^2 - k}\,\theta^1.$$

Bao–Shen showed that Randers spaces $(\mathbb{S}^3, F_k = \alpha_k + \beta_k)$ equipped with the Busemann–Hausdorff volume measure m_{BH} are of constant flag curvature 1 and zero S-curvature, Thus, $\mathrm{Ric}_N = \mathrm{Ric} = 2$. Moreover the diameter $d = \pi$ ([Sh1]). Consequently, for any constant $c > 0$, Randers spaces $(\mathbb{S}^3, \tilde{F}_k = cF_k, m_{BH})$ are of positive constant flag curvature c^{-2} and zero S-curvature. Moreover the diameter $d = c\pi$ and the first eigenvalue $\lambda_1 = \frac{3}{2}c^{-2}$ by Theorem 8.4.2. Obviously, $\tilde{F}_k$ is reversible if $k = 1$ and nonreversible if $k > 1$.

It is interesting to determine the geometric or topologic structure of compact Finsler measure spaces (M, F, m) with constant flag curvature one and zero S-curvature (especially, when $d = \pi$). If F is reversible and $m = m_{BH}$, Kim–Yim ([KY]) proved that such a manifold (M, F, m_{BH}) is Riemannian and it is the standard Riemannian sphere $\mathbb{S}^n$ of constant sectional curvature 1 when $d = \pi$. It is open for the other cases.

8.5 Gradient Estimates for Eigenfunctions

Since the idea to derive the gradient estimates for the eigenfunctions is similar to that for the harmonic functions, we mainly present the related results and the ideas of their proofs without detailed proofs. The readers refer to [Xia7] for more details. However, we shall give a complete proof of the global gradient estimate for eigenfunctions on forward complete Finsler spaces. In particular, we obtain the global gradient estimate for harmonic functions.

Let (M, F, m) be a compact Finsler measure space. Assume that u is a positive eigenfunction for $\mathbf{\Delta}$ associated to λ, i.e., $\mathbf{\Delta} u = -\lambda u$ (in the weak sense). In particular, $\mathbf{\Delta} u = -\lambda u$ pointwise on M_u. Then, in the same way as in the proof of Lemma 7.4.1,

$$F(\nabla u)\mathbf{\Delta}^{\nabla u}(F(\nabla u)) \geq \frac{1}{N-1}g_{\nabla u}\left(\nabla^{\nabla u}F(\nabla u), \nabla^{\nabla u}F(\nabla u)\right)$$

$$+ Ric_N(\nabla u) - \lambda F^2(\nabla u) \tag{8.5.1}$$

holds on M_u. On the other hand, let $f(x) = F(\nabla \log u)$. By following the proof of Theorem 7.4.1, (7.4.9) becomes

$$\mathbf{\Delta}^{\nabla u} f = u^{-1}\mathbf{\Delta}^{\nabla u}F(\nabla u) - 2u^{-2}g_{\nabla u}(\nabla^{\nabla u}F(\nabla u), \nabla u) + 2f^3 + \lambda f \leq 0. \tag{8.5.2}$$

Plugging (8.5.1) into (8.5.2) also yield (7.4.10), where the terms including the eigenvalue λ are canceled. Thus, one obtains the following estimate.

Theorem 8.5.1. *Let (M, F, m) be an n-dimensional compact Finsler measure space without boundary or with a convex boundary. Assume that $Ric_N \geq K$ for some $N \in (n, \infty)$ and $K \in \mathbb{R}$, and u is an positive eigenfunction of the Finsler Laplacian associated to the eigenvalue λ, i.e., $\mathbf{\Delta} u = -\lambda u$ in M (in the weak sense) with a Neumann boundary condition $\nabla u \in T_x(\partial M)$ if $\partial M \neq \emptyset$. Then $K \leq 0$ and*

$$F(x, \nabla(\log u)) \leq \sqrt{-(N-1)K}. \tag{8.5.3}$$

Similarly, if u is a positive eigenfunction for $\mathbf{\Delta}$ associated to λ on a complete and noncompact Finsler measure space (M, F, m), then the functions $v := \log u$ and $f := F^2(x, \nabla v)$ as in the proof of Theorem 7.4.2 satisfy

$$\nabla v = \frac{1}{u}\nabla u, \quad \mathbf{\Delta} v = -f - \lambda, \quad (\mathbf{\Delta} v)^2 \geq f^2 \tag{8.5.4}$$

on $M_v = M_u$. Thus, Corollary 7.1.2 implies that, for any $\eta \in W_0^{1,2}(B_{2R}) \cap L^\infty(B_{2R})$, we still have (7.4.13), i.e.,

$$\int_M d\eta(\nabla^{\nabla v} f) dm \leq \int_M \eta \left(2df(\nabla v) - 2Ric_N(\nabla v) - \frac{2f^2}{N} \right) dm, \quad (8.5.5)$$

where the terms including the eigenvalue λ are canceled. Following the same arguments after (7.4.13) in the proof of Theorem 7.4.2 yields the following.

Theorem 8.5.2. *Let (M, F, m) be an $n(\geq 2)$-dimensional forward complete and noncompact Finsler measure space equipped with a uniformly convex and uniformly smooth Finsler metric F and a smooth measure m. Assume that $Ric_N \geq -K$ for some $N \in [n, \infty)$ and $K \geq 0$. Let u be a positive eigenfunction corresponding to the eigenvalue λ, i.e.,*

$$\Delta u = -\lambda u \qquad (8.5.6)$$

in the weak sense in a forward geodesic ball $B_{2R}^+(x_0)$. Then there exists a positive constant $C = C(N, \kappa, \kappa^*)$ depending on N, the uniform constants κ and κ^*, such that

$$\sup_{x \in B_R^+(x_0)} \{F(x, \nabla \log u(x)), F(x, \nabla(-\log u(x)))\} \leq C \frac{1 + \sqrt{K}R}{R}. \quad (8.5.7)$$

In particular, $F(x, \nabla \log u(x))$ and $F(x, \nabla \log(-u(x)))$ are bounded on M.

As an application, one obtains the following Harnack inequality for eigenfunctions by a standard argument.

Corollary 8.5.1. *Let (M, F, m), Ric_N be as in Theorem 8.5.2 and u be a positive eigenfunction in a geodesic ball $B_{2R}^+(x_0) \subset M$. Then there exists a positive constant $C = C(N, \kappa, \kappa^*)$ such that*

$$\sup_{x \in B_R^+(x_0)} u(x) \leq e^{C(1+\sqrt{K}R)} \inf_{x \in B_R^+(x_0)} u(x).$$

If $K = 0$, then we have a uniform positive constant $c = c(N, \kappa, \kappa^*)$ independent of R such that

$$\sup_{x \in B_R^+(x_0)} u(x) \leq c \inf_{x \in B_R^+(x_0)} u(x).$$

Moreover, if $Ric_N \geq 0$, then taking $R \to \infty$ in (8.5.7) yields that u is a positive constant. But any nonzero constant function on M cannot be a solution to $\Delta u = -\lambda u$ (in the weak sense) if $\lambda \neq 0$. Thus, we obtain the following nonexistence result.

Corollary 8.5.2. *Let (M, F, m) be a forward complete and noncompact Finsler measure space with nonnegative weighted Ricci curvature Ric_N. Then there does not admit a positive eigenfunction corresponding to the nonzero eigenvalue on M.*

Based on the local gradient estimate and the volume growth estimate of a geodesic ball (see Proposition 6.4.2), we have the following global gradient estimate for eigenfunctions.

Theorem 8.5.3. *Let (M, F, m) be an $n(\geq 2)$-dimensional forward complete and noncompact Finsler measure space equipped with a uniformly convex and uniformly smooth Finsler metric F and a smooth measure m. Assume that $Ric_N \geq -K$ for some $N \in [n, \infty)$ and $K > 0$. Let u be a positive eigenfunction on M corresponding to the first eigenvalue λ_1. Then*

$$\sup_{x \in M} \left\{ F(x, \nabla \log u(x)), F(x, \nabla(-\log u(x))) \right\} \leq \chi, \qquad (8.5.8)$$

where χ is the largest positive root of the equation

$$\chi^2 - \Lambda \sqrt{(N-1)K}\chi + \lambda_1 = 0, \qquad (8.5.9)$$

where Λ is the reversibility of F.

Proof. Observe that (8.5.9) has two positive roots and the value χ is the bigger one, which is well defined by Theorem 8.2.1. As in the proof of Theorem 8.5.2, let $v = \log u$ and $f = F^2(x, \nabla v)$. Then $\nabla v = \frac{1}{u}\nabla u$ and $\Delta v = -f - \lambda_1$ on $M_v = M_u$ as in (8.5.4). It suffices to consider the case when $f > 0$. To obtain the global estimate, we need to estimate $L_v(f)$ in a more refined way than (7.1.12).

Choose a local orthonormal basis $\{e_1, \ldots, e_n\}$ with respect to $g_{\nabla v}$ at $x \in M_v$ such that $\nabla v = F(\nabla v)e_1$. Then $v_1 = F(\nabla v) = f^{1/2} > 0$ and $v_i = 0 (2 \leq i \leq n)$. Differentiating $f = v_1^2$ yields $f^{-1}df(\nabla v) = 2v_{1;1}$ and $f^{-1/2}df(e_j) = 2v_{1;j}(\forall j)$, where $v_{i;j}$ stand for the covariant derivatives of v with respect to the Levi-Civita connection of $g_{\nabla v}$. Thus, $\Delta v = -f - \lambda_1$ on M_v can be rewritten as

$$\sum_{i=2}^{n} v_{i;i} + \frac{1}{2}f^{-1}df(\nabla v) - S(\nabla v) + f + \lambda_1 = 0.$$

Hence,

$$
\begin{aligned}
\|\nabla^2 v\|^2_{HS(\nabla v)} &= \sum_{i,j=1}^n v_{i;j}^2 \geq \sum_{j=1}^n v_{1;j}^2 + \frac{1}{n-1}\left(\sum_{i=2}^n v_{ii}\right)^2 \\
&= \frac{1}{n-1}\left(f + \frac{1}{2}f^{-1}df(\nabla v) + \lambda_1 - S(\nabla v)\right)^2 + \frac{1}{4}f^{-1}\sum_{j=1}^n df(e_j)^2 \\
&\geq \frac{1}{N-1}\left(f + \frac{1}{2}f^{-1}df(\nabla v) + \lambda_1\right)^2 - \frac{1}{N-n}S^2(\nabla v) \\
&\quad + \frac{1}{4}f^{-1}\|\nabla^{\nabla v}f\|^2_{HS(\nabla v)},
\end{aligned}
$$

where we used $(a-b)^2 \geq \frac{a^2}{1+\delta} - \frac{b^2}{\delta}$ with $\delta = (N-n)/(n-1) > 0$ in the last inequality. Plugging this into (7.1.11), and using $Ric_N \geq -K$ and (2.1.5) yield

$$
\begin{aligned}
L_v(f) &\geq \frac{2}{N-1}\left(f + \Lambda^{-1}\sqrt{(N-1)K}f^{1/2} + \lambda_1\right) \\
&\quad \cdot \left(f - \Lambda\sqrt{(N-1)K}f^{1/2} + \lambda_1\right) - \frac{2(N-2)}{N-1}df(\nabla v) \\
&\quad + \frac{2}{N-1}\lambda_1 f^{-1}df(\nabla v).
\end{aligned}
\tag{8.5.10}
$$

Note that the above inequality holds only on M_v.

Next, let w be the largest positive root of the equation

$$
w - \Lambda\sqrt{(N-1)K}w^{\frac{1}{2}} + \lambda_1 = 0.
\tag{8.5.11}
$$

In fact, $\sqrt{w}$ exactly corresponds to χ, where χ is the largest positive root of (8.5.9). For any $\delta > 0$, consider the nonnegative function $\hat{f} = (f - (w + \delta))^+$. We denote by $\Omega := \{f \geq w + \delta\} \subset M_v$. Then $0 < w + \delta \leq f \leq c(N, K, \kappa, \kappa^*)$ on Ω by Theorem 8.5.2. Since $L_v(f) = L_v(\hat{f})$ and

$$
|df(\nabla v)| \leq \sqrt{f}\sqrt{g^{ij}(\nabla v)f_i f_j} \leq \sqrt{\tilde{\kappa}f}F(\nabla f),
$$

by (8.5.10), there exist positive constants $c_i = c_i(N, K, \kappa, \kappa^*, \delta)(i = 1, 2)$ such that

$$
L_v(\hat{f}) \geq c_1\left(f - \Lambda\sqrt{(N-1)K}f^{1/2} + \lambda_1\right) - c_2 F^*(d\hat{f}) \quad \text{on } \Omega.
\tag{8.5.12}
$$

Observe that

$$
f - \Lambda\sqrt{(N-1)K}f^{1/2} + \lambda_1 \geq c_3 \hat{f}
\tag{8.5.13}
$$

for some positive constant $c_3 = c_3(N, K, \Lambda, \delta)$. In fact, (8.5.13) clearly holds when $f = w$ for any choices of c_3. On the other hand, we view both sides of (8.5.13) as functions of f, and the derivative of the left side is given by

$$\frac{1}{2} f^{-\frac{1}{2}} \left(2f^{1/2} - \Lambda \sqrt{(N-1)K} \right).$$

Note that $\lambda_1 \leq \left(\frac{\Lambda \sqrt{(N-1)K}}{2} \right)^2$ by Theorem 8.2.1. Hence, $w \geq \frac{1}{4}\Lambda^2(N-1)K$ by (8.5.11). From this and $f > w + \delta$, we get $2f^{1/2} - \Lambda\sqrt{(N-1)K} \geq c(N, K, \Lambda, \delta) > 0$. Thus, (8.5.13) is true on Ω by choosing $0 < c_3 < c$. Consequently,

$$L_v(\hat{f}) \geq c_4 \hat{f} - c_2 F^*(d\hat{f}) \quad \text{on } \Omega, \tag{8.5.14}$$

where $c_4 = c_4(N, K, \kappa, \kappa^*, \delta)$ is a positive constant. Further, we claim that (8.5.14) holds on M in the weak sense. In fact, we have $Z(\hat{f}) = 0$ for any nonzero vector field Z on $\partial\Omega$, i.e., $g_{\nabla\hat{f}}(\nabla\hat{f}, Z) = 0$, which means that $\tilde{\nu} = \frac{\nabla\hat{f}}{F(\nabla\hat{f})} = \frac{\nabla f}{F(\nabla f)}$ is a normal vector field on $\partial\Omega$ with respect to $g_{\nabla f}$. However, $\tilde{\nu}$ points inward. To apply the stokes theorem, we need a normal vector field pointing outward. Actually, $\nu = \frac{\nabla(-f)}{F(\nabla(-f))}$ is a normal vector field pointing outward with respect to $g_{\nabla(-f)}$ on $\partial\Omega$. Its dual is $J^{-1}(\nu) = -\frac{df}{F(\nabla(-f))}$. Note that $F(\nabla(-f)) \neq F(\nabla f)$ in general. Hence, for any nonnegative function $\varphi \in W_0^{1,p}(M)$, one obtains from Lemma 2.4.1 that

$$\int_M \hat{f} L_v(\varphi) dm = \int_\Omega \hat{f} L_v(\varphi) dm = \int_\Omega \varphi L_v(\hat{f}) dm + \int_{\partial\Omega} \hat{f} g_\nu(\nu, \nabla^{\nabla v} \varphi) dm_\nu$$

$$- \int_{\partial\Omega} \varphi g_\nu(\nu, \nabla^{\nabla v} \hat{f}) dm_\nu, \tag{8.5.15}$$

where we used $d\hat{f}(\nabla^{\nabla v}\varphi) = d\varphi(\nabla^{\nabla v}\hat{f})$ and dm_ν is the volume measure on $\partial\Omega$ induced from $dm|_\Omega$ by ν. Note that $\hat{f} = 0$ on $\partial\Omega$ and

$$-g_\nu(\nu, \nabla^{\nabla v}\hat{f}) = F^{-1}(\nabla(-f)) df(\nabla^{\nabla v} f)$$

$$\geq F^{-1}(\nabla(-f)) \|\nabla^{\nabla v} f\|_{HS(\nabla v)}^2 \geq 0. \tag{8.5.16}$$

Consequently, combining (8.5.14)–(8.5.16) together yields

$$\int_M \hat{f} L_v(\varphi) dm \geq \int_\Omega \varphi L_v(\hat{f}) dm \geq \int_\Omega \varphi \left(c_4 \hat{f} - c_2 F^*(d\hat{f}) \right) dm$$

$$= \int_M \varphi \left(c_4 \hat{f} - c_2 F^*(d\hat{f}) \right) dm. \tag{8.5.17}$$

This proves the claim.

Finally, we prove that $\hat{f} \equiv 0$ on M based on (8.5.17). In other words, $f \leq w$ by the arbitrariness of δ. Equivalently, $F(\nabla \log u) \leq \chi$.

Let $\varphi = \eta^2 \hat{f}^t$ be a cut-off function on M as in the proof of Theorem 7.4.2 for some constant $t \geq 1$. Plugging this into (8.5.17) yields

$$- \int_\Omega d\hat{f}(\nabla^{\nabla v}(\eta^2 \hat{f}^t)) dm \geq \int_M \eta^2 \hat{f}^t \left(c_4 \hat{f} - c_2 F^*(d\hat{f}) \right) dm. \tag{8.5.18}$$

By (7.4.15)–(7.4.16) and the boundedness of f, (8.5.18) implies that

$$c_4 \int_M \eta^2 \hat{f}^{t+1} dm \leq c_2 \int_\Omega \eta^2 \hat{f}^t F^*(d\hat{f}) dm + c_5 \int_\Omega \eta \hat{f}^t F^*(d\hat{f}) F(\nabla \eta) dm$$

$$- c_6 t \int_\Omega \eta^2 \hat{f}^{t-1} F^{*2}(d\hat{f}) dm$$

for some positive constants $c_i = c_i(N, K, \kappa, \kappa^*, \delta)(i = 5, 6)$. Thus, for any $0 < \epsilon < c_4$, we have

$$c_4 \int_M \eta^2 \hat{f}^{t+1} dm \leq \epsilon \int_\Omega \eta^2 \hat{f}^{t+1} dm + \frac{c_2^2}{4\epsilon} \int_\Omega \eta^2 \hat{f}^{t-1} F^{*2}(d\hat{f}) dm$$

$$+ \epsilon \int_\Omega \hat{f}^{t+1} F^2(\nabla \eta) dm + \frac{c_5^2}{4\epsilon} \int_\Omega \eta^2 \hat{f}^{t-1} F^{*2}(d\hat{f}) dm$$

$$- c_6 t \int_\Omega \eta^2 \hat{f}^{t-1} F^{*2}(d\hat{f}) dm. \tag{8.5.19}$$

Choose t such that $c_2^2 + c_5^2 = 4\epsilon c_6 t$. Therefore,

$$(c_4 - \epsilon) \int_M \eta^2 \hat{f}^{t+1} dm \leq \epsilon \int_M \hat{f}^{t+1} F^2(\nabla \eta) dm. \tag{8.5.20}$$

We choose the test functions $\eta_k = -1$ on $B_k^+(x)$ and 0 outside $B_{k+1}^+(x)$ as in the proof of Theorem 8.2.1. Then we have $F(\nabla \eta_k) \leq 1$, where k is a positive integer. Thus, we have

$$\int_{B_{k+1}^+(x)} \hat{f}^{t+1} dm \geq \left(\frac{c_4 - \epsilon}{\epsilon}\right) \int_{B_k^+(x)} \hat{f}^{t+1} dm \geq \left(\frac{c_4 - \epsilon}{\epsilon}\right)^k \int_{B_1^+(x)} \hat{f}^{t+1} dm.$$

Consequently, either $\hat{f} \equiv 0$ or

$$\int_{B_R^+(x)} \hat{f}^{t+1} dm \geq c e^{R \log \frac{c_4 - \epsilon}{\epsilon}}$$

for all $R \geq 1$ and some positive constant c independent of ϵ. However, $\hat{f}$ is bounded and $m(B_R^+(x)) \leq C e^{R\sqrt{(N-1)K}}$ for $R \geq 1$ by Proposition 6.4.2. This leads to a contradiction if ϵ is sufficiently small. So, $\hat{f} \equiv 0$. For the estimate of $F(x, \nabla(-\log u))$, the same arguments also work. The proof is finished. □

A direct consequence of Theorem 8.5.3 gives a global gradient estimate for positive harmonic functions.

Corollary 8.5.3. *Let* (M, F, m) *and* Ric_N *be as in Theorem* 8.5.3. *If* u *is a positive harmonic function, then*

$$\sup_{x \in M} \{F(x, \nabla \log u(x)), F(x, \nabla(-\log u(x)))\} \leq \Lambda \sqrt{(N-1)K}, \quad (8.5.21)$$

where Λ *is the reversibility of* F.

Remark 8.5.1. For the Finsler $p(> 1)$-Laplacian, we also have the corresponding results as in Theorems 8.5.2–8.5.3. We refer to [Xia7] for more details. In [Xia7], the author further studied the rigidity of the Finsler space (M, F, m) when the first p-eigenvalue attains its upper bound $\left(\Lambda\sqrt{(N-1)K}/p\right)^p$.

Chapter 9

Heat Flow on Finsler Manifolds

In this chapter, we shall discuss the nonlinear heat flow on Finsler manifolds introduced by Ohta-Sturm ([OS1]). The main goal is to introduce the existence and the local regularity of solutions for the nonlinear heat flow $\partial_t u = \Delta u$, and then give Li–Yau's type gradient estimates on compact or complete noncompact Finsler manifolds with weighted Ricci curvature bounded below.

9.1 Existence and Regularity

Let $\mathbf{B}$ be a Banach space with the norm $\| \cdot \|$. Denote by $L^2([0,T],\mathbf{B})$ the space of all measurable functions $u : [0,T] \to \mathbf{B}$ with

$$\|u\|_{L^2([0,T],\mathbf{B})} := \left(\int_0^T \|u\|^2 dt \right)^{\frac{1}{2}} < \infty$$

and by $C([0,T],\mathbf{B})$ the space of all continuous functions $u : [0,T] \to \mathbf{B}$ with

$$\|u\|_{C([0,T],\mathbf{B})} := \max_{t \in [0,T]} \|u(t)\| < \infty.$$

Let (M,F,m) be a Finsler measure space. Recall that $H^1(M) = W^{1,2}(M)$ and $H_0^1(M) = W_0^{1,2}(M)$. Likewise, $H^{-1}(M)$ means the dual Banach space of $H_0^1(M)$.

Definition 9.1.1. We say that a function u on $[0,T] \times M$, $T > 0$ (T maybe the infinity), is a global (super, sub) solution to the heat equation

$\partial_t u = \boldsymbol{\Delta} u$ (with the Dirichlet boundary condition) if it satisfies

(i) $u \in L^2\left([0,T], H_0^1(M)\right) \cap H^1\left([0,T], H^{-1}(M)\right)$.
(ii) For almost every $t \in (0,T)$ and $0 \le \phi \in C_0^\infty(M)$ (or, equivalently, for every $\phi \in H_0^1(M)$), it holds that

$$\int_M \phi \partial_t u \, dm \; (\ge, \le) = - \int_M d\phi(\nabla u) dm. \tag{9.1.1}$$

Note that $H_0^1(M) \subset L^2(M) \subset H^{-1}(M)$. Then $u \in L^2\left([0,T], H_0^1(M)\right)$ means that

$$\int_0^T \int_M F^*(du)^2 dm \, dt < \infty$$

and $u \in L^2\left([0,T], H^{-1}(M)\right)$ means that the weak derivative $\partial_t u$ in $H^{-1}(M)$ exists for almost every $t \in (0,T)$ in the sense that

$$\int_0^T \int_M \varphi(\partial_t u) dm \, dt = - \int_0^T \int_M (\partial_t \varphi) u \, dm \, dt$$

for all $\varphi \in C_0^\infty((0,T) \times M)$ and $\int_0^T \|u\|_{H^{-1}}^2 dt < \infty$. Moreover, we remark that the condition (i) implies that $u \in C\left([0,T], L^2(M)\right)$ (see [Ev], p287). Equivalently, we could require that (9.1.1) holds for all $\phi \in L^2\left([0,T], H_0^1(M)\right)$ and a.e. $t \in [0,T]$. If M is compact, then every global solution u to the heat equation is mass preserving, that is, $\int_M u(t,x) dm = \int_M u(0,x) dm$ holds for all t. Indeed, choosing $\phi \equiv 1$ as a test function yields the claim.

Global solutions are constructed as gradient curves of the energy functional $\mathcal{E}(u) = \int_M F^2(\nabla u) dm$ on $u \in H_{loc}^1(M)$ in the Hilbert space $L^2(M)$. As for the regularity, since the Finsler Laplacian is locally uniformly elliptic thanks to the strong convexity of F, the classical theory of partial differential equations applies. We summarize the existence and regularity properties as follows. The readers refer to [Oh4] (also [OS1], [GS]) if necessary.

Theorem 9.1.1. *Assume that $\Lambda < \infty$.*

(1) *For each initial datum $u_0 \in H_0^1(M)$ and $T > 0$, there exists a unique global solution $u(t,x)$ for $t \in [0,T]$ to the heat equation, and the distributional Laplacian $\Delta u(t,x)$ is absolutely continuous with respect to m for all $t \in (0,T)$.*

(2) *Let $u(t,x) \in H_0^1(M)$ be a global solution to the heat equation. Then one can take the continuous version of u, and it enjoys the H_{loc}^2-regularity in x as well as the $C^{1,\beta}$-regularity in both t and x on $(0,\infty) \times M$. Moreover, $\partial_t u$ lies in $H_{loc}^1(M) \cap C(M)$, and further in $H_0^1(M)$ if F has finite uniform smoothness constant κ.*

We remark that the Finsler metric F enjoys the local uniformly elliptic property in the sense of (8.1.22). Thus, the usual elliptic regularity yields that u is C^∞ on $\cup_{t>0}(\{t\} \times M_{u(t,x)})$. We also can study the regularity of local solutions to the heat equation.

Definition 9.1.2 ([OS1]). Given an open subset $\Omega \subset M$ and an open interval $I \subset \mathbb{R}$, we say that a function u on $I \times \Omega$ is a local (super, sub) solution to the heat equation $\partial_t u = \mathbf{\Delta} u$ on $I \times \Omega$ if $u \in L_{loc}^2(I \times \Omega)$ with $F^*(du) \in L_{loc}^2(I \times \Omega)$, and for any $\phi \in C_0^\infty(I \times \Omega)$ (or, equivalently, for any $\phi \in H_0^1(I \times \Omega)$),

$$\int_I \int_\Omega u \partial_t \phi \, dm dt \ (\leq, \ \geq) = \int_I \int_\Omega d\phi(\nabla u) dm dt. \qquad (9.1.2)$$

A function u being a local solution to the heat equation implies that $c_1 u + c_2$ is a local solution for every $c_1 \in \mathbb{R}^+$ and $c_2 \in \mathbb{R}$. In general, it will not imply that $-u$ is a local solution. In fact, u is a local (resp., global) solution to $\partial_t u = \mathbf{\Delta} u$ if and only if $-u$ is a local (resp., global) solution to $\partial_t u = \overleftarrow{\mathbf{\Delta}} u$.

Example 9.1.1. Let $\|\cdot\|$ be any smooth, strictly convex Minkowski norm on $\mathbb{R}^n$. Put $F(x,\cdot) = \|\cdot\|$ for all $x \in \mathbb{R}^n$ and choose m as the Legesgue measure. Then for each fixed $y \in \mathbb{R}^n$, the function

$$u(t,x) = t^{-n/2} \exp\left(-\frac{\|y-x\|^2}{4t}\right) \qquad (9.1.3)$$

is a local solution to the heat equation $\partial_t u = \mathbf{\Delta} u$ on $\mathbb{R}^+ \times \mathbb{R}^n$. Note that (9.1.3) is C^2 in the space variable at $x = y$ if and only if $\|\cdot\|$ is a Hilbert norm. More generally, $u(t,x) = f(t,\|y-x\|)$ is a local solution to the heat equation for each smooth function $f : \mathbb{R}^+ \times \mathbb{R}^+ \to \mathbb{R}$ satisfying $\partial_r f(t,r) \leq 0$ and

$$\partial_r^2 f(t.r) + \frac{n-1}{r} \partial_r f(t,r) = \partial_t f(t,r), \quad \partial_r f(t,0) = 0. \qquad (9.1.4)$$

If f satisfies $\partial_r f(t,r) \geq 0$ and (9.1.4), then the function $u(t,x) = f(t,\|x-y\|)$ is a local solution to the heat equation. If $\|\cdot\|$ is even a norm (i.e., if in addition it is symmetric), then the latter holds without any restriction on the sign of $\partial_r f(t, r)$.

Write $dm = e^{-V} dx$ and $J = \mathcal{L}^{-1}$, the inverse of the Legendre transformation $\mathcal{L}$. Assume that the logarithmic derivative $-V(x) = \log[dm/dx]$ of the measure m is Lipschitz continuous in x. For each $x_0 \in M$, there exists a local coordinate system $(x^i)_{i=1}^n$ in a neighborhood $\mathcal{U}$ of x_0 such that $\bar{\mathcal{U}}$ is compact. Thus, there is a positive constant C such that $|\partial V / \partial x^k| \leq C$ and

$$\left| J^i(x, \xi) \right| \leq CF^*(x, \xi) \leq C'|\xi|, \quad \left| \frac{\partial J^i(x, \xi)}{\partial x^j} \right| \leq CF^*(x, \xi) \leq C'|\xi|$$

$$(9.1.5)$$

for almost $x \in \mathcal{U}$ and all $\xi \in T^*M$ (also see (8.1.19)–(8.1.22) in Section 8.1).

Theorem 9.1.2 ($C^{1,\beta}$**-Regularity, [OS1]**). *Assume that the inverse J of the Legendre transformation as well as the logarithmic density of the measure m are differentiable in x as specified in (9.1.5). Then every continuous local solution to the heat equation $\partial_t u = \Delta u$ on $I \times \Omega$ is $C^{1,\beta}$ in t and x.*

Remark 9.1.1. If F is a smooth Finsler structure and if the logarithmic density of the measure m is smooth, then local solutions u of the heat equation $\partial_t u = \Delta u$ are C^∞ in t and x outside the set $\{(t, x) | du(t, x) = 0\}$. On this set, however, the solutions will not be C^2. See Example 9.1.1.

9.2 L^p-Uniqueness

In this section, we consider the uniqueness of nonnegative $L^p (p > 0)$ subsolutions to the heat equation on forward complete Finsler measure spaces ([Xia11]). For the case when $p > 1$, similar to the case of nonnegative subharmonic function, uniqueness is automatic for any forward complete Finsler measure spaces as indicated by the following result.

Theorem 9.2.1. *Let (M, F, m) be a forward complete Finsler measure space. If $u(t, x)$ is a nonnegative global subsolution to the heat equation, i.e., $\partial_t u \leq \Delta u$ (in the weak sense), on $[0, \infty) \times M$ with*

$$\lim_{t \to 0} \int_M u^p(t, x) dm = 0, \quad \int_M u^p(t, x) dm < \infty$$

for all $t > 0$ and $p > 1$, then $u(t, x) \equiv 0$ for all $x \in M$ and $t > 0$.

Proof. Let $\psi(x)$ be a cut-off function defined by (8.2.8) with $0 \le \psi \le 1$ and $F^*(-d\psi) \le 1/R$ a.e., on $B_{2R}^+(x_0)$. Since $\lim_{t\to 0} \int_M u^p(t, x)dm = 0$, we have

$$0 \le \lim_{t\to 0} \int_M \psi^2 u^p dm \le \lim_{t\to 0} \int_M u^p dm = 0,$$

which means that $\lim_{t\to 0} \int_M \psi^2 u^p dm = 0$. For any $T > 0$, choosing $\phi := \psi^2 u^{p-1}$ as a test function in (9.1.1) yields

$$-\int_0^T \int_M d(\psi^2 u^{p-1})(\nabla u)dmdt \ge \int_0^T \int_M \psi^2 u^{p-1}\frac{\partial u}{\partial t}dmdt$$

$$\ge \frac{1}{p}\int_0^T \frac{\partial}{\partial t}\left(\int_M \psi^2 u^p dm\right)dt$$

$$= \frac{1}{p}\int_M \psi^2 u^p(T, x)dm$$

by the assumption. On the other hand, the LHS of above inequality is equal to

$$-2\int_0^T \int_M \psi u^{p-1}d\psi(\nabla u)dmdt - (p-1)\int_0^T \int_M \psi^2 u^{p-2}F^2(\nabla u)dmdt$$

$$\le \frac{2}{p-1}\int_0^T \int_M u^p F^{*2}(-d\psi)dmdt - \frac{2(p-1)}{p^2}\int_0^T \int_M \psi^2 F^2(\nabla u^{p/2})dmdt,$$

where we used the inequality $-d\psi(\nabla u) \le F^*(-d\psi)F(\nabla u)$. Therefore,

$$\int_M \psi^2 u^p(T, x)dm + \frac{2(p-1)}{p}\int_0^T \int_M \psi^2 F^2(\nabla u^{p/2})dmdt$$

$$\le \frac{2p}{(p-1)R^2}\int_0^T \int_M u^p dmdt.$$

Letting $R \to \infty$, one obtains

$$\int_M u^p(T, x)dm = 0, \quad \text{and} \quad \int_0^T \int_M F^2(\nabla u^{p/2})dmdt = 0,$$

which implies that $u(T, x) = 0$ and $\nabla u^{p/2} = 0$. Hence, $u(t, x) \equiv 0$. $\qquad\square$

Now, we establish the local L^p mean value inequality to study the uniqueness of nonnegative subsolutions to the heat equation. Let us first consider the case when $p = 2$.

Proposition 9.2.1. *Let (M, F, m) be an n-dimensional forward complete Finsler measure space with finite reversibility Λ. Fix $R > 0$, assume that (7.2.6) holds. If u is a nonnegative local subsolution of the heat equation in the cylinder $Q := (s - r^2, s + \varepsilon) \times B_r^+(x_0)$ for any $s \in \mathbb{R}$ and $0 < r < R$, where ε is a positive constant, then there are constants $\nu > 2$, and $c = c(N, \nu, \Lambda) > 0$ such that for any $0 < \delta < \delta' \leq 1$ we have*

$$\sup_{Q_\delta} u^2 \leq \frac{e^{c(1+r\sqrt{K})}}{(\delta' - \delta)^{2+\nu} r^2 m(B_r^+(x_0))} \int_{Q_{\delta'}} u^2 dmdt, \qquad (9.2.1)$$

where $Q_\delta := (s - \delta r^2, s] \times B_{\delta r}^+(x_0)$ and $Q_{\delta'} := (s - \delta' r^2, s] \times B_{\delta' r}^+(x_0)$. In particular, when $Ric_N \geq -K$ for some $N \in [n, \infty)$ and $K > 0$, (9.2.1) holds.

Proof. For the sake of simplicity, we denote $B_\bullet := B_\bullet^+(x_0)$ in the following. Since u is a local subsolution to the heat equation in Q, we have $u \in L^2_{loc}(Q)$ with $F^*(du) \in L^2_{loc}(Q)$ and

$$\int_{s-r^2}^t \int_{B_r} (\phi \partial_t u + d\phi(\nabla u)) dmdt \leq 0 \qquad (9.2.2)$$

for any $t \in (s - \delta r^2, s]$ and nonnegative function $\phi \in W_0^{1,2}(Q)$. Let $\lambda = \lambda(t) \in C_0^\infty(\mathbb{R})$ and $\varphi = \varphi(x) \in C_0^\infty(B_r)$ be cut-off functions. Replacing ϕ with $u\varphi^2\lambda^2$ in (9.2.2) and using the Cauchy–Schwarz inequality yield

$$\int_{s-r^2}^t \int_{B_r} \lambda^2\varphi^2 F^2(\nabla u) dmdt + \int_{s-r^2}^t \int_{B_r} \lambda^2\varphi^2 u \partial_t u dmdt$$

$$\leq -2 \int_{s-r^2}^t \int_{B_r} \lambda^2\varphi u d\varphi(\nabla u) dmdt$$

$$\leq 2 \int_{s-r^2}^t \int_{B_r} \lambda^2\varphi u F^*(-d\varphi) F(\nabla u) dmdt$$

$$\leq \frac{1}{3} \int_{s-r^2}^t \int_{B_r} \lambda^2\varphi^2 F^{*2}(du) dmdt + 3 \int_{s-r^2}^t \int_{B_r} \lambda^2 u^2 F^{*2}(-d\varphi) dmdt,$$

where we used $F(\nabla u) = F^*(du)$. Thus,

$$\int_{s-r^2}^t \int_{B_r} \lambda^2\varphi^2 \partial_t(u^2) dmdt + \frac{4}{3} \int_{s-r^2}^t \int_{B_r} \lambda^2\varphi^2 F^{*2}(du) dmdt$$

$$\leq 6 \int_{s-r^2}^t \int_{B_r} \lambda^2 u^2 F^{*2}(-d\varphi) dmdt.$$

Note that $F^*(d(\varphi u)) \leq u F^*(d\varphi) + \varphi F^*(du)$. Hence,

$$\int_{s-r^2}^t \int_{B_r} \frac{\partial}{\partial t} (\lambda \varphi u)^2 \, dm dt + \int_{s-r^2}^t \int_{B_r} \lambda^2 F^{*2}(d(\varphi u)) \, dm dt$$

$$\leq 2 \int_{s-r^2}^t \int_{B_r} \lambda \lambda' \varphi^2 u^2 \, dm dt + \int_{s-r^2}^t \int_{B_r} \lambda^2 \varphi^2 \partial_t(u^2) \, dm dt$$

$$+ \frac{4}{3} \int_{s-r^2}^t \int_{B_r} \lambda^2 \varphi^2 F^{*2}(du) u^2 \, dm dt + 4 \int_{s-r^2}^t \int_{B_r} \lambda^2 F^{*2}(d\varphi) \, dm dt$$

$$\leq 2 \int_{s-r^2}^t \int_{B_r} \lambda u^2 \left[\lambda' \varphi^2 + 5\Lambda^2 \lambda F^{*2}(d\varphi) \right] dm dt. \tag{9.2.3}$$

Now, we choose $\lambda(t)$ on $\mathbb{R}$ with $0 \leq \lambda \leq 1$, $\lambda = 0$ in $(-\infty, s - \delta'r^2)$, $\lambda = 1$ in $(s - \delta r^2, +\infty)$ and $|\lambda'(t)| \leq \frac{1}{[(\delta'-\delta)r]^2}$, and $\varphi = \varphi(x)$ on B_r defined as in (7.2.9) with $F^*(-d\varphi) \leq \frac{1}{(\delta'-\delta)r}$ and $F^*(d\varphi) \leq \Lambda F^*(-d\varphi) \leq \frac{\Lambda}{(\delta'-\delta)r}$ a.e. on $B_{\delta'r}$. Denote $I_\delta := (s - \delta r^2, s]$ and $I_{\delta'} := (s - \delta'r^2, s]$. Taking the supremum with respect to t on both sides of (9.2.3) yields

$$\sup_{t \in I_\delta} \left\{ \int_{B_r} (\varphi u)^2 \, dm \right\} + \int_{I_\delta \times B_r} F^{*2}(d(\varphi u)) \, dm dt \leq \frac{12\Lambda^2}{(\delta'-\delta)^2 r^2} \int_{Q_{\delta'}} u^2 \, dm dt. \tag{9.2.4}$$

On the other hand, by Hölder's inequality and Soblev's inequality (7.2.6), we have

$$\int_{B_r} (u\varphi)^{2(1+\frac{2}{\nu})} \, dm$$

$$\leq \left(\int_{B_r} (u\varphi)^{\frac{2\nu}{\nu-2}} \, dm \right)^{\frac{\nu-2}{\nu}} \cdot \left(\int_{B_r} (u\varphi)^2 \, dm \right)^{\frac{2}{\nu}}$$

$$\leq \left(A \int_{B_r} [F^{*2}(d(u\varphi)) + r^{-2}(u\varphi)^2] \, dm \right) \cdot \left(\int_{B_r} (u\varphi)^2 \, dm \right)^{\frac{2}{\nu}} \tag{9.2.5}$$

for some $\nu > 2$, where $A := e^{c_0(1+r\sqrt{K})} r^2 m(B_r)^{-2/\nu}$, here $c_0 = c_0(N, \Lambda)$ is a positive constant. From (9.2.4)–(9.2.5), one obtains

$$\int_{Q_\delta} u^{2(1+\frac{2}{\nu})} \, dm dt$$

$$= \int_{Q_\delta} (u\varphi)^{2(1+\frac{2}{\nu})} \, dm dt$$

$$\leq \left(A \int_{I_\delta \times B_r} [F^{*2}(d(u\varphi)) + r^{-2}(u\varphi)^2] dmdt \right) \left(\sup_{t \in I_\delta} \int_{B_r} (u\varphi)^2 dm \right)^{2/\nu}$$

$$\leq A \left(\frac{12\Lambda^2}{((\delta'-\delta)r)^2} \int_{Q_{\delta'}} u^2 dmdt \right)^{1+\frac{2}{\nu}}.$$

For any $\tau \geq 1$, it is easy to see that u^τ is also a nonnegative subsolution of the heat equation. Let $\chi := 1 + \frac{2}{\nu}$. The above inequality implies that

$$\int_{Q_\delta} u^{2\chi\tau} dmdt \leq A \left(\frac{12\Lambda^2}{((\delta'-\delta)r)^2} \int_{Q_{\delta'}} u^{2\tau} dmdt \right)^\chi. \qquad (9.2.6)$$

For any $0 < \delta < \delta' \leq 1$, let $\delta_0 = \delta', \delta_{i+1} = \delta_i - \frac{\delta'-\delta}{2^{i+1}}$ and $\tau = \chi^i > 1$ on $B_{\delta_i r}$ for $i = 0, 1, \dots$. Applying (9.2.6) for $\delta' = \delta_i$, $\delta = \delta_{i+1}$ and $\tau = \chi^i$, we get

$$\int_{Q_{\delta_{i+1}}} u^{2\chi^{i+1}} dmdt \leq A \left(\frac{(12\Lambda^2)4^{i+1}}{((\delta'-\delta)r)^2} \int_{Q_{\delta_i}} u^{2\chi^i} dmdt \right)^\chi.$$

By iteration, one obtains

$$\left(\int_{Q_{\delta_{i+1}}} u^{2\chi^{i+1}} dmdt \right)^{\frac{1}{\chi^{i+1}}}$$

$$\leq C^{\sum j\chi^{1-j}} (A)^{\sum \chi^{-j}} [(\delta'-\delta)r]^{-2\sum \chi^{1-j}} \cdot \int_{Q_{\delta'}} u^2 dmdt, \qquad (9.2.7)$$

in which $\sum$ denotes the summation on j from 1 to $i + 1$ and C is a positive constant depending on Λ. Since $\sum_{j=1}^\infty \chi^{1-j} = 1 + \frac{\nu}{2}$, $\sum_{j=1}^\infty \chi^{-j} = \frac{\nu}{2}$ and $\sum_{j=1}^\infty j\chi^{1-j}$ converges, (9.2.7) implies (9.2.1) by taking $i \to \infty$. This finishes the proof. $\qquad \square$

Based on Proposition 9.2.1, we have the following local L^p mean value inequality.

Theorem 9.2.2. *Let (M, F, m) be an n-dimensional forward complete Finsler measure space with finite reversibility Λ. Fix $R > 0$, assume that (7.2.6) holds. If u is a nonnegative L^p subsolution to the heat equation in the cylinder $Q = (s - r^2, s + \varepsilon) \times B_r^+(x_0)$ for any $s \in \mathbb{R}$ and $0 < r < R$, where ε is a positive constant, then there are constants $\nu > 2$ and $c = c(N, p, \nu, \Lambda) > 0$ such that*

$$\sup_{Q_\delta} u^p \leq \frac{e^{c(1+r\sqrt{K})}}{(\delta'-\delta)^{2+\nu} r^2 m(B_r)} \int_{Q_{\delta'}} u^p dmdt \qquad (9.2.8)$$

for any $0 < p \leq 2$ and $0 < \delta < \delta' \leq 1$, where Q_δ and $Q_{\delta'}$ are as in Proposition 9.2.1. In particular, when $Ric_N \geq -K(K > 0)$ for some $N \in [n, \infty)$, (9.2.8) holds.

Proof. The case when $p = 2$ has been proved in Proposition 9.2.1. It suffices to prove (9.2.8) for $0 < p < 2$. We denote $B_\bullet := B_\bullet^+(x_0)$ as in the proof of Proposition 9.2.1. For any $0 < \delta < \delta'' \leq \delta' \leq 1$, it follows from the proof of Proposition 9.2.1 that there are positive constants $\nu = \nu(N) > 2$ and $c = c(N, \nu, \Lambda)$ such that

$$\sup_{Q_\delta} u^2 \leq e^{c(1+r\sqrt{K})}(\delta'' - \delta)^{-2-\nu} r^{-2} m(B_r)^{-1} \int_{Q_{\delta''}} u^2 \, dm \, dt.$$

On the other hand,

$$\int_{Q_{\delta''}} u^2 \, dm \, dt \leq \sup_{Q_{\delta''}} u^{2-p} \int_{Q_{\delta''}} u^p \, dm \, dt \leq \left(\sup_{Q_{\delta''}} u^2 \right)^{1-p/2} \int_{Q_{\delta'}} u^p \, dm \, dt.$$

Thus,

$$\sup_{Q_\delta} u^2 \leq e^{c(1+r\sqrt{K})}(\delta'' - \delta)^{-2-\nu} r^{-2} m(B_r)^{-1} \left(\sup_{Q_{\delta''}} u^2 \right)^{1-p/2} \int_{Q_{\delta'}} u^p \, dm \, dt.$$

$$(9.2.9)$$

Let $\mu = 1 - p/2 > 0$ and

$$\mathcal{M}(\delta) := \sup_{Q_\delta} u^2, \quad \mathcal{R} = e^{c(1+r\sqrt{K})} r^{-2} m(B_r)^{-1} \int_{Q_{\delta'}} u^p \, dm \, dt.$$

Let $\delta_0 = \delta$ and $\delta_i = \delta_{i-1} + \frac{\delta' - \delta}{2^i}$ for $i = 1, 2, \ldots$. Applying (9.2.9) for $\delta = \delta_{i-1}$ and $\delta'' = \delta_i$ yields

$$\mathcal{M}(\delta_{i-1}) \leq \mathcal{R} 2^{i(2+\nu)}(\delta' - \delta)^{-2-\nu} \mathcal{M}(\delta_i)^\mu.$$

By iterating, we get

$$\mathcal{M}(\delta_0) \leq \mathcal{R}^{\sum \mu^{i-1}} 2^{(2+\nu) \sum i\mu^{i-1}} (\delta' - \delta)^{(-2-\nu) \sum \mu^{i-1}} \mathcal{M}(\delta_j)^{\mu^j},$$

$$(9.2.10)$$

in which $\sum$ denotes the summation on i from 1 to j. Obviously, $\lim_{j \to \infty} \delta_j = \delta'$ and $\lim_{j \to \infty} \mu^j = 0$. Moreover, $\sum_{i=1}^\infty \mu^{i-1} = 2/p$ and $\sum_{i=1}^\infty i\mu^{i-1}$

converges. Letting $j \to \infty$ on the both sides of (9.2.10), there is a positive constant $c = c(N, \nu, p)$ such that

$$\sup_{Q_\delta} u^p \le \mathcal{M}(\delta_0)^{p/2} \le e^{c(1+r\sqrt{K})}(\delta' - \delta)^{-2-\nu}r^{-2}m(B_r)^{-1} \int_{Q_{\delta'}} u^p dm dt.$$

The proof is finished. $\square$

If $u(t, x)$ is independent of the time t, then it is a subharmonic function on (M, F, m). Theorem 9.2.2 implies the following local mean value inequality for L^p subharmonic functions, which essentially coincides with (7.2.13).

Corollary 9.2.1. *Let (M, F, m) be an n-dimensional forward complete Finsler measure space with finite reversibility Λ. Assume that $Ric_N \ge -K$ for some $N \in [n, \infty)$ and $K > 0$. If u is a nonnegative L^p subharmonic function on $B_r := B_r^+(x_0)$, then there are constants $\nu > 2$ and $c = c(N, p, \nu, \Lambda) > 0$ such that*

$$\sup_{B_{\delta r}} u^p \le \frac{e^{c(1+r\sqrt{K})}}{(\delta' - \delta)^{2+\nu}m(B_r)} \int_{B_{\delta' r}} u^p dm \qquad (9.2.11)$$

for any $0 < p \le 2$ and $0 < \delta < \delta' \le 1$.

As an application of Theorem 9.2.2, we have the following Liouville property.

Corollary 9.2.2. *Let (M, F, m) be an n-dimensional forward complete and noncompact Finsler measure space with finite reversibility Λ. If $Ric_N \ge 0$ for some $N \in [n, \infty)$, then every nonnegative subsolutions in $L^p(\mathbb{R} \times M)(0 < p \le 2)$ to the heat equation on $\mathbb{R} \times M$ is identically zero.*

Proof. By Theorem 9.2.2 and the assumption, we have

$$u(t, x_0) \le \sup_{Q_\delta} u^p \le \frac{c}{(\delta' - \delta)^{2+\nu}r^2 m(B_r)} \int_{Q_{\delta'}} u^p dm dt \qquad (9.2.12)$$

for $t \in I_\delta$ and $0 < \delta < \delta' \le 1$. Since $Ric_N \ge 0$ and $\Lambda < \infty$, the volume of a geodesic ball grows linearly with respect to its radius by Theorem 6.4.2. Note that $0 \le u(t, x) \in L^p(\mathbb{R} \times M)$ by the assumption. Letting $\delta \to 0$ and then $r \to \infty$ on both sides of (9.2.12) yields $u(s, x_0) = 0$. Since $s \in \mathbb{R}$ and $x_0 \in M$ are arbitrary, we have $u(t, x) \equiv 0$ on $\mathbb{R} \times M$. $\square$

9.3 Li-Yau's Gradient Estimates

In the celebrated paper [LY], Li and Yau proved the local gradient estimate for positive solutions of the heat equation on a complete Riemannian manifold (M, g) with Ricci curvature bounded from below. Li–Yau's gradient estimates for the parabolic heat equation are fundamental to study the global analysis on manifolds ([SY]). In this section, we shall give Li–Yau's type estimates for the the nonlinear heat flow on Finsler manifolds. Because of the nonlinearity, the approaches to study Li–Yau's type estimates for the nonlinear heat equation are essentially different from those for the linear heat equation in Riemannian case.

9.3.1 *Compact Case*

Li–Yau's estimate for the nonlinear heat flow in Finsler geometry was first studied by Ohta–Sturm in [OS2]. They gave Li–Yau's type estimate on a compact Finsler manifold without boundary via establishing the Bochner–Weitzenböck formula and using the maximum principle for the linearized operator of Finsler Laplacian. In fact, this estimate is also true on a compact Finsler manifold with a convex boundary ∂M if $\partial M \neq \emptyset$ and u is C^1 on $[0, \infty) \times \partial M$. Precisely, we have the following result.

Theorem 9.3.1. *Let (M, F, m) be an n-dimensional compact Finsler measure space without boundary or with a convex boundary. Assume that $Ric_N \geq -K$ for some $N \in [n, \infty)$ and $K \geq 0$ and u is a globally positive solution $u : [0, \infty) \times M \to \mathbb{R}$ to $u_t = \Delta u$ (in the weak sense) with the Neumann boundary condition, i.e., $\nabla u \in T(\partial M)$ for any $t \in [0, \infty)$ if $\partial M \neq \emptyset$ and $u \in C^1([0, \infty) \times \partial M)$. Then for any constant $\alpha > 1$ and $t > 0$, we have*

$$F^2(\nabla(\log u)) - \alpha(\log u)_t \leq N\alpha^2 \left(\frac{1}{2t} + \frac{K}{4(\alpha - 1)} \right). \qquad (9.3.1)$$

For its proof, the readers refer to that of Theorem 4.4 in [OS2] when $\partial M = \emptyset$. For the case when $\partial M \neq \emptyset$, it suffices to prove that (9.3.1) holds on ∂M. This proof is similar to that of Theorem 9.3.2 below. We omit it here. We proceed along the line of Theorem 5.3.5 in [Da] and obtain the following Harnack inequality.

Corollary 9.3.1. *Under the same assumptions as in Theorem 9.2.1, we have*

$$u(t_1, x_1) \leq u(t_2, x_2) \cdot \left(\frac{t_2}{t_1} \right)^{\alpha N/2} \exp\left(\frac{\alpha d_F(x_2, x_1)^2}{4(t_2 - t_1)} + \frac{\alpha K N(t_2 - t_1)}{4(\alpha - 1)} \right)$$

for all $\alpha > 1$, $0 < t_1 < t_2 \leq T$ and $x_1, x_2 \in M$.

Proof. Replacing u by $u + \varepsilon$ if necessary, we may assume without loss of generality that u is positive. Let $\gamma(\tau) = \exp_{x_2}((t_2 - \tau)y)$ for $\tau \in [t_1, t_2]$ be the reverse curve of minimal geodesic from $x_2 = \gamma(t_2)$ to $x_1 = \gamma(t_1)$ with suitable $y \in T_{x_2}M$. Then obviously $F(-\dot\gamma(\tau)) = d_F(x_2, x_1)/(t_2 - t_1)$ for all τ. We also put $v := \log u$,

$$\Theta := \frac{\alpha d_F(x_2, x_1)^2}{4(t_2 - t_1)^2} + \frac{\alpha KN}{4(\alpha - 1)}, \quad \sigma(\tau) := v(\tau, \gamma(\tau)) + \frac{\alpha N}{2}\log\tau + \Theta\tau.$$

Then we have by Theorem 9.2.1,

$$\sigma'(\tau) = dv(\dot\gamma) + v_\tau + \frac{\alpha N}{2\tau} + \Theta$$

$$\geq -F(\nabla v)F(-\dot\gamma) + \frac{F(\nabla v)^2}{\alpha} - N\alpha\left(\frac{1}{2\tau} + \frac{K}{4(\alpha - 1)}\right) + \frac{\alpha N}{2\tau} + \Theta$$

$$\geq -F(\nabla v)\frac{d_F(x_2, x_1)}{t_2 - t_1} + \frac{F(\nabla v)^2}{\alpha} + \frac{\alpha d_F(x_2, x_1)^2}{4(t_2 - t_1)^2} \geq 0.$$

Hence, we obtain

$$u(t_1, x_1) \cdot t_1^{\alpha N/2}e^{\Theta t_1} = e^{\sigma(t_1)} \leqslant e^{\sigma(t_2)} = u(t_2, x_2) \cdot t_2^{\alpha N/2}e^{\Theta t_2},$$

which proves the claim. $\qquad\qquad\square$

Eestimate (9.3.1) is usually called the *Davies' estimate* ([Da]), in which α is a constant greater than 1. However, it tells nothing if α is not constant. To get a more general Li–Yau's type estimate, we introduce the following notations.

For any constant $K \geq 0$, $N \in [n, \infty)$ and any positive C^1 function $a(t)$ satisfying

(A1) for all $t > 0$, $a'(t) > 0$, $\lim_{t\to 0+} a(t) = 0$ and $\lim_{t\to 0+} \frac{a(t)}{a'(t)} = 0$;

(A2) for any $T > 0$, $\frac{a'^2}{a}$ is continuous and integrable on the interval $[0, T]$.

Let

$$\alpha(t) := 1 + \frac{2K}{a}\int_0^t a(s)ds, \qquad\qquad (9.3.2)$$

$$\varphi(t) := \frac{NK}{2} + \frac{NK^2}{2a}\int_0^t a(s)ds + \frac{N}{8a}\int_0^t \frac{a'^2(s)}{a(s)}ds. \qquad (9.3.3)$$

It is easy to check that $\alpha(t)$ and $\varphi(t)$ satisfy the following ODEs:

$$(\alpha - 1)' + (\log a)'(\alpha - 1) - 2K = 0,$$

$$\varphi' + (\log a)'\varphi - \frac{N}{8}[2K + (\log a)']^2 = 0 \qquad (9.3.4)$$

with $\lim_{t \to 0+} \alpha(t) = 1$ and $\lim_{t \to 0+} \varphi(t) = \infty$. Obviously, $\alpha(t) \geq 1$ and $\varphi(t) \geq 0$. It follows from (9.3.2) that $\alpha(t) \equiv 1$ if and only if $K = 0$. In this case,

$$\varphi_1(t) = \frac{N}{8a} \int_0^t \frac{a'^2(s)}{a(s)} ds. \qquad (9.3.5)$$

Theorem 9.3.2. *Let (M, F, m) be an n-dimensional compact Finsler measure space without boundary or with a convex boundary. Assume that $Ric_N \geq -K$ for some $N \in [n, \infty)$ and $K \geq 0$ and u is a globally positive solution $u : [0, \infty) \times M \to \mathbb{R}$ to $\partial_t u = \mathbf{\Delta} u$ (in a weak sense) with the Neumann boundary condition, i.e., $\nabla u \in T(\partial M)$ for any $t \in [0, \infty)$ if $\partial M \neq \emptyset$ and $u \in C^1([0, \infty) \times \partial M)$. Then*

$$\sup_{x \in M} \left\{ F^2(\nabla(\log u)) - \alpha(t)(\log u)_t, \ F^2(\nabla(-\log u)) + \alpha(t)(\log u)_t \right\} \leq \varphi(t)$$

$$(9.3.6)$$

for any $t > 0$, where $\alpha(t)$ and $\varphi(t)$ are respectively given by (9.3.2) and (9.3.3). In particular, if $Ric_N(M) \geq 0$, then

$$\sup_{x \in M} \left\{ F^2(\nabla(\log u)) - (\log u)_t, \ F^2(\nabla(-\log u)) + (\log u)_t \right\} \leq \varphi_1 \quad (9.3.7)$$

for any $t > 0$.

To prove Theorem 9.3.2, we need some lemmas as follows. Let

$$v(t, x) := \log u(t, x), \qquad f := F^2(\nabla v) - \alpha(t)v_t - \varphi(t).$$

Then $v \in H^2(M) \cap C^{1,\beta}([0, T] \times M)$, $\mathbf{\Delta} v \in H^1(M)$ and $f \in H^1(M) \cap C^\beta([0, T] \times M)$ by the regularity of u. Moreover, $g_{\nabla v} = g_{\nabla u}$ and hence $\nabla v = \frac{1}{u}\nabla u$, $\mathbf{\Delta} v = \mathbf{\Delta}^{\nabla u} v$ on $M_u(= M_v)$. For the sake of simplicity, we write $(\cdot)_t$ to represent the usual derivative of $(\cdot)$ with respect to t instead of $\partial_t(\cdot)$ in the following.

Lemma 9.3.1. *The following identities hold on* $(0,T) \times M$ *in the weak sense.*

(i) $\boldsymbol{\Delta} v + F^2(\nabla v) = v_t$;

(ii) $\boldsymbol{\Delta}^{\nabla u} v_t + 2dv_t(\nabla v) = \partial_t(v_t)$;

(iii) $\boldsymbol{\Delta}^{\nabla u} f + 2df(\nabla v) - f_t = 2\|\nabla^2 v\|^2_{HS(\nabla v)} + 2Ric_\infty(\nabla v) + \alpha' v_t + \varphi'$,
where $\|\cdot\|_{HS(\nabla v)}$ *stands for the Hilbert–Schmidt norm with respect to* $g_{\nabla v}$.

Proof. It is easy to check (i). Now, we prove (ii). Observe that

$$\partial_t(F^2(\nabla v)) = \partial_t \left(g^{ij}(\nabla v) \frac{\partial v}{\partial x^i} \frac{\partial v}{\partial x^j} \right)$$

$$= 2g^{ij}(\nabla v) \frac{\partial^2 v}{\partial x^i \partial t} \frac{\partial v}{\partial x^j} - 2C_{ijk}(\nabla v)\partial_t(\nabla^k v)\nabla^i v \nabla^j v$$

$$= 2d(v_t)(\nabla v), \tag{9.3.8}$$

on $(0,T) \times M_v$, where $\nabla v = (\nabla^i v)\frac{\partial}{\partial x^i}$ in local coordinates. Obviously, (9.3.8) a.e. holds on $(0,T) \times M$ (cf. the arguments in the paragraph after (9.3.13)). Thus, for any $\phi \in H^1_0([0,T] \times M)$,

$$-\int_0^T \int_M \phi_t F^2(\nabla v) dm dt = \int_0^T \int_M \phi \partial_t(F^2(\nabla v)) dm dt$$

$$= 2\int_0^T \int_M \phi d(v_t)(\nabla v) dm dt. \tag{9.3.9}$$

Also, it is easy to see that $\partial_t(\boldsymbol{\Delta} v) = \boldsymbol{\Delta}^{\nabla u} v_t$ in the weak sense. From this and (i), we get

$$\int_0^T \int_M \left\{ -d\phi(\nabla^{\nabla u} v_t) + 2\phi d(v_t)(\nabla v) + \phi_t v_t \right\} dm dt$$

$$= \int_0^T \int_M \left\{ -d\phi(\nabla^{\nabla u} v_t) + \phi_t \left(v_t - F^2(\nabla v) \right) \right\} dm dt$$

$$= -\int_0^T \int_M \left\{ d\phi(\nabla^{\nabla u} v_t) + \phi \partial_t(\boldsymbol{\Delta} v) \right\} dm dt$$

$$= -\int_0^T \int_M \left\{ d\phi(\nabla^{\nabla u} v_t) + \phi \boldsymbol{\Delta}^{\nabla u} v_t \right\} dm dt$$

$$= -\int_0^T \int_M \text{div}(\phi \nabla^{\nabla u} v_t) dm dt = 0.$$

(iii) From (9.3.9) and (i)–(ii), one obtains, for each $\phi \in H_0^1((0,T) \times M)$,

$$\int_0^T \int_M \left\{ -d\phi(\nabla^{\nabla u} f) + 2\phi df(\nabla v) + f\phi_t \right\} dmdt$$

$$= \int_0^T \int_M \left\{ -d\phi(\nabla^{\nabla u}(F^2(\nabla v))) + 2\phi d(F^2(\nabla v))(\nabla v) - 2\phi d(v_t)(\nabla v) \right.$$

$$\left. + \phi(\alpha' v_t + \varphi') \right\} dmdt$$

$$= \int_0^T \int_M \left\{ -d\phi(\nabla^{\nabla u}(F^2(\nabla v))) - 2\phi d(\mathbf{\Delta} v)(\nabla v) + \phi(\alpha' v_t + \varphi') \right\} dmdt$$

$$= \int_0^T \int_M \phi \left\{ 2\|\nabla^2 v\|^2_{HS(\nabla v)} + 2Ric_\infty(\nabla v) + \alpha' v_t + \varphi' \right\} dmdt,$$

$$(9.3.10)$$

in which the last equality follows from Corollary 7.1.2. This finishes the proof. $\qquad \square$

Lemma 9.3.2. *Given an $N \in [n, \infty)$, we have*

$$\int_0^T \int_M \left\{ -d\phi(\nabla^{\nabla u} f) + 2\phi df(\nabla v) + f\phi_t \right\} dmdt$$

$$\geq \int_0^T \int_M \phi \left\{ \frac{2}{N}(\mathbf{\Delta} v)^2 + 2Ric_N(\nabla v) + \alpha' v_t + \varphi' \right\} dmdt \quad (9.3.11)$$

for any nonnegative function $\phi \in H_0^1((0,T) \times M)$.

Proof. By Cauchy–Schwarz's inequality and the elementary inequality $(a+b)^2 \geq \frac{a^2}{1+\delta} - \frac{b^2}{\delta}$ for any $\delta > 0$, we have

$$\|\nabla^2 v\|^2_{HS(\nabla v)} \geq \frac{1}{n}\left(\mathrm{tr}_{\nabla v}\, \nabla^2 v\right)^2 = \frac{1}{n}\left(\mathbf{\Delta} v + S(\nabla v)\right)^2 \geq \frac{(\mathbf{\Delta} v)^2}{N} - \frac{S^2(\nabla v)}{N-n},$$

which implies that

$$\mathrm{Ric}_\infty(\nabla v) + \|\nabla^2 v\|^2_{HS(\nabla v)} \geq \frac{(\mathbf{\Delta} v)^2}{N} + \mathrm{Ric}_N(\nabla v) \qquad (9.3.12)$$

pointwise on $(0,T) \times M_v$. For any nonnegative $\phi \in H_0^1((0,T) \times M_v)$, it follows from (9.3.12) that

$$\int_0^T \int_{M_v} \phi \left\{ \mathrm{Ric}_\infty(\nabla v) + \|\nabla^2 v\|^2_{HS(\nabla v)} \right\} dmdt$$

$$\geq \int_0^T \int_{M_v} \phi \left\{ \mathrm{Ric}_N(\nabla v) + \frac{(\mathbf{\Delta} v)^2}{N} \right\} dmdt. \qquad (9.3.13)$$

Obviously, the first term on the left (resp., right)-hand side is zero on $M \backslash M_v$. In order to see that the second term on the LHS (resp. RHS) vanishes, let us fix a local coordinate $\{(x^i)\}_{i=1}^n$ and apply Lemma 7.1.3 to the function $h := \partial v / \partial x^i \in H^1_{loc}(M)$ for each t. It implies that $\partial h / \partial x^j = 0$ a.e. on $h^{-1}(0)$ for all $j = 1, 2, \ldots, n$. That is to say, $\partial^2 v / \partial x^i \partial x^j = 0$ for all $i, j = 1, 2, \ldots, n$ on $\bigcap_{i=1}^n \{\partial v / \partial x^i = 0\}$. In particular, $\nabla^2 v = 0$ and $\Delta v = 0$ a.e. on $M \backslash M_v$. Thus, the integrands on the left and right hand sides of (9.3.13) are actually integrated on M. Consequently,

$$\int_0^T \int_M \phi \left\{ \mathrm{Ric}_\infty(\nabla v) + \left\| \nabla^2 v \right\|_{HS(\nabla v)}^2 \right\} dm dt$$

$$\geq \int_0^T \int_M \phi \left\{ \mathrm{Ric}_N(\nabla v) + \frac{(\Delta v)^2}{N} \right\} dm dt. \qquad (9.3.14)$$

Plugging this into (9.3.10) yields (9.3.11).

For the case when $0 \leq \phi \in H^1_0((0,T) \times M)$, set

$$\phi_k := \min\{\phi, k^2 F^2(\nabla v)\}, \quad k \in \mathbb{N}.$$

Then $\phi_k \in H^1_0((0,T) \times M_v)$ and $\lim_{k \to \infty} \phi_k(t,x) = \phi(t,x)$ for all $(t,x) \in (0,T) \times M_v$. Hence,

$$\int_0^T \int_{M_v} \phi_k \left\{ \mathrm{Ric}_\infty(\nabla v) + \left\| \nabla^2 v \right\|_{HS(\nabla v)}^2 \right\} dm dt$$

$$\geq \int_0^T \int_{M_v} \phi_k \left\{ \mathrm{Ric}_N(\nabla v) + \frac{(\Delta v)^2}{N} \right\} dm dt.$$

As $k \to \infty$, we have (9.3.13). By the same arguments as in the paragraph after (9.3.13), we still have (9.3.14). Thus, we obtain (9.3.11) from (9.3.10) and (9.3.14). $\qquad \square$

Proof of Theorem 9.3.2. Let $v = \log u$ and $f := F^2(\nabla v) - \alpha(t)v_t - \varphi(t)$ as before. Given any C^1 function $\mu = \mu(t)$ on $[0,T]$, it is obvious that $(\Delta v)^2 \geq -2\mu\Delta v - \mu^2$. From Lemma 9.3.2, we have

$$\int_0^T \int_M \left\{ -d\phi(\nabla^{\nabla u} f) + 2\phi df(\nabla v) + f\phi_t \right\} dm dt$$

$$\geq \int_0^T \int_M \phi \left\{ 2\mathrm{Ric}_N(\nabla v) - \frac{4\mu}{N}\Delta v - \frac{2}{N}\mu^2 + \alpha' v_t + \varphi' \right\} dm dt \qquad (9.3.15)$$

for any nonnegative function $\phi \in H_0^1((0,T) \times M)$. By (9.3.15) and Lemma 9.3.1, one obtains

$$\int_0^T \int_M \{-d\phi(\nabla^{\nabla u} f) + 2\phi df(\nabla v) + f\phi_t\} \, dmdt$$

$$\geq \int_0^T \int_M \phi \left\{ 2\mathrm{Ric}_N(\nabla v) - \frac{4\mu}{N}(v_t - F^2(\nabla v)) - \frac{2\mu^2}{N} + \alpha' v_t + \varphi' \right\} dmdt$$

$$\geq \int_0^T \int_M \phi \left\{ \left(\frac{4\mu}{N} - 2K \right) F^2(\nabla v) - \left(\frac{4\mu}{N} - \alpha' \right) v_t + \varphi' - \frac{2\mu^2}{N} \right\} dmdt.$$

$$(9.3.16)$$

Note that the above inequality holds for any C^1 function $\mu = \mu(t)$. We choose μ such that $\frac{4\mu}{N} - 2K = (\log a)' > 0$, where $a(t)$ satisfies (A1)–(A2). On the other hand, by the assumptions, $\alpha(t)$ and $\varphi(t)$ satisfy (9.3.4). Thus, we have

$$\frac{4\mu}{N} - \alpha' = \alpha \left(\frac{4\mu}{N} - 2K \right), \quad \varphi' + \left(\frac{4\mu}{N} - 2K \right) \varphi - \frac{2\mu^2}{N} = 0, \quad (9.3.17)$$

with $\lim_{t \to 0+} \alpha(t) = 1$ and $\lim_{t \to 0+} \varphi(t) = \infty$. In particular, $\alpha(t) \equiv 1$ if and only if $K = 0$ from (9.3.17)$_1$. In fact, if $K = 0$ and $\alpha > 1$, then $-\alpha' = (\alpha - 1)(\log a)' > 0$. Thus, $\alpha(t) < \lim_{t \to 0+} \alpha(t) = 1$ for $t > 0$. This is impossible. In this case, $\varphi = \varphi_1$ given by (9.3.5). Plugging (9.3.17) into (9.3.16) yields

$$\int_0^T \int_M \{-d\phi(\nabla^{\nabla u} f) + 2\phi df(\nabla v) + f\phi_t\} \, dmdt \geq \int_0^T \int_M \phi(\log a)' f dmdt.$$

$$(9.3.18)$$

Since M is compact, we may assume that (t_0, x_0) is the maximum point of f on $[0,T] \times M$. We claim that $f(t_0, x_0) \leq 0$ by (9.3.18), which implies the conclusion.

Case 1. $\partial M = \emptyset$ and $x_0 \in \mathrm{int}(M)$. Note that $\lim_{t \to 0+} \varphi(t) = \infty$, which implies that $f(0, x_0) = -\infty$. If $f(t_0, x_0) \leq 0$, the assertion of Theorem 9.3.2 is obvious. Thus, we may assume that $t_0 > 0$ and $f(t_0, x_0) > 0$, which implies that there is a neighborhood $Q := (t_0 - \delta, t_0 + \delta) \times B_\delta^+(x_0)$ of (t_0, x_0) in which $f(t, x) > 0$. Note that $(\log a)' > 0$ on $(0, T]$ by the assumption on $a(t)$. Hence, f is a strict subsolution to the linear and divergence type parabolic operator $\Delta^{\nabla u} f + 2df(\nabla v) - f_t$ in Q. By the maximum principle, $f(t_0, x_0) < \max_{\partial Q} f$, which contradicts the maximality of $f(t_0, x_0)$. This proves the claim.

Case 2. $\partial M \neq \emptyset$. We only need consider the case when $x_0 \in \partial M$ since the arguments are the same as Case 1 when $x_0 \in \text{int}(M)$. By arguments as in Case 1, we may assume that $t_0 > 0$. If $\nabla u(t_0, x_0) = 0$, then $\nabla v(t_0, x_0) = 0$. Thus, $x_0 \in M \backslash M_v$. Similar to the proof of Lemma 9.3.2, $\mathbf{\Delta} v(t_0, \cdot) = 0$ a.e. on $M \backslash M_v$. Hence, $v_t(t_0, \cdot) = 0$ a.e. in $M \backslash M_v$ from Lemma 9.3.1(i). In this case, $f(t_0, x_0) = -\varphi(t_0) < 0$ and nothing needs to be proved. Now, we assume that $(t_0, x_0) \in (0, T] \times M_u$ and $f(t_0, x_0) > 0$. There exists a neighborhood $Q := (t_0 - \delta, t_0 + \delta) \times (B_\delta^+(x_0) \cap \bar{M})$ of (t_0, x_0) in which $f(t, x) > 0$, where $\bar{M}$ is the closure of M. Without loss of generality, we assume that ν is the unit normal vector field of ∂M pointing outward with respect to F. Then there is another normal vector field $\nu_{\nabla u(t, \cdot)}$ of ∂M for each $t \in (0, T]$ such that $g_{\nabla u}(\nu_{\nabla u}, Y) = 0, g_{\nabla u}(\nu_{\nabla u}, \nu_{\nabla u}) = 1$ and $g_\nu(\nu, \nu_{\nabla u}) > 0$, where $Y \in \{t\} \times T(\partial M)$. Note that $\nu_{\nabla u}$ depends on the time t as well as the spacial variable x. Then

$$g_\nu(\nu, X) = 0 \Leftrightarrow X \in (0, T] \times T(\partial M) \Leftrightarrow g_{\nabla u}(\nu_{\nabla u}, X) = 0$$

by Proposition 2.4.1. Since $\nu_{\nabla u}$ points outward to ∂M, by taking normal derivative of f with respect to $\nu_{\nabla u}$, we have $df(\nu_{\nabla u}) \geq 0$. On the other hand, the Neumann boundary condition $\nabla u(t, x) \in T_x(\partial M)$ for each t means that

$$g_\nu(\nu, \nabla u(t, x)) = 0, \quad \text{equivalently,} \quad g_{\nabla u}(\nu_{\nabla u}, \nabla u)(t, x) = 0,$$

$$(9.3.19)$$

The latter is equivalent to $du(\nu_{\nabla u}) = 0$. Differentiating the first equation of (9.3.19) with respect to t yields

$$g_\nu(\nu, \partial_t(\nabla u)) = 0, \quad \text{equivalently,} \quad g_{\nabla u}(\nu_{\nabla u}, \partial_t(\nabla u)) = 0.$$

$$(9.3.20)$$

Differentiating the second equation of (9.3.19) with respect to t yields

$$0 = \partial_t \left(g_{\nabla u}(\nu_{\nabla u}, \nabla u) \right) = g_{\nabla u} \left(\partial_t(\nu_{\nabla u}), \nabla u \right) + g_{\nabla u} \left(\nu_{\nabla u}, \partial_t(\nabla u) \right).$$

$$(9.3.21)$$

Combining (9.3.20) with (9.3.21) gives $g_{\nabla u}(\nabla u, \partial_t(\nu_{\nabla u})) = 0$, i.e., $du(\partial_t(\nu_{\nabla u})) = 0$. Therefore,

$$d(u_t)(\nu_{\nabla u}) = \partial_t(du(\nu_{\nabla u})) - du(\partial_t(\nu_{\nabla u})) = 0.$$

Moreover, in local coordinates, we have

$$\frac{\partial}{\partial x^i}\left(\frac{F^2(\nabla u)}{2}\right) = g_{\nabla u}\left(\nabla u, D^{\nabla u}_{\partial/\partial x^i}(\nabla u)\right) = g_{\nabla u}\left(D^{\nabla u}_{\nabla u}(\nabla u), \frac{\partial}{\partial x^i}\right).$$

by the symmetry of $\nabla^2 u$. From these, one obtains

$$\begin{aligned}
df(\nu_{\nabla u}) &= d\left(F^2(\nabla v) - \alpha(t)v_t - \varphi(t)\right)(\nu_{\nabla u})\\
&= d\left(u^{-2}F^2(\nabla u) - \alpha u^{-1}u_t\right)(\nu_{\nabla u})\\
&= 2u^{-2}g_{\nabla u}\left(D^{\nabla u}_{\nu_{\nabla u}}\nabla u, \nabla u\right) = 2u^{-2}g_{\nabla u}\left(D^{\nabla u}_{\nabla u}\nabla u, \nu_{\nabla u}\right).
\end{aligned}$$

By the convexity of ∂M, we have $g_\nu(\nu, D^{\nabla u}_{\nabla u}\nabla u) \leq 0$, equivalently, $g_{\nabla u}\left(\nu_{\nabla u}, D^{\nabla u}_{\nabla u}\nabla u\right) \leq 0$. Consequently, $df(\nu_{\nabla u})(t_0, x_0) \leq 0$ and hence $df(\nu_{\nabla u})(t_0, x_0) = 0$. Obviously, the tangent derivative of f at (t_0, x_0) vanishes due to its maximality. Consequently, $df(t_0, x_0) = 0$. The rest proof follows that of Case 1.

For the estimate of $F^2(\nabla(-\log u)) + \alpha(t)(\log u)_t$, the same arguments as above work for $v = -\log u$. The proof is completed. $\qquad\square$

Remark 9.3.1. In fact, using the linearized heat semigroup approach, we can prove that (9.3.6)–(9.3.7) also hold on a forward complete Finsler manifold (M, F, m) satisfying (2.1.4) under the assumption that $\mathrm{Ric}_N \geq K$ for some $N \in [n, \infty)$ and $K \in \mathbb{R}$. The readers refer to [Xia10] for more details.

When $\mathrm{Ric}_N \geq 0$, i.e., $K = 0$, equivalently, $\alpha(t) \equiv 1$, letting $a(t) = t^2(t > 0)$ in (9.3.2)–(9.3.3), (9.3.6) is reduced to

$$F^2(\nabla \log u) - (\log u)_t \leq \frac{N}{2t},$$

which is sharp in Riemannian case since the equality holds for the heat kernel on the Euclidean space $\mathbb{R}^n$ ([LY]). However, (9.3.6) is not sharp when $K \neq 0$, even in Riemannian case. It generalizes the related results in Riemannian geometry. For example, take $a(t) = \sinh^2(Kt)$ in (9.3.2)–(9.3.3), we obtain the estimate (9.3.6) with

$$\alpha(t) = 1 + \frac{\sinh(Kt)\cosh(Kt) - Kt}{\sinh^2(Kt)}, \quad \varphi(t) = \frac{NK}{2}(1 + \coth(Kt)).$$

In Riemannian case, this estimate was due to Li–Xu ([LX]).

As an application of Theorem 9.3.2, one obtains the Harnack inequality as follows.

Corollary 9.3.2. *Under the same assumptions as in Theorem 9.3.2, we have*

$$u(t_1, x_1) \le u(t_2, x_2) \exp\left(\frac{d_F^2(x_2, x_1)}{4(t_2 - t_1)^2} \int_{t_1}^{t_2} \alpha(t)dt + \int_{t_1}^{t_2} \frac{\varphi(t)}{\alpha(t)}dt\right) \quad (9.3.22)$$

for all $0 < t_1 < t_2 < \infty$ and $x_1, x_2 \in M$.

Proof. The proof is similar to that of Corollary 9.3.1. Replacing u by $u + \varepsilon$ if necessary, we may assume without loss of generality that u is positive. Let $\gamma(t) = \exp_{x_2}((t_2 - t)y)$ for $t \in [t_1, t_2]$ be the reverse curve of the minimal geodesic from $x_2 = \gamma(t_2)$ to $x_1 = \gamma(t_1)$ with suitable $y \in T_{x_2}M$. Then $F(-\dot{\gamma}(t)) = d_F(x_2, x_1)/(t_2 - t_1)$ for all t . We also put $v := \log u$,

$$\Theta := \frac{d_F(x_2, x_1)^2}{4(t_2 - t_1)^2}, \quad \sigma(t) := v(t, \gamma(t)) + \int_0^t \frac{\varphi(s)}{\alpha(s)}ds + \Theta \int_0^t \alpha(s)ds.$$

Then we have by Theorem 9.3.2,

$$\sigma'(t) = dv(\dot{\gamma}) + v_t + \frac{\varphi}{\alpha} + \Theta\alpha(t)$$

$$\ge -F(\nabla v)F(-\dot{\gamma}) + \frac{F(\nabla v)^2}{\alpha} - \frac{\varphi}{\alpha} + \frac{\varphi}{\alpha} + \Theta\alpha(t)$$

$$\ge -F(\nabla v)\frac{d_F(x_2, x_1)}{t_2 - t_1} + \frac{F(\nabla v)^2}{\alpha} + \frac{\alpha d_F(x_2, x_1)^2}{4(t_2 - t_1)^2} \ge 0.$$

Hence, we obtain

$$v(t_1, x_1) \le v(t_2, x_2) + \Theta \int_{t_1}^{t_2} \alpha(t)dt + \int_{t_1}^{t_2} \frac{\varphi(t)}{\alpha(t)}dt,$$

which implies the claim. $\qquad\qquad\square$

9.3.2 *Complete and Noncompact Case*

In this section, we give Li-Yau's gradient estimates to the heat equation on a forward complete and noncompact Finsler manifold with Ric_N bounded from below.

To state our results, let $\varphi_\varsigma(t)$ be a positive C^1 function on $\mathbb{R}$ defined by

$$\varphi_\varsigma(t) := \begin{cases} \frac{N\alpha^2}{2\varsigma}, & t \leq 0, \\ \frac{N\alpha^2}{2(t+\varsigma)}, & 0 < t \leq \frac{\alpha-1}{K} - \varsigma, \\ \frac{NK\alpha^2}{4(\alpha-1)}\left(1 + e^{2-\frac{2K}{\alpha-1}(t+\varsigma)}\right), & t > \frac{\alpha-1}{K} - \varsigma, \end{cases} \quad (9.3.23)$$

where ς is a positive constant less than $\frac{\alpha-1}{K}$. For $t > 0$, $\varphi(t) = \lim_{\varsigma \to 0} \varphi_\varsigma(t)$ given by

$$\varphi(t) = \begin{cases} \frac{N\alpha^2}{2t}, & 0 < t \leq \frac{\alpha-1}{K}, \\ \frac{NK\alpha^2}{4(\alpha-1)}\left(1 + e^{2-\frac{2K}{\alpha-1}t}\right), & t > \frac{\alpha-1}{K} \end{cases} \quad (9.3.24)$$

with $\lim_{t \to 0+} \varphi(t) = +\infty$.

Theorem 9.3.3. *Let (M, F, m) be an n-dimensional forward complete and noncompact Finsler measure space equipped with a uniformly convex and uniformly smooth Finsler metric F and a smooth measure m. Fix $R > 0$ and $s \in \mathbb{R}$, assume that $\text{Ric}_N \geq -K$ in a forward geodesic ball $B_{2R} := B_{2R}^+(x_0) \subset M$ for some $N \in [n, \infty)$ and $K \geq 0$, and u is a locally positive solution to $u_t = \Delta u$ (in the weak sense) on $(s - R^2, s + \epsilon) \times B_{2R}$, where ϵ is a positive constant. Then there exist constants $\nu > 2$ and $C = C(\kappa, \kappa^*, N, \nu) > 0$ such that*

$$\sup_{I_\tau \times B_R} \left\{ F^2(\nabla \log u) - \alpha(\log u)_t - \varphi_\varsigma,\ F^2(\nabla(-\log u)) + \alpha(\log u)_t - \varphi_\varsigma \right\}$$

$$\leq \frac{C\alpha^2}{(1-\tau)^{3+\nu}} \left(\frac{1 + \sqrt{K}R}{R^2}\right), \quad (9.3.25)$$

for any constants $\alpha > 1$ and $0 < \tau < 1$, where $I_\tau = (s - \tau R^2, s]$.

To prove Theorem 9.3.3, we are careful of the regularity for the distributional time derivative u_t of the solution u to the heat equation. In fact, it is only Hölder continuous in t since u_t lies in $H^1(B_{2R}) \cap C^\beta(B_{2R})$. we have no sufficient regularity for u_t to argue in the sequel. To overcome this difficulty, we consider the regularization u_ε of u defined by

$$u_\varepsilon(t, x) = J_\varepsilon * u(t, x) = \int_I J_\varepsilon(t-s)u(s, x)ds, \quad t \in I_\varepsilon = \{t \in I | \text{dist}(t, \partial I) > \varepsilon\}$$

for any $\varepsilon > 0$ and $x \in B_{2R}$, where J_ε is a mollifier, which is nonnegative, belongs to $C_0^\infty(\mathbb{R})$ with (i) $J_\varepsilon(t) = 0$ for $|t| \geq \varepsilon$ and (ii) $\int_\mathbb{R} J_\varepsilon(t)dt = 1$. For example, we may take

$$J_\varepsilon(t) = \frac{1}{\varepsilon} \begin{cases} k \exp\left(\frac{\varepsilon^2}{t^2 - \varepsilon^2}\right), & \text{if } |t| < \varepsilon, \\ 0, & \text{if } |t| \geq \varepsilon, \end{cases} \tag{9.3.26}$$

where $k > 0$ is chosen such that (ii) is satisfied. Then $u_\varepsilon(t, x)$ is smooth in $t \in I_\varepsilon$, and $\lim_{\varepsilon \to 0} u_\varepsilon(t, x) = u(t, x)$ for a.e. $t \in I_\varepsilon$ and uniformly on each compact subsets of I for each $x \in B_{2R}$ ([Ad], [Ev]). In particular, if we take $I = (0, T)$ for any $0 < T < \infty$, then $I_\varepsilon = (\varepsilon, T - \varepsilon)$.

Lemma 9.3.3. *Let u be a local solution to the heat equation on $I \times B_{2R}$. Then, for any $\varepsilon > 0$, we have*

(i) $\lim_{\varepsilon \to 0}(u_\varepsilon)_t = u_t$, $\lim_{\varepsilon \to 0} \nabla u_\varepsilon = \nabla u$ and $\lim_{\varepsilon \to 0} F(\nabla u_\varepsilon) = F(\nabla u)$ on $I_\varepsilon \times B_{2R}$;

(ii) *If $u \in L^2(I \times B_{2R})$ with $F^*(du) \in L^2(I \times B_{2R})$, then $u_\varepsilon \in L^2(I_\varepsilon \times B_{2R})$ with $F^*(du_\varepsilon) \in L^2(I_\varepsilon \times B_{2R})$. Further, $(u_\varepsilon)_t \in L^2(I_\varepsilon \times B_{2R})$ and hence $u_\varepsilon \in H^1(I_\varepsilon \times B_{2R}) \cap C^\infty(\mathbb{R}) \cap H^2(B_{2R})$.*

(iii) *u_ε is a local solution to the heat equation on $I_\varepsilon \times B_{2R}$.*

Proof. (i) It is clear that $(u_\varepsilon)_t$ converges to u_t as $\varepsilon \to 0$. Since $\int_{|t-s|<\varepsilon} J_\varepsilon(t-s)ds = 1$, we have

$$|du_\varepsilon(t, x) - du(t, x)| = \left| \int_{|t-s|<\varepsilon} J_\varepsilon(t-s)(du(s, x) - du(t, x))ds \right|$$

$$\leq \int_{|t-s|<\varepsilon} J_\varepsilon(t-s)\left| du(s, x) - du(t, x) \right| ds$$

$$\leq \sup_{|t-s|<\varepsilon} |du(s, x) - du(t, x)|,$$

which means that $\lim_{\varepsilon \to 0} du_\varepsilon = du$ and hence $M_u = M_{u_\varepsilon}$. Thus, we have $\lim_{\varepsilon \to 0} \nabla u_\varepsilon = \nabla u$ since the Lengendre transformation J is a diffeomorphism on the slit tangent bundle. Note that the norm $F(x, y)$ and the Euclidean norm $|y|$ are equivalent locally for any $y \in T_x M$ by (8.1.19). From this and Minkowski's inequality (see Lemma 1.1.2), we have

$$|F(\nabla u_\varepsilon) - F(\nabla u)| \leq \max\{F(\nabla u_\varepsilon - \nabla u), F(\nabla u - \nabla u_\varepsilon)\} \to 0 \ (\text{as } \varepsilon \to 0),$$

which implies that $\lim_{\varepsilon \to 0} F(\nabla u_\varepsilon) = F(\nabla u)$, equivalently, $\lim_{\varepsilon \to 0} F^*(du_\varepsilon) = F^*(du)$.

(ii) Observe that

$$u_\varepsilon(t, x) = \int_I J_\varepsilon(t - s)u(s, x)ds$$

$$\leq \left(\int_I J_\varepsilon(t - s)ds \right)^{1/2} \left(\int_I J_\varepsilon(t - s)u^2(s, x)ds \right)^{1/2}$$

for any $t \in I_\varepsilon$. Since $\int_{|t-s|<\varepsilon} J_\varepsilon(t - s)ds = 1$, by Fubini's Theorem, this inequality implies that

$$\int_{I_\varepsilon} \int_{B_{2R}} u_\varepsilon^2(t, x)dmdt \leq \int_I J_\varepsilon(t - s)dt \int_{B_{2R}} dm \int_I u^2(s, x)ds$$

$$\leq \int_I \int_{B_{2R}} u^2(s, x)dmds < \infty,$$

which means that $u_\varepsilon \in L^2(I_\varepsilon \times B_{2R})$. Moreover, since $F^*(du_\varepsilon) \leq F^*(du_\varepsilon - du) + F^*(du)$, we have

$$\int_{I_\varepsilon} \int_{B_{2R}} F^{*2}(du_\varepsilon)dmdt \leq \int_{I_\varepsilon} \int_{B_{2R}} \left\{ 2F^{*2}(du_\varepsilon - du) + 2F^{*2}(du) \right\} dmdt$$

which implies that $F^*(du_\varepsilon) \in L^2(I_\varepsilon \times B_{2R})$. Consequently, $u_\varepsilon \in H^1(I_\varepsilon \times B_{2R})$. Further, the assertion of $(u_\varepsilon)_t$ follows from:

$$(u_\varepsilon(t, x))_t = \lim_{\delta \to 0} \frac{u_\varepsilon(t + \delta, x) - u_\varepsilon(t, x)}{\delta}$$

and the regularity of u_ε.

(iii) By the assumption, we have (i)–(ii) and $u_t = \Delta u$ (in a weak sense). Moreover, we remark that $J_\varepsilon \in C_0^\infty(\mathbb{R})$ and $\partial_t(J_\varepsilon(t - s)) = -\partial_s(J_\varepsilon(t - s))$ (cf. (9.3.26)). Thus, for any $\phi = \phi(t, x) \in H_0^1(I_\varepsilon \times B_{2R})$, applying Fubini's Theorem yields

$$\int_{I_\varepsilon} \int_{B_{2R}} \phi(t, x)\partial_t u_\varepsilon(t, x)dmdt$$

$$= \int_{I_\varepsilon} \int_I \int_{B_{2R}} \phi(t, x)\partial_t(J_\varepsilon(t - s))u(s, x)dmdsdt$$

$$= -\int_{I_\varepsilon} \int_I \int_{B_{2R}} \partial_s\left(\phi(t, x)J_\varepsilon(t - s)\right) u(s, x)dmdsdt$$

$$= -\int_{I_\varepsilon} \int_I \int_{B_{2R}} J_\varepsilon(t - s)d\phi(\nabla u(s, x))dmdsdt$$

$$= -\int_{I_\varepsilon} \int_{B_{2R}} d\phi(\nabla u_\varepsilon(t, x))dmds,$$

where we used (9.1.2) with the test function $\phi(t,x)J_\varepsilon(t-s)$ in the third equality. The above equality implies that u_ε is a local solution to the heat equation on $I_\varepsilon \times B_{2R}$. □

Lemma 9.3.3 shows that if u is a locally positive solution to the heat equation on $I \times B_{2R}$, then, for any small $\delta > 0$, $u_\varepsilon + \delta \in H^1(I_\varepsilon \times B_{2R}) \cap C^\infty(I_\varepsilon) \cap H^2(B_{2R})$, which is a locally positive solution to the heat equation on $I_\varepsilon \times B_{2R}$. Let $v_{\varepsilon,\delta} := \log(u_\varepsilon + \delta)$ and $f_{\varepsilon,\delta} := F^2(\nabla v_{\varepsilon,\delta}) - \alpha(t)\partial_t(v_{\varepsilon,\delta}) - \varphi(t)$, where $\alpha = \alpha(t)$ and $\varphi = \varphi(t)$ are C^1 functions on $\mathbb{R}$. Then $v_{\varepsilon,\delta} \in H^1(I_\varepsilon \times B_{2R}) \cap C^\infty(I_\varepsilon) \cap H^2(B_{2R})$ with $\lim_{(\varepsilon,\delta) \to (0,0)} v_{\varepsilon,\delta} = v$ and $f_{\varepsilon,\delta} \in L^2(I_\varepsilon \times B_{2R}) \cap C^\infty(I_\varepsilon) \cap H^1(B_{2R})$ with $\lim_{(\varepsilon,\delta) \to (0,0)} f_{\varepsilon,\delta} = f$ by Lemma 9.3.3(i)–(ii), here $f := F^2(\nabla v) - \alpha v_t - \varphi$. Without loss of generality, we assume that $u \in H^1(I \times B_{2R}) \cap C^\infty(I) \cap H^2(B_{2R})$ and hence $f \in L^2(I \times B_{2R}) \cap C^1(I) \cap H^1(B_{2R})$ in the following. Otherwise we use $u_\varepsilon + \delta$ and f_ε instead of u and f respectively and then taking approximating arguments. Similar to the proofs of Lemmas 9.3.1–9.3.2, we have the following local version.

Lemma 9.3.4. *Let u be a positive solution to the heat equation on $I \times B_{2R}$ and $v = \log u$. Then*

(i) $\mathbf{\Delta} v + F^2(\nabla v) = v_t$ *in the distributional sense that*

$$\int_{B_{2R}} \{-d\phi(\nabla v) + \phi F^2(\nabla v)\}\, dm = \int_{B_{2R}} \phi v_t dm \qquad (9.3.27)$$

for each $t \in I$ and any $\phi \in H_0^1(B_{2R})$.

(ii) $\mathbf{\Delta}^{\nabla u} v_t + 2 dv_t(\nabla v) = \partial_t(v_t)$ *in the distributional sense that*

$$\int_I \int_{B_{2R}} \{-d\phi(\nabla^{\nabla u} v_t) + 2\phi dv_t(\nabla v)\}\, dm dt = \int_I \int_{B_{2R}} \phi \partial_t(v_t)\, dm dt$$

$$(9.3.28)$$

for any $\phi \in H^1(I \times B_{2R}) \cap H_0^1(B_{2R})$.

(iii) *For any C^1 functions $\alpha(t), \varphi(t)$ defined on $\mathbb{R}$ and $N \in [n,\infty)$, $f := F^2(\nabla v) - \alpha(t)v_t - \varphi(t)$ satisfies*

$$\int_I \int_{B_{2R}} \{-d\phi(\nabla^{\nabla u} f) + 2\phi df(\nabla v) - \phi f_t\}\, dm dt$$

$$\geq \int_I \int_{B_{2R}} \phi \left(\frac{2}{N}(\mathbf{\Delta} v)^2 + 2\mathrm{Ric}_N(\nabla v) + \alpha' v_t + \varphi'\right) dm dt,$$

$$(9.3.29)$$

where $\phi \in H^1(I \times B_{2R}) \cap H_0^1(B_{2R})$ is a nonnegative function.

Based on Lemma 9.3.4, one obtains the following result.

Proposition 9.3.1. *Assume that $Ric_N \geq -K$ on a forward geodesic ball B_{2R} for some $N \in [n, \infty)$ and $K \geq 0$, and u is a positive solution to the heat equation on $I \times B_{2R}$. Then, for any constant $\alpha > 1$, $f_\varsigma = F^2(\nabla v) - \alpha v_t - \varphi_\varsigma(t)$ satisfies*

$$\int_I \int_{B_{2R}} \left\{ -d\phi \left(\nabla^{\nabla u} f_\varsigma \right) + 2\phi df_\varsigma(\nabla v) - \phi(f_\varsigma)_t \right\} dm dt$$

$$\geq \int_I \int_{B_{2R}} \phi \left\{ \frac{2}{N\alpha^2} f_\varsigma^2 + \frac{4(\alpha-1)}{N\alpha^2} F^2(\nabla v) f_\varsigma + \frac{4\varphi_\varsigma}{N\alpha^2} f_\varsigma \right\} dm dt,$$

$$(9.3.30)$$

where $\phi \in H^1(I \times B_{2R}) \cap H_0^1(B_{2R})$ is a nonnegative function and $\varphi_\varsigma(t)$ is defined by (9.3.23).

Proof. For the sake of simplicity, we write f and φ instead of f_ς and φ_ς in the following. It follows from Lemma 9.3.4 that

$$\int_I \int_{B_{2R}} \left\{ -d\phi \left(\nabla^{\nabla u} f \right) + 2\phi df(\nabla v) - \phi f_t \right\} dm dt$$

$$\geq \int_I \int_{B_{2R}} \phi \left(\frac{2}{N} (\Delta v)^2 - 2K F^2(\nabla v) + \varphi' \right) dm dt \qquad (9.3.31)$$

for any nonnegative $\phi \in H^1(I \times B_{2R}) \cap H_0^1(B_{2R})$. By Lemma 9.3.4(i), we have

$$\Delta v = v_t - F^2(\nabla v) = -\frac{f}{\alpha} - \frac{\alpha-1}{\alpha} F^2(\nabla v) - \frac{1}{\alpha}\varphi$$

in the distributional sense. Therefore,

$$\frac{2}{N}(\Delta v)^2 - 2K F^2(\nabla v) + \varphi'$$

$$= \frac{2}{N} \left(-\frac{f}{\alpha} - \frac{\alpha-1}{\alpha} F^2(\nabla v) - \frac{1}{\alpha}\varphi \right)^2 - 2K F^2(\nabla v) + \varphi'$$

$$= \frac{2}{N} \left\{ \frac{f^2}{\alpha^2} + \frac{2(\alpha-1)}{\alpha^2} F^2(\nabla v) f + \frac{2\varphi}{\alpha^2} f \right\} + \frac{2(\alpha-1)^2}{N\alpha^2} F^4(\nabla v)$$

$$+ \left(\frac{4(\alpha-1)}{N\alpha^2} \varphi - 2K \right) F^2(\nabla v) + \varphi' + \frac{2\varphi^2}{N\alpha^2} \qquad (9.3.32)$$

Since $F^2(\nabla v)$ is nonnegative, by an elementary argument, we find that

$$\frac{2(\alpha-1)^2}{N\alpha^2}F^4(\nabla v) + \left(\frac{4(\alpha-1)}{N\alpha^2}\varphi - 2K\right)F^2(\nabla v) + \varphi' + \frac{2\varphi^2}{N\alpha^2} \geq 0$$

(9.3.33)

if and and only if one of the following cases holds:

(1) $\varphi' + \frac{2\varphi^2}{N\alpha^2} \geq 0$, and $\frac{4(\alpha-1)}{N\alpha^2}\varphi - 2K \geq 0$;
(2) the discriminant is nonpositive, i.e.,

$$\left(\frac{4(\alpha-1)}{N\alpha^2}\varphi - 2K\right)^2 - \frac{8(\alpha-1)^2}{N\alpha^2}\left(\varphi' + \frac{2\varphi^2}{N\alpha^2}\right) \leq 0,$$

equivalently,

$$\varphi' + \frac{2K}{\alpha-1}\varphi - \frac{N\alpha^2K^2}{2(\alpha-1)^2} \geq 0. \tag{9.3.34}$$

From the assumption on φ_ς, it is easy to check that $\varphi = \varphi_\varsigma$ satisfies (1) or (2). Thus, (9.3.33) is always true. Consequently, (9.3.30) follows from (9.3.31)–(9.3.33). $\qquad\square$

Now, we give the proof of Theorem 9.3.3 based on Proposition 9.3.1.

Proof of Theorem 9.3.3. Let $I = (s - R^2, s + \epsilon)$. Since u is a locally positive solution to $u_t = \Delta u$ on $I \times B_{2R}$, by Lemma 9.3.3 and the regularization arguments in the paragraph preceding to Lemma 9.3.4, we may assume that $u \in C^{1,\beta}(I \times B_{2R}) \cap C^\infty(I) \cap H^2(B_{2R})$. Let $v = \log u$ and $f_\varsigma = F^2(\nabla v) - \alpha v_t - \varphi_\varsigma$ as before. Then v has the same regularity as u on $Q := I \times B_{2R}$. In particular, both u and v are C^∞ on $\cup_{t \in I}(\{t\} \times (M_u \cap B_{2R}))$. Further, $f_\varsigma \in L^2(Q) \cap C^1(I) \cap H^2(B_{2R})$. If $f_\varsigma \leq 0$, then the conclusion is obvious. Now we show (9.3.25) for $f_\varsigma > 0$. To avoid the burdensome notations, we simply write f and φ instead of f_ς and φ_ς in the following.

Note that $df(\nabla v) \leq F^*(df)F(\nabla v)$. By the inequality $ax^2 - bx \geq -\frac{b^2}{4a}(a > 0)$, we get

$$\frac{4(\alpha-1)}{N\alpha^2}fF^2(\nabla v) - 2df(\nabla v) \geq \frac{4(\alpha-1)}{N\alpha^2}fF^2(\nabla v) - 2F^*(df)F(\nabla v)$$

$$\geq -\frac{N\alpha^2}{4(\alpha-1)}f^{-1}F^{*2}(df)$$

for any $t \in I$. Thus, Proposition 9.3.1 implies that

$$
\int_I \int_{B_{2R}} d\phi \left(\nabla^{\nabla u} f \right) dm dt + \int_I \int_{B_{2R}} \phi f_t dm dt
$$

$$
\leq \int_I \int_{B_{2R}} \phi \left(\frac{N\alpha^2}{4(\alpha-1)} f^{-1} F^{*2}(df) - \frac{2}{N\alpha^2} f^2 \right) dm dt
$$

for all nonnegative $\phi \in H^1(Q) \cap H_0^1(B_{2R})$. Let $\phi = \lambda \eta^2 f^\ell \in H^1(Q) \cap C^\infty(\mathbb{R}) \cap H_0^1(B_{2R})$, where $\eta \in C_0^\infty(B_{2R})$ and $\lambda = \lambda(t) \in C^\infty(\mathbb{R})$ are nonnegative functions, and $\ell > 1$, a constant to be determined later. Then

$$
\int_I \int_{B_{2R}} \lambda d(\eta^2 f^\ell) \left(\nabla^{\nabla u} f \right) dm dt
$$

$$
+ \int_I \int_{B_{2R}} \frac{1}{\ell+1} \left\{ \frac{\partial}{\partial t} \left(\lambda \eta^2 f^{\ell+1} \right) - \lambda' \eta^2 f^{\ell+1} \right\} dm dt
$$

$$
\leq \int_I \int_{B_{2R}} \lambda \eta^2 f^\ell \left(\frac{N\alpha^2}{4(\alpha-1)} f^{-1} F^{*2}(df) - \frac{2}{N\alpha^2} f^2 \right) dm dt. \qquad (9.3.35)
$$

Since F satisfies the uniform convexity and uniform smoothness, we have by (2.1.6)

$$
\tilde{\kappa}^* F^2(x, \nabla f) \leq df(\nabla^{\nabla v} f) = g^{ij}(x, \nabla v) f_i f_j \leq \tilde{\kappa} F^2(x, \nabla f).
$$

From this and the Cauchy–Schwarz inequality, we get

$$
|d\eta(\nabla^{\nabla v} f)| = |df(\nabla^{\nabla v} \eta)| \leq \tilde{\kappa} F(x, \nabla \eta) F(x, \nabla f).
$$

Thus,

$$
\int_{B_{2R}} d(\eta^2 f^\ell) \left(\nabla^{\nabla v} f \right) dm
$$

$$
= \int_{B_{2R}} \left\{ 2\eta f^\ell d\eta \left(\nabla^{\nabla v} f \right) + \ell \eta^2 f^{\ell-1} df \left(\nabla^{\nabla v} f \right) \right\} dm
$$

$$
\geq \int_{B_{2R}} \left\{ \ell \tilde{\kappa}^* \eta^2 f^{\ell-1} F^2(\nabla f) - 2\tilde{\kappa} \eta f^\ell F(\nabla f) F(\nabla \eta) \right\} dm
$$

$$
\geq \int_{B_{2R}} \left\{ \frac{\ell \tilde{\kappa}^*}{2} \eta^2 f^{\ell-1} F^2(\nabla f) - \frac{2\tilde{\kappa}^2}{\ell \tilde{\kappa}^*} f^{\ell+1} F^2(\nabla \eta) \right\} dm,
$$

where we used the elementary inequality $2ab \leq a^2 + b^2$ to the second term $2\tilde{\kappa} \eta f^\ell F(\nabla f) F(\nabla \eta)$ in the last inequality. Plugging the above inequality into (9.3.35) yields

$$(\ell+1) \int_I \int_{B_{2R}} \left(\frac{\ell\tilde{\kappa}^*}{2} - \frac{N\alpha^2}{4(\alpha-1)} \right) \lambda\eta^2 f^{\ell-1} F^2(\nabla f) dmdt$$

$$+ \int_I \frac{\partial}{\partial t} \left(\int_{B_{2R}} \lambda\eta^2 f^{\ell+1} dm \right) dt$$

$$\leq \int_I \int_{B_{2R}} \left\{ \lambda'\eta^2 f^{\ell+1} + 4\tilde{\kappa}^2(\tilde{\kappa}^*)^{-1}\lambda f^{\ell+1} F^2(\nabla\eta) \right.$$

$$\left. - (\ell+1)\frac{2\lambda\eta^2}{N\alpha^2} f^{\ell+2} \right\} dmdt. \tag{9.3.36}$$

Since $\alpha(>1)$ is a constant, it is obvious that there is an $\ell \geq \ell_1 := \max\left\{1, \frac{\alpha-1+N\alpha^2}{(2\tilde{\kappa}^*-1)(\alpha-1)}\right\}$ such that $\frac{\ell\tilde{\kappa}^*}{2} - \frac{N\alpha^2}{4(\alpha-1)} \geq \frac{\ell+1}{4}$. Thus, by (9.3.36), one obtains

$$\int_I \int_{B_{2R}} \lambda F^{*2} \left(d(\eta f^{\frac{\ell+1}{2}}) \right) dmdt + \int_I \frac{\partial}{\partial t} \left(\int_{B_{2R}} \lambda\eta^2 f^{\ell+1} dm \right) dt$$

$$\leq \int_I \int_{B_{2R}} \left\{ \lambda'\eta^2 f^{\ell+1} + c_1\lambda f^{\ell+1} F^2(\nabla\eta) - (\ell+1)\frac{2\lambda\eta^2}{N\alpha^2} f^{\ell+2} \right\} dmdt. \tag{9.3.37}$$

where $c_1 := 1 + \frac{4\tilde{\kappa}^2}{\tilde{\kappa}^*}$.

For any $0 < \tau < \tau' < 1$, we denote $I_\tau = (s - \tau R^2, s]$ and $I_{\tau'} = (s - \tau'R^2, s]$. Choose $\lambda \in C^\infty(\mathbb{R})$ with $0 \leq \lambda \leq 1$ such that λ is 1 on $(s - \tau R^2, \infty)$ and 0 on $(-\infty, s - \tau'R^2)$. Since the solution u to the heat equation is local, (9.3.37) still holds on $(s-R^2, t) \times B_{2R}$ for any $t \in I_\tau$. From this, taking the supremum on both sides of this inequality with respect to t yields

$$\int_{Q_\tau} \lambda F^{*2} \left(d(\eta f^{\frac{\ell+1}{2}}) \right) dmdt + \sup_{I_\tau} \left(\int_{B_{2R}} \lambda\eta^2 f^{\ell+1} dm \right)$$

$$\leq \int_{Q_{\tau'}} \left\{ |\lambda'|\eta^2 f^{\ell+1} + c_1\lambda f^{\ell+1} F^2(\nabla\eta) - 2(\ell+1)\frac{\lambda\eta^2}{N\alpha^2} f^{\ell+2} \right\} dmdt, \tag{9.3.38}$$

where $Q_\tau := I_\tau \times B_{2R}$ and $Q_{\tau'} := I_{\tau'} \times B_{2R}$.

On the other hand, let $\chi := 1 + \frac{2}{\nu}$ for some $\nu > 2$. By Hölder's inequality and Soblev's inequality (7.2.6), one obtains

$$\int_{B_{2R}} \lambda^\chi \eta^{2\chi} f^{(\ell+1)\chi} dm$$

$$\leq \left(\int_{B_{2R}} \left(\lambda \eta^2 f^{\ell+1} \right)^{\frac{\nu}{\nu-2}} dm \right)^{\frac{\nu-2}{\nu}} \cdot \left(\int_{B_{2R}} \lambda \eta^2 f^{\ell+1} dm \right)^{\frac{2}{\nu}}$$

$$\leq 4AR^2 \left(\int_{B_{2R}} \lambda \eta^2 f^{\ell+1} dm \right)^{\frac{2}{\nu}}$$

$$\times \left(\int_{B_{2R}} \lambda F^{*2} \left(d \left(\eta f^{\frac{\ell+1}{2}} \right) \right) dm + (2R)^{-2} \int_{B_{2R}} \lambda \eta^2 f^{\ell+1} dm \right),$$

where $A := e^{c_0(1+2\sqrt{K}R)} m \left(B_{2R} \right)^{-\frac{2}{\nu}}$. From this, we have

$$\int_{Q_\tau} \lambda^\chi \eta^{2\chi} f^{(\ell+1)\chi} dm dt = \int_{I_\tau} dt \int_{B_{2R}} \lambda^\chi \eta^{2\chi} f^{(\ell+1)\chi} dm$$

$$\leq 4AR^2 \left(\int_{Q_\tau} \lambda F^{*2} \left(d \left(\eta f^{\frac{\ell+1}{2}} \right) \right) dm dt + \sup_{I_\tau} \int_{B_{2R}} \lambda \eta^2 f^{\ell+1} dm \right)^\chi.$$

Therefore, by (9.3.38),

$$\left(\int_{Q_\tau} \lambda^\chi \eta^{2\chi} f^{(\ell+1)\chi} dm dt \right)^{\frac{1}{\chi}} \leq (4AR^2)^{\frac{1}{\chi}} \int_{Q_{\tau'}} \left\{ |\lambda'| \eta^2 f^{\ell+1} + c_1 \lambda f^{\ell+1} F^2(\nabla \eta) \right.$$

$$\left. - 2(\ell+1) \frac{\lambda \eta^2}{N\alpha^2} f^{\ell+2} \right\} dm dt. \qquad (9.3.39)$$

To complete the proof of Theorem 9.3.3, we need the following key lemma.

Lemma 9.3.5. *Under the same assumptions and notations as in Theorem 9.3.3, there exist positive constants $c_0 = c_0(\kappa, \kappa^*, N)$ and $C_0 = C_0(\kappa, \kappa^*, N)$ such that for $\ell_0 = c_0(1 + 2\sqrt{K}R)$ and $\ell_1 = \chi(1 + \ell_0)$, where $\chi = 1 + \frac{2}{\nu}$ for some $\nu > 2$, we have $f \in L^{\ell_1}(I_\tau \times B_{R_1})$ with*

$$\|f\|_{L^{\ell_1}(I_\tau \times B_{R_1})} \leq \frac{C_0 \alpha^2}{(\tau' - \tau)^3} \left(\frac{1 + \sqrt{K}R}{R^2} \right) \left(R^2 m(B_{2R}) \right)^{\frac{1}{\ell_1}} \qquad (9.3.40)$$

for any $0 < \tau < \tau' < 1$, where $R_1 := R + \frac{(\tau'-\tau)R}{2}$.

Proof. Taking $\ell = \ell_0 = c_0(1 + 2\sqrt{K}R)$ in (9.3.39) (we always may choose a large R such that $\ell_0 \geq \max\left\{1, \frac{N\alpha^2 + \alpha - 1}{(2\tilde{\kappa}^* - 1)(\alpha - 1)}\right\}$), we get

$$
\left(\int_{Q_\tau} \lambda^\chi \eta^{2\chi} f^{(\ell_0+1)\chi} dmdt\right)^{\frac{1}{\chi}}
$$

$$
\leq (4AR^2)^{\frac{1}{\chi}} \int_{Q_{\tau'}} \left\{ |\lambda'|\eta^2 f^{\ell_0+1} + c_1 \lambda f^{\ell_0+1} F^2(\nabla\eta) \right.
$$

$$
\left. - 2(\ell_0 + 1)\frac{\lambda\eta^2}{N\alpha^2} f^{\ell_0+2} \right\} dmdt. \tag{9.3.41}
$$

To estimate the first and second terms of the RHS in (9.3.41), we choose $\lambda(t) = \tilde{\lambda}(t)^{\ell_0+2} \in C^\infty(\mathbb{R})$, where $\tilde{\lambda} \in C^\infty(\mathbb{R})$ satisfying $0 \leq \tilde{\lambda} \leq 1$, $\tilde{\lambda} = 1$ on $(s - \tau R^2, \infty)$, $\tilde{\lambda} = 0$ on $(-\infty, s - \tau' R^2)$, and $|\tilde{\lambda}'| \leq \frac{c'}{(\tau'-\tau)^2 R^2}$ for some positive constant c'. Thus, $(\tau' - \tau)^2 R^2 |\lambda'| \leq 3c'\ell_0 \lambda^{\frac{\ell_0+1}{\ell_0+2}}$. Similarly, let $\eta(z) = \psi(z)^{\ell_0+2}$ with $\psi(z) = \tilde{\psi}(d_F(x,z)) \in C_0^\infty(B_{2R})$ satisfying

$$
0 \leq \tilde{\psi} \leq 1, \quad \tilde{\psi} = 1 \quad \text{in } [0, R_1], \quad |\tilde{\psi}'| \leq \frac{c''}{(\tau'-\tau)R}
$$

for some positive constant c''. Note that $F^*(d(d_F(x,\cdot))) = 1$ a.e. in B_{2R}. Thus, ψ satisfies $0 \leq \psi \leq 1$, $\psi = 1$ in B_{R_1} and $F^*(d\psi) \leq \frac{c''\Lambda}{(\tau'-\tau)R}$. Since F is uniformly convex and uniformly smooth, we have $1 \leq \Lambda \leq \min\{\sqrt{\kappa}, \sqrt{1/\kappa^*}\}$. Hence, $F^*(d\psi) \leq \frac{c_2}{(\tau'-\tau)R}$, where $c_2 = c_2(\kappa, \kappa^*)$. Thus, $0 \leq \eta \leq 1$ and $c_1(\tau' - \tau)^2 R^2 F^{*2}(d\eta) \leq c_3 \ell_0^2 \eta^{\frac{2(\ell_0+1)}{\ell_0+2}}$ for some positive constant $c_3 = c_3(\kappa, \kappa^*)$. In the following, we always denote the positive constants c_i as $c_i = c_i(\kappa, \kappa^*)(4 \leq i \leq 9)$ only depending on κ, κ^*.

For the first term of the RHS in (9.3.41), by Hölder's inequality and weighted Young's inequality $ab \leq \epsilon \frac{a^p}{p} + \epsilon^{-\frac{q}{p}} \frac{b^q}{q}$ for any $a, b \geq 0$ and $\epsilon > 0$, where $\frac{1}{p} + \frac{1}{q} = 1$ and $p > 1$, one obtains

$$
\int_{Q_{\tau'}} |\lambda'|\eta^2 f^{\ell_0+1} dmdt
$$

$$
\leq \frac{3c'\ell_0}{(\tau'-\tau)^2 R^2} \int_{Q_{\tau'}} \left(\frac{\lambda\eta^2}{\alpha^2}\right)^{\frac{\ell_0+1}{\ell_0+2}} f^{\ell_0+1} \cdot \alpha^{\frac{2(\ell_0+1)}{\ell_0+2}} \eta^{\frac{2}{\ell_0+2}} dmdt
$$

$$
\leq \frac{3c'\ell_0}{(\tau'-\tau)^2 R^2} \left(\int_{Q_{\tau'}} \frac{\lambda\eta^2}{\alpha^2} f^{\ell_0+2} dmdt\right)^{\frac{\ell_0+1}{\ell_0+2}}
$$

$$\cdot \left(m(B_{2R}) \int_{I_{\tau'}} \alpha^{2(\ell_0+1)} dt \right)^{\frac{1}{\ell_0+2}}$$

$$\leq \frac{1}{N}(\ell_0 + 1) \int_{Q_{\tau'}} \frac{\lambda \eta^2}{\alpha^2} f^{\ell_0+2} dm dt$$

$$+ (c_4 N)^{\ell_0+1} \left(\frac{1}{(\tau'-\tau)^2 R^2} \right)^{\ell_0+2} m(B_{2R}) \int_{I_{\tau'}} \alpha^{2(\ell_0+1)} dt, \quad (9.3.42)$$

where c_4 is a positive constant only depending on c'.

Similarly, for the second term of the RHS in (9.3.41), by Hölder's and weighted Young's inequalities again, one obtains

$$c_1 \int_{Q_{\tau'}} \lambda f^{\ell_0+1} F^{*2}(d\eta) dm dt$$

$$\leq \frac{c_4 \ell_0^2}{(\tau'-\tau)^2 R^2} \int_{Q_{\tau'}} \left(\frac{\lambda \eta^2}{\alpha^2} \right)^{\frac{\ell_0+1}{\ell_0+2}} f^{\ell_0+1} \cdot \alpha^{\frac{2(\ell_0+1)}{\ell_0+2}} \lambda^{\frac{1}{\ell_0+2}} dm dt$$

$$\leq \frac{c_4 \ell_0^2}{(\tau'-\tau)^2 R^2} \left(\int_{Q_{\tau'}} \frac{\lambda \eta^2}{\alpha^2} f^{\ell_0+2} dm dt \right)^{\frac{\ell_0+1}{\ell_0+2}}$$

$$\cdot \left(m(B_{2R}) \int_{I_{\tau'}} \alpha^{2(\ell_0+1)} dt \right)^{\frac{1}{\ell_0+2}}$$

$$\leq \frac{1}{N}(\ell_0 + 1) \int_{Q_{\tau'}} \frac{\lambda \eta^2}{\alpha^2} f^{\ell_0+2} dm dt$$

$$+ (c_5 N)^{\ell_0+1} \left(\frac{\ell_0}{(\tau'-\tau)^2 R^2} \right)^{\ell_0+2} m(B_{2R}) \int_{I_{\tau'}} \alpha^{2(\ell_0+1)} dt. \quad (9.3.43)$$

Plugging (9.3.42)–(9.3.43) into (9.3.41) yields

$$\left(\int_{I_\tau \times B_{R_1}} f^{(\ell_0+1)\chi} dm dt \right)^{\frac{1}{\chi}} \leq \left(\int_{Q_\tau} \lambda^\chi \eta^{2\chi} f^{(\ell_0+1)\chi} dm dt \right)^{\frac{1}{\chi}}$$

$$\leq (4AR^2)^{\frac{1}{\chi}} (c_6 N)^{\ell_0+1} \left(\frac{\ell_0}{(\tau'-\tau)^2 R^2} \right)^{\ell_0+2} m(B_{2R}) \int_{I_{\tau'}} \alpha^{2(\ell_0+1)} dt$$

$$\leq \frac{\ell_0 e^{\ell_0}}{(\tau'-\tau)^2} \left(\frac{c_7 N \ell_0}{(\tau'-\tau)^2 R^2} \right)^{\ell_0+1} \left(R^2 m(B_{2R}) \right)^{\frac{1}{\chi}} \alpha^{2(\ell_0+1)}.$$

Note that $\ell_0^{\frac{1}{\ell_0+1}} \le e$. Taking the $\frac{1}{\ell_0+1}$-th power on both sides of the above inequality, we have

$$\|f\|_{L^{\ell_1}(I_\tau \times B_{R_1})} \le \frac{c_8 N\alpha^2}{(\tau'-\tau)^3}\left(\frac{\ell_0}{R^2}\right)\left(R^2 m(B_{2R})\right)^{\frac{1}{\ell_1}},$$

which gives (9.3.40). $\square$

Now, we go back to prove Theorem 9.3.3. It follows from (9.3.39) that

$$\left(\int_{Q_\tau} \lambda^\chi \eta^{2\chi} f^{(\ell+1)\chi} dmdt\right)^{\frac{1}{\chi}}$$

$$\le (4AR^2)^{\frac{1}{\chi}}\left\{\int_{Q_{\tau'}} |\lambda'|\eta^2 f^{\ell+1} dmdt + c_1 \int_{Q_{\tau'}} \lambda f^{\ell+1} F^2(\nabla\eta) dmdt\right\}.$$

$$(9.3.44)$$

Let ℓ_0, ℓ_1 and I_τ be those given in Lemma 9.3.5 and $\ell_{k+1} = \chi\ell_k$ $(k \ge 1)$. We choose $\tau_{k+1} = \tau_k - \frac{\tau'-\tau}{2^k}, \tau_1 = \tau'$, and $\lambda_k \in C_0^\infty(\mathbb{R})$ with $0 \le \lambda_k \le 1$ such that λ_k is 1 on $(s-\tau_{k+1}R^2, \infty)$ and 0 on $(-\infty, s-\tau_k R^2)$, $|\lambda'_k| \le \tilde{c}'\frac{2^k}{(\tau'-\tau)^2 R^2}$. Moreover, we choose $R_k = R + \frac{(\tau'-\tau)R}{2^k}$ and $\eta_k \in C_0^\infty(B_{R_k})$ with $0 \le \eta_k \le 1$ such that

$$\eta_k = 1 \quad \text{in } B_{R_{k+1}}, \qquad F^*(x, d\eta_k) \le \tilde{c}''\frac{2^k}{(\tau'-\tau)R}, \qquad k = 1, 2, \ldots.$$

where $\tilde{c}'$ and $\tilde{c}'' = \tilde{c}''(\Lambda)$ are positive constants. Taking $\ell + 1 = \ell_k$, $\tau = \tau_{k+1}, \tau' = \tau_k, \lambda = \lambda_k$ and $\eta = \eta_k$ in (9.3.44), one obtains

$$\|f\|_{L^{\ell_{k+1}}(I_{\tau_{k+1}} \times B_{R_{k+1}})}$$

$$\le \left(4e^{\ell_0}R^2 m(B_{2R})^{-\frac{2}{\nu}}\right)^{\frac{1}{\ell_{k+1}}}\left(\frac{c_9(2^k+4^k)}{(\tau'-\tau)^2 R^2}\right)^{\frac{1}{\ell_k}}\|f\|_{L^{\ell_k}(I_{\tau_k} \times B_{R_k})}.$$

Note that $\sum_{k=1}^\infty \frac{1}{\ell_k} = \frac{1}{\ell_1}(1+\frac{\nu}{2}) = \frac{\nu}{2(\ell_0+1)}, \sum_{k=1}^\infty \frac{1}{\ell_{k+1}} = \frac{\nu}{2\ell_1}$ and $\sum_{k=1}^\infty \frac{k}{\ell_k} = \frac{\nu(\nu+2)}{4(\ell_0+1)}$. By the standard Moser's iteration and using Lemma 9.3.5 (in which τ is arbitrary), we get

$$\|f\|_{L^\infty(I_\tau \times B_R)} = 4^{\sum \frac{k}{\ell_k}}\left(4e^{\ell_0}R^2\right)^{\sum \frac{1}{\ell_{k+1}}} m(B_{2R})^{-\frac{2}{\nu}\sum \frac{1}{\ell_{k+1}}}$$

$$\times \left(\frac{2c_9}{(\tau'-\tau)^2 R^2}\right)^{\sum \frac{1}{\ell_k}}\|f\|_{L^{\ell_1}(I_{\tau_1} \times B_{R_1})}$$

$$\leq c_{10} \left(R^2 m(B_{2R}) \right)^{-\frac{1}{\ell_1}} (\tau' - \tau)^{-\frac{\nu}{\ell_1}} \|f\|_{L^{\ell_1}(I_{\tau'} \times B_{R_1})}$$

$$\leq \frac{C\alpha^2}{(\tau' - \tau)^{3+\nu}} \cdot \frac{1 + \sqrt{K}R}{R^2}$$

for some positive constants $c_{10} = c_{10}(\kappa, \kappa^*, \nu)$ and $C = C(\kappa, \kappa^*, N, \nu)$, where the sums in the first equality are taken on k from 1 to ∞. Letting $\tau' \to 1$, we get (9.3.25) for F.

The same arguments as above also work if we use $-v$ instead of v. Thus, one obtains the estimate for $F^2(\nabla(-\log u)) + \alpha(t)(\log u)_t$. This finishes the proof. $\qquad\square$

If the solution u to the heat equation is independent of t, then u is a Finsler harmonic function. In this case, Theorem 9.3.3 is reduced to Theorem 7.4.2.

Letting $\tau \to 0$ and then $R \to \infty$ in (9.3.25) yield the global gradient estimate of u. For a locally positive solution u to the heat flow on $\mathbb{R}^+ \times M$, where $\mathbb{R}^+$ is the set of positive real numbers, we may extend this flow such that u is a locally positive solution on $\mathbb{R} \times M$. Such an extension always exists. Thus, we obtain a locally positive solution $\tilde{u}$ to the heat flow on $\mathbb{R} \times M$ and hence get the local estimate of $\tilde{u}$ by Theorem 9.3.2. Restricting this estimate to $\mathbb{R}^+ \times M$ and then letting $\varsigma \to 0$, we can get the required estimate. Summing up, we obtain the following result.

Corollary 9.3.3. *Let (M, F, m), φ_ς and α be the same as in Theorem 9.3.2. Assume that $Ric_N \geq -K$ for some $N \in [n, \infty)$ and $K \geq 0$ on M.*

If u is a positive solution to the heat flow on $\mathbb{R} \times M$, then

$$\sup_{x \in M} \left\{ F^2(\nabla(\log u)) - \alpha(\log u)_t, \ F^2(\nabla(-\log u)) + \alpha(\log u)_t \right\} \leq \varphi_\varsigma$$

$$(9.3.45)$$

for any $t \in \mathbb{R}$.

If u is a positive solution to the heat flow on $\mathbb{R}^+ \times M$, then

$$\sup_{x \in M} \left\{ F^2(\nabla(\log u)) - \alpha(\log u)_t, \ F^2(\nabla(-\log u)) + \alpha(\log u)_t \right\} \leq \varphi$$

$$(9.3.46)$$

for any $t > 0$, where φ is given by (9.3.24).

Note that $\varphi(t) \leq \frac{NK\alpha^2}{2(\alpha-1)}$ if $t > \frac{\alpha-1}{K}$ in (9.3.24). The following corollary follows from (9.3.46).

Corollary 9.3.4. *Let (M, F, m) and α be the same as in Theorem 9.3.2. Assume that $Ric_N \geq -K$ for some $N \in [n, \infty)$ and $K \geq 0$ on M. If u is a positive solution to the heat flow on $\mathbb{R}^+ \times M$, then*

$$\sup_{x \in M} \left\{ F^2(\nabla(\log u)) - \alpha(\log u)_t, \ F^2(\nabla(-\log u)) + \alpha(\log u)_t \right\}$$

$$\leq \max \left\{ \frac{N\alpha^2}{2t}, \frac{NK\alpha^2}{2(\alpha - 1)} \right\}$$

for any $t > 0$.

Similar to the proof of Corollaries 9.3.3, we can obtain the following Harnack inequality from Corollary 9.3.4.

Corollary 9.3.5. *Under the same assumptions as in Corollary 9.3.4, we have*

$$u(t_1, x_1) \leq u(t_2, x_2) \exp\left(\frac{\alpha d_F^2(x_2, x_1)}{4(t_2 - t_1)} \right.$$

$$\left. + \max \left\{ \frac{N\alpha}{2} \log\left(\frac{t_2}{t_1} \right), \frac{NK\alpha}{2(\alpha - 1)}(t_2 - t_1) \right\} \right) \quad (9.3.47)$$

for all $0 < t_1 < t_2 < \infty$ and $x_1, x_2 \in M$.
In particular, if $Ric_N \geq 0$, then

$$u(t_1, x_1) \leq u(t_2, x_2) \left(\frac{t_2}{t_1} \right)^{\frac{N}{2}} \exp\left(\frac{d_F^2(x_2, x_1)}{4(t_2 - t_1)} \right).$$

In this case, $\alpha = 1$.

Remark 9.3.2. In [Xia10], we further develop the linearized heat semi-group approach. Using this technique, we obtain a generalized Li–Yau's inequality on a complete Finsler measure space under the assumptions that F satisfies (2.1.4) and $Ric_N \geq K$ for some $N \in [n, \infty)$ and $K \in \mathbb{R}$. This inequality includes Davies' one (9.3.1) and the gradient estimates in Theorem 9.3.2 as special cases. The readers refer to [Xia10] for more details.

Chapter 10

Appendix: Sobolev Spaces on Compact Finsler Manifolds

Let (M, F, m) be a Finsler measure space. Recall that the class $L^p(M)$ is defined in terms of the manifold structure of M (i.e., independent of the choice of F) and the Lebesgue space $(L^p(M), \|\cdot\|_p)(p \geq 1)$ is a Banach space, where $\|u\|_p = \left(\int_M |u|^p dm\right)^{1/p}$ is the norm of $L^p(M)$. For any open set $\Omega \subset M$, let

$$W^{1,p}(\Omega) := \left\{ u \in L^p(\Omega) \mid \int_\Omega [F^*(du)]^p dm < \infty \right\}$$

and

$$\|u\|_{\Omega,1,p} := \left(\int_\Omega |u|^p dm\right)^{1/p} + \left(\int_\Omega [F^*(du)]^p dm\right)^{1/p} \tag{10.0.1}$$

for $u \in W^{1,p}(\Omega)$. Then $\|\cdot\|_{\Omega,1,p}$ is a (positively homogeneous) norm by the Minkowski inequality $F^*(\xi + \eta) \leq F^*(\xi) + F^*(\eta)$ for any $\xi, \eta \in T^*(M)$(Lemma 1.1.2). We will suppress Ω in (10.0.1) if $\Omega = M$, e.g., $\|u\|_{1,p} = \|u\|_{M,1,p}$. In general, $(W^{1,p}(\Omega), \|u\|_{\Omega,1,p})$ is not a linear space over $\mathbb{R}$. For example, let $\mathbb{B}^n(1) := \{x \in \mathbb{R}^n \mid |x| < 1\}$, $u : \mathbb{B}^n(1) \to \mathbb{R}$ be defined by $u(x) = -\sqrt{1 - |x|}$ and

$$F(x, y) = \frac{\sqrt{|y|^2 - (|x|^2|y|^2 - \langle x, y\rangle^2)}}{1 - |x|^2} + \frac{\langle x, y\rangle}{1 - |x|^2}, \quad x \in \mathbb{B}^n(1), \quad y \in \mathbb{R}^n$$

be the Funk metric. Then $u \in W_0^{1,2}(\mathbb{B}^n(1))$, but $-u \notin W_0^{1,2}(\mathbb{B}^n(1))$ with respect to the Busemann–Hausdorff measure ([KR]). However, if M is compact, then the metrics F and $\overleftarrow{F}$ are equivalent. Consequently, $(W^{1,p}(M), \|\cdot\|_{1,p})$ is a Sobolev space, i.e., a normed linear space with respect to the (positive homogeneous) norm $\|\cdot\|_{1,p}$ from (10.0.1) and the Minkowski inequality. For the sake of convenience, we always assume that (M, F, m) is a compact Finsler space without boundary or with a smooth boundary in the following.

A sequence of functions $\{u_i\}$ in the space $(W^{1,p}(\Omega), \|\cdot\|_{1,p})$ is called a *forward (resp., backward) Cauchy sequence* if for any $\varepsilon > 0$, there exists an integer N such that for any $j > i > N$, we have $\|u_i - u_j\|_{1,p} < \varepsilon$ (resp., $\|u_j - u_i\|_{1,p} < \varepsilon$). A sequence $\{u_i\}$ is said to be *forward convergent* (resp., *backward convergent*) if there is a function $u \in W^{1,p}(\Omega)$ such that $\lim_{i \to \infty} \|u_i - u\|_{1,p} = 0$ (resp., $\lim_{i \to \infty} \|u - u_i\|_{1,p} = 0$). If M is compact, then the forward Cauchy (resp., convergent) sequence is also a backward Cauchy (resp., convergent) sequence and vice versa. In this case, we call it a Cauchy (resp., convergent) sequence.

Let (M, F, m) be a compact Finsler space without boundary, simply say that (M, F, m) is compact. In this case, (8.1.16)–(8.1.19) and (8.1.23) hold on the whole M.

Proposition 10.0.2. *Let (M, F, m) be an n-dimensional compact Finsler measure space. Then $W^{1,p}(M)(p \geq 1)$ is a Banach space with respect to the norm $\|\cdot\|_{1,p}$.*

Proof. Let $\{u_k\}$ be a Cauchy sequence in the normed linear space $(W^{1,p}(M), \|\cdot\|_{1,p})$. Then both $\{u_k\}$ and $\{du_k\}$ are Cauchy sequences in $(L^p(M), \|\cdot\|_p)$ from (10.0.1) and (8.1.19). Since $(L^p(M), \|\cdot\|_p)$ is complete, there exist functions u and u_1 such that $u_k \to u$ and $du_k \to u_1$ in $L^p(M)$ as $k \to \infty$. Assume $p > 1$ and $p' > 1$ with $\frac{1}{p} + \frac{1}{p'} = 1$. For any function $\varphi \in C_0^\infty(M)$, by Hölder inequality, we have

$$\left| \int_M \varphi(u_k - u)dm \right| \leq \int_M |u_k - u|\varphi dm \leq \|\varphi\|_{p'} \|u_k - u\|_p \to 0(k \to \infty),$$

$$(10.0.2)$$

which implies that $\lim_{k \to \infty} \int_M \varphi u_k dm = \int_M \varphi u dm$. Similarly, we have $\lim_{k \to \infty} \int_M \varphi du_k dm = \int_M \varphi u_1 dm$. It follows that

$$\int_M \varphi u_1 dm = \lim_{k \to \infty} \int_M \varphi du_k dm = -\lim_{k \to \infty} \int_M u_k d\varphi dm$$

$$= -\int_M u d\varphi dm = \int_M \varphi du dm,$$

which means that $du = u_1$ in the weak sense on M. Hence, by (8.1.19), we have $u \in W^{1,p}(M)$ and $\lim_{k \to \infty} \|u_k - u\|_{1,p} = 0$. Thus, $W^{1,p}(M)$ is complete for $p > 1$. The case that $p = 1$ is obvious. This completes the proof.   □

Let

$$H^{1,p}(M) := \text{the completion of } \{u \in C^\infty(M) \mid \|u\|_{1,p} < \infty\}$$

with respect to the norm $\|u\|_{1,p}$. From Proposition 10.0.2, one obtains that $H^{1,p}(M) \subset W^{1,p}(M)$. In fact, we have the following result.

Proposition 10.0.3. *Let (M, F, m) be an n-dimensional compact Finsler measure space. Then $H^{1,p}(M) = W^{1,p}(M)$.*

Proof. It suffices to show that $W^{1,p}(M) \subset H^{1,p}(M)$. If $u \in W^{1,p}(M)$ and $\varepsilon > 0$, we shall in fact show that there exists a function $w \in C^\infty(M)$ such that $\|u - w\|_{1,p} < \varepsilon$, which means that $C^\infty(M)$ is dense in $W^{1,p}(M)$. In this case, we have $\|w\|_{1,p} < \infty$. Since M is compact, M can be covered by a finite number of charts $\{U_i, \psi_i\}_{i=1,\dots,N}$ such that (8.1.19) holds for some constant $C > 1$ on each $\overline{U}_i$. Let $dm := \sigma(x)dx$ for some positive smooth function $\sigma(x)$. There are constants C_1, C_2 such that $C_1 \leq \sigma(x) \leq C_2$ since M is compact.

Let $\{f_i\}$ be a smooth partition of unity subordinated to the covering $\{U_i\}$. Then

$$
\begin{aligned}
\|f_i u\|_{U_i, 1, p} &= \left(\int_{U_i} |f_i u|^p dm \right)^{1/p} + \left(\int_{U_i} [F^*(d(f_i u))]^p dm \right)^{1/p} \\
&\leq \left(1 + \sup_{x \in \text{supp}(f_i)} F^*(df_i) \right) \\
&\quad \times \left(\left(\int_{U_i} |u|^p dm \right)^{1/p} + \left(\int_{U_i} F^{*p}(du) dm \right)^{1/p} \right) \\
&\leq \left(1 + \sup_{x \in \text{supp}(f_i)} F^*(df_i) \right) \|u\|_{U_i, 1, p} < \infty, \quad (10.0.3)
\end{aligned}
$$

which means that $f_i u \in W^{1,p}(U_i)$, equivalently, $(f_i u) \circ \psi_i^{-1} \in W^{1,p}(\psi_i(U_i))$. By Lemma 3.16 in [Ad], there are smooth functions $J_{\epsilon_i} * (f_i u) \circ \psi_i^{-1} \in C^\infty(\psi_i(U_i))$ such that

$$\|J_{\epsilon_i} * (f_i u) \circ \psi_i^{-1} - (f_i u) \circ \psi_i^{-1}\|_{U_i', 1, p} < \frac{\varepsilon}{N C_2^{1/p}(1 + C)} \quad (10.0.4)$$

for any $\varepsilon > 0$, where $U_i' := \psi_i(\text{supp}(f_i)) \subset \psi_i(U_i)$, J_{ε_i} are mollifiers and $*$ means a convolution. Let

$$u(x) = \sum_{i=1}^{N} f_i(x)u(x), \quad \text{and} \quad w(x) = \sum_{i=1}^{N} J_{\varepsilon_i} * (f_i u)(x).$$

Then $w(x) \in C^\infty(M)$. By (8.1.19), we have

$$\|w - u\|_{1,p}$$

$$\leq \sum_{i=1}^{N} \|J_{\varepsilon_i} * (f_i u) - (f_i u)\|_{U_i,1,p}$$

$$\leq C_2^{1/p} \sum_{i=1}^{N} \left(\int_{\psi_i(U_i)} |J_{\varepsilon_i} * (f_i u) \circ \psi_i^{-1} - f_i u \circ \psi_i^{-1}|^p dx \right)^{1/p}$$

$$+ C_2^{1/p} C \sum_{i=1}^{N} \left(\int_{\psi_i(U_i)} |d(J_{\varepsilon_i} * (f_i u) \circ \psi_i^{-1}) - d((f_i u) \circ \psi_i^{-1})|^p dx \right)^{1/p}$$

$$\leq (1 + C) C_2^{1/p} \sum_{i=1}^{N} \|J_{\varepsilon_i} * (f_i u) \circ \psi_i^{-1} - (f_i u) \circ \psi_i^{-1}\|_{U_i',1,p} < \varepsilon.$$

This finishes the proof. $\qquad\qquad\qquad\qquad\qquad\qquad\qquad\qquad\qquad\qquad\qquad$ $\square$

Remark 10.0.3. If (M, F) is noncompact, then the norms $\| \cdot \|_{1,p}$ in (10.0.1) defined by F and $\overleftarrow{F}$ are not equivalent. Consequently, $(W^{1,p}(M), \| \cdot \|_{1,p})$ is not Sobolev space. To remedy this case, we define $W^{1,p}(M)$ (resp., $W_0^{1,p}(M)$) by the completion of the space $C^\infty(M)$ (resp., $C_0^\infty(M)$) with respect to the norm (7.1.17), that is,

$$\|u\|_{M,1,p} := \|u\|_{L^p} + \|F^*(du)\|_{L^p} + \|\overleftarrow{F}^*(du)\|_{L^p},$$

where $C_0^\infty(M)$ is the space of smooth functions on M with compact support. Then $(W_0^{1,p}(M), \| \cdot \|_{M,1,p})$ (resp., $W_0^{1,p}(M), \| \cdot \|_{M,1,p})$ is a Banach space.

Let M be a compact manifold endowed with two Finsler metric F, $\tilde{F}$ and two measures m, $\tilde{m}$. Assume that $dm = \sigma(x)dx$ and $d\tilde{m} = \tilde{\sigma}(x)dx$. It is easy to see that there exists a constant $C > 1$ such that

$$C^{-1}F(x,y) \leq \tilde{F}(x,y) \leq CF(x,y), \quad \text{and} \quad C^{-1}\sigma(x) \leq \tilde{\sigma}(x) \leq C\sigma(x)$$

on M. This leads to the following result.

Proposition 10.0.4. *If M is compact, then $W^{1,p}(M)$ does not depend on the Finsler metric and the measure m.*

Note that $(L^p(M), |\cdot|_p)$ is reflexive if $p > 1$. One gets the following.

Proposition 10.0.5. *If $p > 1$, then $W^{1,p}(M)$ is reflexive.*

If (M, F, m) is a compact Finsler space with C^1 boundary, then we can choose a finite C^1 atlas (U_i, ψ_i) of $\overline{M}$, each U_i being homeomorphic either to an open set $\psi_i(U_i)$ of $\mathbb{R}^n$ or to a set $\psi_i(U_i) \cap \partial\mathbb{R}^n_+$. Similar to the proofs of Propositions 10.0.2–10.0.3, one obtains that $W^{1,p}(M)$ is also a Banach space with respect to the norm $\|\cdot\|_{1,p}$ and $C^\infty(\overline{M})$ is dense in $(W^{1,p}(M), \|\cdot\|_{1,p})$. Therefore, Propositions 10.0.2–10.0.5 are also true in this case.

Let $C^\alpha(M)$ be the set of continuous functions $u : M \to \mathbb{R}$ for which the norm

$$\|u\|_{C^\alpha} = \sup_{x \neq y \in M} \frac{|u(y) - u(x)|}{d_F(x,y)^\alpha}$$

is finite. If $x, y \in \overline{M}$, the corresponding norm $\|\cdot\|_{C^\alpha}$ is the norm in $C^\alpha(\overline{M})$. The following Sobolev embedding theorem is important to study analysis on Finsler manifolds, in which the compactly embedding theorem is often referred to as the *Rellich–Kondrakov theorem*.

Theorem 10.0.4. *Let (M, F, m) be an n-dimensional compact Finsler measure space without boundary or with C^1 boundary. Then*

(1) *If $1 \le p < n$, then $W^{1,p}(M)$ is continuously embedded in $L^{p^*}(M)$ and compactly embedded in $L^q(M)$ for any $1 \le q < p^*$, where $p^* = \frac{np}{n-p}$.*

(2) *If $p > n$, then $W^{1,p}(M)$ is continuously embedded in $C^\alpha(\overline{M})$ and compactly embedded in $C^\beta(\overline{M})$ for any $0 \le \beta < \alpha$, where $\alpha = 1 - \frac{n}{p}$.*

(3) *If $1 \le p < n$, then under the trace mapping $W^{1,p}(M)$ is continuously embedded in $L^{q^*}(\partial M)$ and compactly embedded in $L^q(\partial M)$ for any $1 \le q < q^*$, where $q^* = \frac{(n-1)p}{n-p}$.*

To prove Theorem 10.0.4, we first recall Sobolev's embedding theorem in $\mathbb{R}^n$.

Lemma 10.0.6. *Let Ω be a bounded domain in $\mathbb{R}^n$ with C^1 boundary.*

(i) *If $1 \le p < n$, the space $W^{1,p}(\Omega)$ is continuously embedded in $L^{p^*}(\Omega)$ and compactly embedded in $L^q(\Omega)$ for any $1 \le q < p^*$, where $p^* = \frac{np}{n-p}$.*

(ii) *If $p > n$, the space $W^{1,p}(\Omega)$ is continuously embedded in $C^\alpha(\overline{\Omega})$ and compactly embedded in $C^\beta(\overline{\Omega})$ for any $0 \le \beta < \alpha$, where $\alpha = 1 - \frac{n}{p}$.*

(iii) *If $1 \leq p < n$, then under the trace mapping the space $W^{1,p}(\Omega)$ is continuously embedded in $L^{q^*}(\partial\Omega)$ and compactly embedded in $L^q(\partial\Omega)$ for any $1 \leq q < q^*$, where $q^* = \frac{(n-1)p}{n-p}$.*

For the proof of Lemma 10.0.6, we refer to [Ad], [GT] and [Le], etc. In (i) and (ii), we do not need the C^1 boundary requirement if $W^{1,p}(\Omega)$ is replaced by $W_0^{1,p}(\Omega)$. In (iii), it is known that trace of $u \in W^{1,p}(\Omega)$ is zero if and only if $u \in W_0^{1,p}(\Omega)$ (see Theorem 15.29 in [Le]).

Lemma 10.0.7. *Let (M, F, m) be an n-dimensional compact Finsler measure space. If $W^{1,1}(M)$ is continuously embedded in $L^{n/(n-1)}(M)$, then $W^{1,p}(M)$ is continuously embedded in $L^{p^*}(M)$ for any $1 \leq p < n$.*

Proof. Let $C_0 > 0$ be a constant such that for any $u \in W^{1,1}(M)$,

$$\left(\int_M |u|^{n/(n-1)} dm \right)^{(n-1)/n} \leq C_0 \int_M (|u| + F^*(du)) dm. \quad (10.0.5)$$

Set $v = |u|^{p^*(n-1)/n}$. Then

$$\left(\int_M |u|^{p^*} dm \right)^{\frac{n-1}{n}} = \left(\int_M |v|^{n/(n-1)} dm \right)^{\frac{n-1}{n}} \leq C_0 \int_M (|v| + F^*(dv)) \, dm. \quad (10.0.6)$$

By Lemma 2.1.1, the reversibility Λ^* of F^* is the reversibility Λ of F, which is bounded by (8.1.23) since M is compact. Thus,

$$F^*(dv) = \frac{p^*(n-1)}{n} F^* \left(\mathrm{sgn}(u)|u|^{p'} du \right) \leq \frac{p^*(n-1)}{n} \Lambda |u|^{p'} F^*(du), \quad (10.0.7)$$

where $p' = \frac{n(p-1)}{n-p}$. Inserting (10.0.7) into (10.0.6) and using Hölder inequality yield

$$\left(\int_M |u|^{p^*} dm \right)^{\frac{n-1}{n}} \leq C_0 \int_M |u|^{\frac{p^*(n-1)}{n}} dm + \frac{C_0 p^*(n-1)\Lambda}{n} \int_M |u|^{p'} F^*(du) dm$$

$$\leq C_0 \left(\int_M |u|^{p^*} dm \right)^{\frac{p-1}{p}} \left(\int_M |u|^p dm \right)^{\frac{1}{p}} + \frac{C_0 p^*(n-1)\Lambda}{n}$$

$$\times \left(\int_M |u|^{p^*} dm \right)^{\frac{p-1}{p}} \left(\int_M [F^*(du)]^p dm \right)^{\frac{1}{p}}, \quad (10.0.8)$$

which implies that

$$\left(\int_M |u|^{p^*} dm \right)^{1/p^*} \leq C_0 \left(1 + \frac{p(n-1)\Lambda}{n} \right) \|u\|_{1,p}.$$

The lemma follows. $\qquad\square$

Proof of Theorem 10.0.4. Since M is compact, M can be covered by a finite number of charts $\{U_i, \psi_i\}_{i=1,\dots,N}$ such that (8.1.19) holds for some constant $C > 1$ on each $\overline{U}_i$. It is worth mentioning that the coordinate chart (U_i, ψ_i) of x will be replaced by $(U_i \cap \partial M, \psi_i)$ if $x \in \partial M$. In this case, $\psi_i(U_i \cap \partial M) = \psi_i(U_i) \cap \mathbb{R}^n_+$. We will use the unified notations $\{U_i, \psi_i\}$ for the sake of simplicity in the following. Let $dm := \sigma(x)dx$. There are constants C_1, C_2 such that $C_1 \leq \sigma(x) \leq C_2$ as in the proof of Proposition 10.0.3. Let $\{f_i\}$ be a smooth partition of unity subordinated to the covering $\{U_i\}$.

(1) By Lemma 10.0.7, it suffices to prove that $W^{1,1}(M) \subset L^{\frac{n}{n-1}}(M)$. For any $u \in W^{1,1}(M)$ and any $1 \leq i \leq N$, we have

$$\int_{U_i} |f_i u|^{\frac{n}{n-1}} dm \leq C_2 \int_{\psi_i(U_i)} |(f_i u) \circ \psi_i^{-1}|^{\frac{n}{n-1}} dx \qquad (10.0.9)$$

and

$$\int_{U_i} F[d(f_i u)] dm \geq C^{-1} C_1 \int_{\psi_i(U_i)} |d[(f_i u) \circ \psi_i^{-1}]| dx. \qquad (10.0.10)$$

By Theorem 2.5 in [Ad], one obtains

$$\left(\int_{\psi_i(U_i)} |(f_i u) \circ \psi_i^{-1}|^{\frac{n}{n-1}} dx \right)^{\frac{n-1}{n}} \leq \frac{1}{2} \int_{\psi_i(U_i)} |d\left((f_i u) \circ \psi_i^{-1} \right)| dx \qquad (10.0.11)$$

for each i. Hence,

$$\left(\int_M |u|^{\frac{n}{n-1}} dm \right)^{\frac{n-1}{n}}$$

$$\leq \sum_{i=1}^N \left(\int_{U_i} |f_i u|^{\frac{n}{n-1}} dm \right)^{\frac{n-1}{n}}$$

$$\leq C_2^{\frac{n-1}{n}} \sum_{i=1}^N \left(\int_{\psi_i(U_i)} |(f_i u) \circ \psi_i^{-1}|^{\frac{n}{n-1}} dx \right)^{\frac{n-1}{n}}$$

$$\leq \frac{1}{2} C_2^{\frac{n-1}{n}} \sum_{i=1}^{N} \left(\int_{\psi_i(U_i)} |d\left((f_i u) \circ \psi_i^{-1}\right)| \, dx \right)$$

$$\leq \frac{1}{2} C_1^{-1} C_2^{\frac{n-1}{n}} \left(\int_M |du| \, dm + \sum_{i=1}^{N} \max_{supp(f_i)} |df_i| \int_M |u| \, dm \right)$$

$$\leq \frac{1}{2} CC_1^{-1} C_2^{\frac{n-1}{n}} \left(1 + \sum_{i=1}^{N} \max_{supp(f_i)} |df_i| \right) \|u\|_{1,1}, \qquad (10.0.12)$$

which means that $W^{1,1} \subset L^{\frac{n}{n-1}}(M)$.

We next prove that the embedding $W^{1,p} \subset L^q(M)$ is compact for $1 \leq q < p^*$ if $1 \leq p < n$. Given a bounded sequence $\{u_k\}$ in $W^{1,p}(M)$, and for any i, let

$$u_k^i = (f_i u_k) \circ \psi_i^{-1}.$$

Obviously, $\{u_k^i\}$ is a bounded sequence in $W_0^{1,p}(\psi_i(U_i))$ for any i. By Lemma 10.0.6, there is a subsequence of $\{u_k^i\}$, also denoted by $\{u_k^i\}$, in $W_0^{1,p}(\psi_i(U_i))$ and a function $u \in L^q(\psi_i(U_i))$ for which $u_k^i \to u$ in $L^q(\psi_i(U_i))$ for all $1 \leq q \leq p^*$. Choose a subsequence, also denoted by $\{u_k\}$, of $\{u_k\}$ such that $\{u_k^i\}$ is the corresponding convergent subsequence in $W_0^{1,p}(\psi_i(U_i))$. Thus, coming back to the manifold by (8.1.8) and the boundedness of $\sigma(x)$ as before, one easily gets that $\{f_i u_k\}$ is a convergent subsequence in $L^q(M)$. Since $u_k = \sum_{i=1}^N (f_i u_k)$, $\{u_k\}$ is a convergent subsequence in $L^q(M)$.

(2) Note that $C^\infty(\overline{M})$ is dense in $W^{1,p}(M)$. We assume that $u \in C^\infty(\overline{M})$. From (8.1.6)–(8.1.7), we have

$$\|f_i u\|_{C^\alpha(U_i)} \leq C^\alpha \|(f_i u) \circ \psi_i^{-1}\|_{C^\alpha(\psi_i(U_i))} \qquad (10.0.13)$$

for all i, where the norm on the right-hand side is the one with respect to the Euclidean norm. Since $(f_i u) \circ \psi_i^{-1} \in W_0^{1,p}(\psi_i(U_i))$ and $W_0^{1,p}(\psi_i(U_i)) \subset C^\alpha(\psi_i(U_i))$ for $p > n$ by Lemma 10.0.6, there exists a constant $C_0 > 0$ such that for any i and any $u \in C^\infty(M)$,

$$\|(f_i u) \circ \psi_i^{-1}\|_{C^\alpha(\psi_i(U_i))} \leq C_0 \|(f_i u) \circ \psi_i^{-1}\|_{\psi_i(U_i),1,p}, \qquad (10.0.14)$$

where we used the equivalence between the norm of $W_0^{1,p}(\psi_i(U_i))$ and the norm of $W^{1,p}(\psi_i(U_i))$. Similar to the proof of (10.0.12), from (8.1.8), (10.0.13)–(10.0.14), and the boundedness of $\sigma(x)$, there exists a positive constant $C' = C'(C_0, C, C_1, C_2, \max_{supp(f_i)} |df_i|, \alpha)$ such that

$$\|f_i u\|_{C^\alpha(U_i)} \leq C^\alpha \|f_i u \circ \psi_i^{-1}\|_{C^\alpha(\psi_i(U_i))} \leq C' \|f_i u\|_{1,p}. \quad (10.0.15)$$

By (10.0.15), we have

$$\|u\|_{C^\alpha(\overline{M})} \leq \sum_{i=1}^N \|f_i u\|_{C^\alpha(U_i)} \leq C' \|u\|_{1,p}, \quad (10.0.16)$$

which implies that the embedding $W^{1,p}(M) \to C^\alpha(\overline{M})$ is continuous.

Let $\mathcal{A}$ be a bounded subset of $W^{1,p}(M)$. If $u \in \mathcal{A}$, then $\|u\|_{1,p} \leq \tilde{C}$ and $\|u\|_{C^\alpha(\overline{M})} \leq \tilde{C} C'$ from (10.0.16) for some positive constant $\tilde{C}$. Thus, $\mathcal{A}$ is a bounded subset of equicontinuous functions of $C^0(\overline{M})$. By Ascoli's Theorem, there exists a sequence $\{u_m\}$ in $\mathcal{A}$ such that $\{u_m\}$ converges to some u in $C^0(\overline{M})$. Clearly, $u \in C^\alpha(\overline{M})$ and $\|u\|_{C^\alpha(\overline{M})} \leq \tilde{C} C'$. Let $v_m := u_m - u$. Then $\|v_m\|_{C^\alpha(\overline{M})} \leq 2\tilde{C} C'$ and for any $0 \leq \beta < \alpha$,

$$
\begin{aligned}
\|v_m\|_{C^\beta(\overline{M})} &= \sup_{x \neq y \in \overline{M}} \frac{|v_m(x) - v_m(y)|}{d_F^\beta(x, y)} \\
&= \sup_{x \neq y \in \overline{M}} \left(\frac{|v_m(x) - v_m(y)|}{d_F^\alpha(x, y)} \right)^{\beta/\alpha} |v_m(x) - v_m(y)|^{1-\beta/\alpha} \\
&\leq (2\tilde{C} C')^{\beta/\alpha} \left(2\|v_m\|_{C^0(\overline{M})} \right)^{1-\beta/\alpha}. \quad (10.0.17)
\end{aligned}
$$

Thus, $v_m \in C^\beta(\overline{M})$ and $\{v_m\}$ converges to 0 in $C^\beta(\overline{M})$ since $\{v_m\}$ converges to 0 in $C^0(\overline{M})$ as m goes to ∞.

(3) The proof is similar to that of Theorem 5.36 in [Ad] with the help of inequalities (8.1.8)–(8.1.12) and compactness of M. We omit it here. $\quad \square$

Bibliography

[Ad] R. A. Adams, *Sobolev Spaces*. Academic Press, San Diego, 1978.

[AZ] H. Akbar-Zadeh, Sur les espaces de Finsler á courbures secionnelles constantes. *Bull. Acad. Roy. Bel. Cl, Sci, 5e Série - Tome*, LXXXIV (1988), 281–322.

[AM] P. L. Antonelli and R. Miron, *Lagrange and Finsler Geometry*. Applications to Physics and Biology, Fund. Theories Phys., 76, Kluwer Academic Publishers, 1996.

[BM] S. Bácsó and M. Matsumoto, On Finsler spaces of Douglas type: A generalization of the notion of Berwald space. *Publ. Math. Debrecen*, 51 (1997), 385–406.

[BCL] K. Ball, E. Carlen and E. Lieb, Sharp uniform convexity and smoothness inequalities for trace norms. *Invent. Math.*, 115 (1994), 463–482.

[BCS] D. Bao, S. S. Chern and Z. Shen, *An Introduction to Riemann-Finsler Geometry*. GTM, 200, Springer-Verlag, New York, 2000.

[BS] D. Bao and Z. Shen, Finsler metrics of constant curvature on the Lie group $\mathbb{S}^3$. *J. London Math. Soc.*, 66 (2002), 453–467.

[BSR] D. Bao, Z. Shen and C. Robles, Zermelo navigation on Riemannian manifolds. *J. Diff Geom.*, 66 (2004), 377–435.

[BL] D. Bakry and M. Ledoux, Sobolev inequalities and Myers diameter theorem for an abstract Markov generator. *Duke Math. J.*, 85(1) (1996), 253–270.

[BQ] D. Bakry and Z. Qian, Some new results on eigenvectors via dimension, diameter, and Ricci curvature. *Adv. Math.*, 155(1) (2000), 98–153.

[BS] V. Balan and P. C. Stavrinos, *Finslerian (α, β)-Metrics in Weak Gravitational Models*. In: Finsler and Lagrange geometries, (eds. M. Anastasiei and P. L. Antonelli), Kluwer Academic Publishers, 2003.

[Ber1] L. Berwald, Untersuchung der krümmung allgemeiner metrischer Räume auf Grund des in ihnen herrschenden parallelismus. *Math. Z.*, 25 (1926), 40–73.

[Ber2] L. Berwald, Parallelübertragung in allgemeinen Räumen. *Atti Congr. Intern. Mat. Bologna*, 4 (1928), 263–270.

[Ber3] L. Berwald, Über die n-dimensionalen geometrien konstanter krümmung, in denen die geraden die kurzesten sind. *Math. Z.*, 30 (1929), 449–469.

[Ber4] L. Berwald, Über Finslersche und Cartansche geometrie IV. Projek-tivkrümmung allgemeiner affiner Räume und Finslersche Räume skalarer Krümmung. *Ann. Math.*, 48 (1947), 755–781.

[BH] N. Bouleau and F. Hirsch, *Dirichlet Forms and Analysis on Wiener Space.* Walter de Gruyter Co. Berlin, 1991.

[Br1] R. Bryant, Finsler structures on the 2-sphere satisfying $K = 1$. *Finsler Geometry, Contemporary Mathematics 196*, Amer. Math. Soc., Providence, RI, 1996, 27–42.

[Br2] R. Bryant, Projectively flat Finsler 2-spheres of constant curvature. *Selecta Math.*, New Series, 3 (1997), 161–204.

[Bu] H. Busemann, Intrinsic area. *Ann. Math.*, 48 (1947), 234–267.

[BK] H. Busemann and J. P. Kelly, *Projective Geometry and Projective Metrics.* Academic Press, New York, 1953.

[Car] E. Cartan, *Les espaces de Finsler.* Actualites 79, Paris, 1934.

[CE] J. Cheeger and D. Ebin, *Comparison Theorems in Riemannian Geometry.* North Holland/American Elsevier, 1975.

[Che] S. Cheng, Eigenvalue comparison theorems and its geometric applications. *Math. Z.*, 143 (1975), 289–297.

[CY] S. Cheng and S.-T. Yau, Differential equations on Riemannian manifolds and their geometric applications. *Commu. Pure Appl. Math.*, 28(3)(1975), 333–354.

[CMS] X. Cheng, X. Mo and Z. Shen, On the flag curvature of Finsler metrics of scalar curvature. *J. London Math. Soc.*, 68(2) (2003), 762–780.

[CS] X. Cheng and Z. Shen, Randers metrics of scalar flag curvature. *J. Aust. Math. Soc.*, 87 (2009), 359–370.

[CnS] X. Cheng and Z. Shen, *Finsler Geometry: An Approach via Randers Spaces.* Springer-Verlag, 2012.

[Ch1] S. S. Chern, On the Euclidean connections in a Finsler space, *Proc. National Acad. Soc*, 29 (1943), 33–37; or Selected Papers, vol. I, 107–111, Springer, 1989.

[Ch2] S. S. Chern, On Finsler geometry. *C. R. Acad. Sc. Paris*, 314 (1992), 757–761.

[ChS] S. S. Chern and Z. Shen, *Riemann-Finsler Geometry.* World Scientific, Singapore, 2005.

[Da] E. B. Davies, *Heat Kernels and Spectral Theory.* Cambridge University Press, Cambridge, 1989.

[Dei] A. Deicke, Über die Finsler-Raume mit $A_i = 0$. *Arch. Math.*, 4 (1953), 45–51.

[Ev] L. C. Evans, *Partial Differential Equations.* Graduate Studies in Mathematics, 19, American Mathematical Society, Providence, R. I., 1998.

[FX] Y. Feng and Q. Xia, On a class of locally projectively flat Finsler metrics II. *Diff. Geom. Appl.*, 62 (2019), 39–59.

[GS] Y. Ge and Z. Shen, Eigenvalues and eigenfunctions of metric measure manifolds. *Proc. London Math. Soc.*, 82(3) (2001), 725–746.

[GT] G. Gilbarg and N. S. Trudinger, *Elliptic Partial Differential Equations of Second Order*. Second edition. Grundlehren der Mathematischen Wissenschaften, 224, Springer, Berlin-New York, 1983.

[GM] E. Guo and X. Mo, *The Geometry of Spherically Symmetric Finsler Metrics*. Springer, 2018.

[Ham] G. Hamel, Über die Geometrien in denen die Geraden die Kürzesten sind. *Math. Ann.*, 57 (1903), 231–264.

[HS] H. Hrimiuc and H. Shimada, On the L-duality between Finsler and Hamilton manifolds. *Nonlinear World*, 3 (1996), 613–641.

[Ic1] Y. Ichijyo, Finsler spaces modeled on a Minkowski space. *J. Math. Kyoto Univ.*, 16 (1976), 639–652.

[Ic2] Y. Ichihyo, On special Finsler connections with vanishing hv-curvature tensor. *Tensor, N. S.*, 32 (1978), 146–155.

[KY] C.-W Kim and J.-W Yim, Finsler manifolds with positive constant flag curvature. *Geom. Dedicata*, 98 (2003), 47–56.

[KR] A. Kristály and I. J. Rudas, Elliptic problems on the ball endowed with Funk-type metrics. *Nonlinear Anal.*, 119 (2015), 199–208.

[KSYZ] A. Kristály, Z. Shen, L. Yuan and W. Zhao, Nonlinear spectrums of Finsler manifolds. *Math. Z.*, 300 (2022), 81–123.

[La1] G. Landsberg, Über die totalkrümmung, *Jahresberichte der deut Math.* Ver. 16 (1907), 36–46.

[La2] G. Landsberg, Über die Krümmung in der variationsrechung, *Math. Ann.*, 65 (1908), 313–349.

[Le] G. Leoni, *A First Course in Sobolev Spaces*, Graduate Studies in Mathematics. 105, Amer. Math. Soc., 2009.

[LX] J. Li and X. Xu, Differential Harnack inequalities on Riemannian manifolds I: Linear heat equation. *Adv. Math.*, 226(5) (2011), 4456–4491.

[LS] P. Li and R. Schoen, L^p and mean value properties of subharmonic functions on Riemannian manifolds. *Acta Math.*, 153 (1984), 279–301.

[LY] P. Li and S.-T. Yau, On the parabolic kernel of the Schrödinger operator, *Acta Math.*, 156(3-4) (1986), 153–201.

[Li] X. Li, Liouville theorems for symmetric diffusion operators on complete Riemannian manifolds. *J. Math. Pures Appl.*, 84 (2005), 1295–1361.

[Lic] A. Lichnerowicz, Geometrie des groupes de transforamtions. In: Travaux et Recherches Mathemtiques, Vol. III, Dunod, Paris, 1958.

[Lieb] Gray M. Lieberman, Boundary regularity for solutions of degenerate elliptic equations. *Nonlinear Anal.*, 12(11) (1988), 1203–1219.

[Ma1] M. Matsumoto, On C-reducible Finsler spaces. *Tensor (N. S.)*, 24 (1972), 29–37.

[Ma2] M. Matsumoto, On Finsler spaces with Randers metric and special forms of important tensors. *J. Math. Kyoto Univ.*, 14 (1974), 477–498.

[Ma3] M. Matsumoto, Projective changes of Finsler metrics and projectively flat Finsler spaces. *Tensor N.S.*, 34 (1980), 303–315.

[Ma4] M. Matsumoto, *Foundations of Finsler Geometry and Special Finsler Spaces*. Kaiseisha Press, Japan, 1986.

[MH]   M. Matsumoto and S. Hojo, A conclusive theorem on C-reducible Finsler spaces. *Tensor, N. S.*, 32 (1978), 225–230.

[MS]   X. Mo and Z. Shen, On negatively curved Finsler manifolds of scalar curvature. *Canad. Math. Bull.*, 48(1) (2005), 112–120.

[Ob]   M. Obata, Certain conditions for a Riemannian manifold to be a sphere. *J. Math. Soc. Japan*, 14 (1962), 333–340.

[Oh1] S. Ohta, Finsler interpolation inequalities, *Calc. Var. PDE*, 36 (2009), 211–249.

[Oh2] S. Ohta, Uniform convexity and smoothness, and their applications in Finsler geometry. *Math. Ann.*, 343 (2009), 669–699.

[Oh3] S. Ohta, Some functional inequalities on non-reversible Finsler manifolds. *Proc. Indian Acad. Sci. (Math. Sci.)*, 127(5) (2017), 833–855.

[Oh4] S, Ohta, *Comparison Finsler Geometry*. Springer Monographs in Mathematics, 2021.

[OS1] S. Ohta and K.-T. Sturm, Heat flow on Finsler manifolds. *Commu. Pure Appl. Math.*, 62 (2009), 1386–1433.

[OS2] S. Ohta and K.-T. Sturm, Bochner-Weitzenböck formula and Li-Yau estimates on Finsler manifolds. *Adv. Math.*, 252 (2014), 429–448.

[Ok]   T. Okada, On models of projectively flat Finsler spaces of constant negative curvature. *Tensor, N. S.*, 40 (1983), 117–123.

[Pa]   D. G. Pavlov (Ed.), *Space-time structure*. Collected papers, TETRU, 2006.

[Qz]   Z. Qian, Estimates for weighted volumes and applications. *Quart. J. Math.*, Oxford Ser., 48(2) (1997), 235–242.

[Ran] G. Randers, On an asymmetric metric in the four-space of general relativity. *Phys. Rev.*, 59 (1941), 195–199.

[Rap] A. Rapcsak, Über die bahntreuen Abbildungen metrischer Rdume. *Publ. Math. Debrecen*, 8 (1961), 285–290.

[RSS] M. Roman, H. Shimada and V. S. Sabau, On β-change of the Antonelli-Shimada ecological metric. *Tensor (N.S.)*, 65 (2004), 65–73.

[Run] H. Rund, *Differential Geometry of Finsler Spaces*. GMW 101, Springer, 1959.

[SY]   R. Schoen and S.-T. Yau, *Lectures on Differential Geometry*. International Press, Cambridge, MA, 1994.

[SS]   Y. Shen and Z. Shen, *Introduction to Modern Finsler Geometry*. World Scientific Publishing, Singapore, 2016.

[Sh1] Z. Shen, *Lectures on Finsler Geometry*. World Scientific Publishing, Singapore, 2001.

[Sh2] Z. Shen, *Differential Geometry of Spray and Finsler Spaces*. Kluwer Academic Publishers, 2001.

[Sh3] Z. Shen, Volume comparison and its applications in Riemann-Finsler geometry. *Adv. Math.*, 128 (1997), 306–328.

[Sh4] Z. Shen, Projectively flat Finsler metrics of constant flag curvature. *Trans. Amer. Math. Soc.*, 355(4) (2003), 1713–1728.

[Sh5] Z. Shen, Nonpositively curved Finsler manifolds with constant S-curvature. *Math. Z.*, 249 (2005), 625–339.

[Sh6] Z. Shen, On projectively flat (α, β)-metrics. *Canad. Math. Bull.*, 52(1) (2009), 132–144.

[Sh7] Z. Shen, On some non-Riemannian quantities in Finsler geometry. *Canad. Math. Bull.*, 56 (2013), 184–193.

[ShY] Z. Shen and G.C. Yildirim, A characterization of Randers metrics of scalar flag curvature. In: Survey Geometric Analysis and Relativity, ALM, 23 (2012), 330–343.

[Shi] H. Shimada, On Finsler spaces with the metric $L = \sqrt[m]{a_{i_1 i_2 \cdots i_m}(x) y^{i_1} y^{i_2} \cdots y^{i_m}}$. *Tensor N.S.*, 33 (1979), 365–372.

[Sz] Z. Szabó, Ein Finslerscher Raum ist gerade dann skalarer Krümmung, wenn seine Weyl sche projektivkrümmüng verschwindet. *Acta Sci. Math. (Szeged)*, 39 (1977), 163–168.

[Th] A. C. Thompson, Minkowski geometry, In: Encyclopedia of Math. and Its Applications. Vol. 63, Cambridge Univ. Press, Cambridge, 1996.

[To] P. Tolksdorf, Regularity for a more general class of quasilinear elliptic equations. *J. Diff. Eq.*, 51(1) (1984), 126–150.

[Va] D. Valtorta, Sharp estimate on the first eigenvalue of the p-Laplacian. *Nonliear Anal.*, 75 (2012), 4974–4994.

[WX] G. Wang and C. Xia, A sharp lower bound for the first eigenvalue on Finsler manifolds. *Ann. I. H. Poincaré-AN*, (30) (2013), 983–996.

[Wh] J. H. C. Whitehead, Convex regions in the geometry of paths. *Quart. J. Math. Oxford Ser.*, 3 (1932), 33–42.

[Wu1] B. Wu, A global rigidity theorem for weakly Landsberg manifolds. *Sci. China Ser. A: Math.*, 50(5) (2007), 609–614.

[Wu2] B. Wu, Comparison theorems in Finsler geometry with weighted curvature bounds and related results. *J. Korean Math. Soc.*, 52(3) (2015), 603–624.

[WuX] B. Wu and Y. Xin, Comparison theorems in Finsler geometry and their applications. *Math. Ann.*, 337 (2007), 177–196.

[Xc] C. Xia, Local gradient estimate for harmonic functions on Finsler manifolds. *Calc. Var. PDE*, 51 (2014), 849–865.

[Xia1] Q. Xia, Some results on the non-Riemannian quantity H of a Finsler metric. *Internat. J. Math.*, 22(7) (2011), 925–936.

[Xia2] Q. Xia, On Kropina metrics of scalar flag curvature. *Diff. Geom. Appl.*, 31 (2013), 393–404.

[Xia3] Q. Xia, On a class of Finsler metrics of scalar flag curvature. *Results Math.*, 71 (2017), 483–507.

[Xia4] Q. Xia, A sharp lower bound for the first eigenvalue on Finsler manifolds with nonnegative weighted Ricci curvature. *Nonlinear Anal.*, 117 (2015), 189–199.

[Xia5] Q. Xia, Sharp spectral gap for the Finsler p-Laplacian. *Sci. China Math.*, 62(8) (2019), 1615–1944.

[Xia6] Q. Xia, Geometric and functional inequalities on Finsler manifolds. *J. Geom. Anal.*, 30 (2020), 3099–3148.

[Xia7] Q. Xia, Local and global gradient estimates for Finsler p-harmonic functions. *Commu. Anal. Geom.*, 30(2) (2022), 451–500.

[Xia8] Q. Xia, Some L^p Liouville theorems on Finsler measure spaces. *Diff. Geom. Appl.*, 87 (2023), 101987, 15pp.

[Xia9] Q. Xia, Li-Yau's estimates on Finsler manifolds. *J. Geom. Anal.*, 33: 49 (2023).

[Xia10] Q. Xia, Linearized heat semigroup and Li-Yau's inequalities on Finsler measure spaces. preprint.

[Xia11] Q. Xia, Uniqueness of L^p subsolutions to the heat equation on Finsler measure spaces. *Canad. Math. Bull.*, 67(1) (2024), 166–175.

[Yau] S.-T. Yau, Some function theoretic properties of complete Riemannian manifolds and their applications to geometry. *Indiana Univ. Math. J.*, 25(7) (1976), 659–670.

[YH1] S. Yin and Q. He, The first eigenvalue of Finsler p-Laplacian. *Diff. Geom. Appl.*, 35 (2014), 30–49.

[YH2] S. Yin and Q. He, The first eigenfunctions and eigenvalue of the p-Laplacian on Finsler manifolds. *Sci. China Math.*, 59 (2016), 1769–1794.

[YO] R. Yoshikawa and K. Okubo, Kropina spaces of constant curvature. *Tensor, N. S.*, 68 (2007), 190–203.

[Yu] C. Yu, Deformations and Hilbert's fourth problem. *Math. Ann.*, 365 (2016), 1379–1408.

[YZ] C. Yu and H. Zhu, On a new class of Finsler metrics. *Diff. Geom. Appl.*, 29 (2011), 244–254.

[ZX] F. Zhang and Q. Xia, Some Liouville-type theorems for harmonic functions on Finsler manifolds. *J. Math. Anal. Appl.*, 417 (2014), 979–995.

[ZY] J. Zhong and H. Yang, On the estimate of the first eigenvalue of a compact Riemannian manifold. *Sci. Sin. Ser. A*, 27(12) (1984), 1265–1273.

[Zh] L. Zhou, Spherically symmetric Finsler metrics in $\mathbb{R}^n$. *Publ. Math. Debrecen*, 80(1-2) (2012), 67–77.